Aqueous Microwave Assisted Chemistry
Synthesis and Catalysis

Edited by

Vivek Polshettiwar
National Risk Management Research Laboratory, U.S. Environmental Protection Agency, Cincinnati, Ohio, US
Heterogeneous and Nano-Catalysis Laboratories, KAUST Catalysis Centre (KCC), Thuwal, KSA

Rajender S. Varma
National Risk Management Research Laboratory, U.S. Environmental Protection Agency, Cincinnati, Ohio, US

RSCPublishing

RSC Green Chemistry No. 7

ISBN: 978-1-84973-038-9
ISSN: 1757-7039

A catalogue record for this book is available from the British Library

© Royal Society of Chemistry 2010

All rights reserved

Apart from fair dealing for the purposes of research for non-commercial purposes or for private study, criticism or review, as permitted under the Copyright, Designs and Patents Act 1988 and the Copyright and Related Rights Regulations 2003, this publication may not be reproduced, stored or transmitted, in any form or by any means, without the prior permission in writing of The Royal Society of Chemistry, or the copyright owner, or in the case of reproduction in accordance with the terms of licences issued by the Copyright Licensing Agency in the UK, or in accordance with the terms of the licences issued by the appropriate Reproduction Rights Organization outside the UK. Enquiries concerning reproduction outside the terms stated here should be sent to The Royal Society of Chemistry at the address printed on this page.

The RSC is not responsible for individual opinions expressed in this work.

Published by The Royal Society of Chemistry,
Thomas Graham House, Science Park, Milton Road,
Cambridge CB4 0WF, UK

Registered Charity Number 207890

For further information see our web site at www.rsc.org

Aqueous Microwave Assisted Chemistry
Synthesis and Catalysis

RSC Green Chemistry

Series Editor:
James H Clark, *Department of Chemistry, University of York, York, UK*
George A Kraus, *Department of Chemistry, Iowa State University, Iowa, USA*

Titles in the Series:
1: The Future of Glycerol: New Uses of a Versatile Raw Material
2: Alternative Solvents for Green Chemistry
3: Eco-Friendly Synthesis of Fine Chemicals
4: Sustainable Solutions for Modern Economies
5: Chemical Reactions and Processes under Flow Conditions
6: Radical Reactions in Aqueous Media
7: Aqueous Microwave Assisted Chemistry: Synthesis and Catalysis

How to obtain future titles on publication:
A standing order plan is available for this series. A standing order will bring delivery of each new volume immediately on publication.

For further information please contact:
Book Sales Department, Royal Society of Chemistry,
Thomas Graham House, Science Park, Milton Road, Cambridge,
CB4 0WF, UK
Telephone: +44 (0)1223 420066, Fax: +44 (0)1223 420247, Email: books@rsc.org
Visit our website at http://www.rsc.org/Shop/Books/

Preface

The demands for green and sustainable synthetic methods in the fields of healthcare and fine chemicals, combined with the pressure to produce these substances expeditiously and in an environmentally benign fashion, pose significant challenges to the synthetic chemical community. Green chemistry can avoid pollution by utilizing techniques that are environmentally friendly by design. One of the best green techniques is the use of microwave (MW) assisted aqueous synthetic protocols.

Fusing the MW technique with water (as a benign reaction medium) can offer an extraordinary synergistic effect with greater potential than these two individual components in isolation. Earlier work by Chris Strauss, Andre Loupy and in recent years by Oliver Kappe, has demonstrated that selective microwave heating can be exploited to develop a high yield protocol and the use of water expedited the MW protocol with better efficiency. We dedicate this book to these pioneering scientists.

This book provides a comprehensive overview of various processes developed using aqueous microwave chemistry and is written for chemists, chemical engineers and researchers in the early stages who want to develop sustainable and green methods. This book cuts across a multi-disciplinary line to achieve a holistic view and chapters are drawn from heterocycle synthesis, metal catalysis, enzyme catalysis, polymer synthesis, nanomaterials synthesis and nanocatalysis.

The book begins with a chapter on the fundamentals of microwave chemistry, including a discussion on thermal–non-thermal MW-effect in aqueous medium. Chapter 2 gives an overview of advances in the field of MW-assisted metal-catalyzed reactions in aqueous media which is complemented by chapter 3 that describes sustainable and economical protocols that have been developed in recent decades for cross-coupling reactions using microwave irradiation. Chapter 4 gives a comprehensive review of bio-active heterocycle synthesis

---

RSC Green Chemistry No. 7
Aqueous Microwave Assisted Chemistry: Synthesis and Catalysis
Edited by Vivek Polshettiwar and Rajender S. Varma
© Royal Society of Chemistry 2010
Published by the Royal Society of Chemistry, www.rsc.org

using aqueous microwave chemistry. Chapter 5, through heuristic discussion, explains the impact of microwave irradiation on enzyme activity in aqueous buffer solutions and aqueous organic solvents. Chapter 6 is devoted to a discussion of advances in microwave-assisted synthesis of polymers in aqueous medium. Chapter 7 is divided into two parts; the first part examines the potential and difficulties related to methods for synthesizing various nanomaterials under microwave irradiation conditions using water; the second introduces the concept of nano-catalysis in water using microwaves, with an interesting discussion on the effect of particle size, shape and morphology on catalytic activity.

Most importantly, each chapter contains representative experimental procedures that will help the reader to quickly replicate some of the experiments and thereby acquire hands-on experience.

This is the only book that exclusively deals with microwave-assisted chemistry conducted in water and represents a significant effort towards green chemistry and should be of interest to both beginner and expert researcher in this field.

We thank the distinguished authors of this book for their scientifically rich, timely and well-organized contributions. Thanks are due to Ms. Lauren Drees of the US EPA, Cincinnati, for her diligent reading of the manuscripts and persuasive adherence to perfection. We also thank the RSC editorial staff for a very thorough and meticulous work and Professor James Clark for his interest in these greener endeavors. Finally, VP thanks his wife Poonam and son Neil for their patience and support during the making of this book.

Vivek Polshettiwar,
Rajender S. Varma

Contents

About the Authors	xii

Chapter 1 Fundamentals of Aqueous Microwave Chemistry 1
Vivek Polshettiwar and Rajender S. Varma

1.1	Introduction		1
1.2	Green Chemistry Approach		2
1.3	Water as Green Solvent		3
1.4	Why Microwaves?		4
	1.4.1	Thermal Effect	4
	1.4.2	Non-thermal Effect	5
	1.4.3	Selectivity towards Water	6
	1.4.4	Selectivity towards Catalyst	7
	1.4.5	Catalyst as Susceptors	7
	1.4.6	Stability of Catalyst	7
1.5	Conclusion		7
References			8

Chapter 2 Metal-catalyzed Reactions in Water under MW Irradiation 10
Victorio Cadierno, Pascale Crochet and
Sergio E. García-Garrido

2.1	Introduction		10
2.2	Addition Reactions		11
2.3	Isomerization Reactions		17
2.4	Substitution Reactions		19
2.5	Cyclization Reactions		28
2.6	Reduction and Oxidation Reactions		38
2.7	Miscellaneous Reactions		44
2.8	Representative Experimental Examples		45
	2.8.1	Addition Reactions	45

RSC Green Chemistry No. 7
Aqueous Microwave Assisted Chemistry: Synthesis and Catalysis
Edited by Vivek Polshettiwar and Rajender S. Varma
© Royal Society of Chemistry 2010
Published by the Royal Society of Chemistry, www.rsc.org

		2.8.2	Isomerization Reactions	46

                2.8.2 Isomerization Reactions   46
                2.8.3 Substitution Reactions   46
                2.8.4 Cyclization Reactions   47
                2.8.5 Reduction and Oxidation Reactions   48
                2.8.6 Miscellaneous Reactions   48
        2.9 Conclusions   49
        Acknowledgements   49
        References   49

Chapter 3   **Microwave-assisted Coupling Reactions in Aqueous Media**   55
            Aziz Fihri and Christophe Len

        3.1 Introduction   55
        3.2 Suzuki–Miyaura Reaction   56
        3.3 Heck Reaction   71
        3.4 Sonogashira Reactions   75
        3.5 Stille Reactions   78
        3.6 Hiyama Reactions   78
        3.7 Cyanation Reactions   79
        3.8 Carbonylation Reactions   80
        3.9 Representative Experimental Syntheses   83
                3.9.1 Synthesis of 4,5-Dimethoxy-2-vinyl-2′-pivaloylamino-biphenyl   83
                3.9.2 Synthesis of 4-Phenyltoluene   83
                3.9.3 Synthesis of m-Carboxycinnamic Acid in Homogeneous Heck Reactions in D_2O 84
                3.9.4 Synthesis of Ethyl 3-(6-Methoxy-2-naphthyl) propanoate   84
                3.9.5 Synthesis of 5-Chloro-2-phenylethynyl-pyridine   84
                3.9.6 Synthesis of 4-Methoxybenzonitrile   85
                3.9.7 Aminocarbonylation and Hydroxycarbonylation   85
        3.10 Conclusion   86
        References   86

Chapter 4   **Microwave-assisted Synthesis of Bio-active Heterocycles in Aqueous Media**   91
            Vivek Polshettiwar and Rajender S. Varma

        4.1 Introduction   91
        4.2 MW-assisted Nitrogen-containing Heterocycle Synthesis in Water   92

	4.3	MW-assisted Synthesis of Oxygen- and Sulfur-containing Heterocycles in Water	106
	4.4	MW-assisted Miscellaneous Reactions in Water	112
	4.5	Representative Experimental Procedures	118
		4.5.1 Synthesis of Dihydropyrimidinones	118
		4.5.2 Synthesis of 2-Aminochromene Derivatives	118
		4.5.3 Synthesis of Dioxane-functionalized Molecules	118
		4.5.4 Synthesis of Sulfonyl-benzothiazole-based Bio-active Compounds	119
		4.5.5 Synthesis of a Series of new β-Aminoketones	119
	4.6	Conclusions	119
	References		119

Chapter 5 Microwave-assisted Enzymatic Reactions in Aqueous Media   **123**
Hua Zhao

5.1	Introduction	123
5.2	Microwave-assisted Enzymatic Reactions in Water (or Aqueous Buffer)	125
	5.2.1 Effect of Microwave Irradiation on Enzyme Activity in Aqueous Solutions	125
	5.2.2 Microwave-assisted Enzymatic Protein Digestion	130
5.3	Microwave-assisted Enzymatic Reactions in Aqueous Solutions of Organic Solvents	132
5.4	Microwave-assisted Enzymatic Reactions in Ionic Liquids	133
5.5	Non-thermal Effect of Microwave Irradiation on Enzymes	136
5.6	Prospects of Microwave Irradiation in Aqueous Phase Biocatalysis	137
5.7	Experimental	137
	5.7.1 Enzymatic Hydrolysis of Starch in Water under Microwave Irradiation	137
	5.7.2 Enzymatic Transglycosylation of Lactose in Hexanol–Water (70:30, v/v) under Microwave Irradiation	138
	5.7.3 Enzymatic Digestion of Proteins (Cytochrome *c*, Myoglobin, Lysozyme, and Ubiquitin)	138
Acknowledgements		138
References		138

Chapter 6	Microwave-assisted Synthesis of Polymers in Aqueous Media	145

Catherine Marestin and Régis Mercier

6.1	Introduction		145
6.2	Radical Polymerization in Aqueous Medium		146
	6.2.1	Free-Radical Polymerization	146
	6.2.2	Radical Polymerization in Dispersed Media	151
	6.2.3	Controlled/Living Radical Polymerization in Dispersed Media	155
6.3	Step-growth Polymerization in Aqueous Media		157
	6.3.1	Synthesis of Poly(ether)s	157
	6.3.2	Synthesis of Poly(amide)s and Poly(imide)s	158
	6.3.3	Synthesis of Polytriazoles	160
	6.3.4	C–C Coupling Polymerizations	160
6.4	Miscellaneous Polymer Synthesis in Aqueous Media		162
	6.4.1	Cationic Polymerization	162
	6.4.2	Polymer Modifications	162
	6.4.3	Solid-phase Peptide Synthesis in Water	169
6.5	Experimental		170
	6.5.1	PMMA Grafting onto a Polysaccharide	170
	6.5.2	PMMA Synthesis in Emulsion Polymerization	170
	6.5.3	Synthesis of PMMA Nanoparticles	171
	6.5.4	Synthesis of Polyacrylamide–Calcium Phosphate Nanocomposites	171
	6.5.5	Nitroxide-assisted Synthesis of Styrene in Miniemulsion	171
	6.5.6	Synthesis of Poly(ether)s	171
	6.5.7	Synthesis of Poly(imide)s	171
	6.5.8	Polymerization by C–C Coupling	172
	6.5.9	Cationic Polymerization	172
	6.5.10	Polymer Modification by Click Chemistry	172
	References		172

Chapter 7	Microwave-assisted Synthesis of Nanomaterials in Aqueous Media	176

Babita Baruwati, Vivek Polshettiwar and Rajender S. Varma

7.1	Introduction		176
7.2	Synthesis of Metal Nanoparticles using Water under Microwave Irradiation		178
	7.2.1	Gold (Au) Nanoparticles	178
	7.2.2	Silver (Ag) Nanoparticles	182
	7.2.3	Palladium (Pd) and Platinum (Pt) Nanoparticles	185
7.3	Synthesis of Metal Oxide Nanoparticles in Aqueous Medium under MW Irradiation Conditions		186

	7.3.1	Synthesis of Titania (TiO$_2$) Nanoparticles	187
	7.3.2	Synthesis of ZnO Nanoparticles	189
	7.3.3	Synthesis of Ferrite Nanoparticles	191
	7.3.4	Synthesis of Quantum Dots in Aqueous Medium under Microwave Conditions	194
7.4	Nanoparticles as Catalysts		195
7.5	Ruthenium Hydroxide Nano-catalyst for Microwave-assisted Hydration of Nitriles in Water		202
7.6	Glutathione-based Nano-organocatalyst for Microwave-assisted Synthesis of Heterocycles in Water		203
7.7	Representative Experimental Procedures		205
	7.7.1	Synthesis of Gold Nanoparticles	205
	7.7.2	Synthesis of Silver Nanoparticles	206
	7.7.3	Synthesis of Different TiO$_2$ Nanoparticles	207
	7.7.4	Synthesis of ZnO Nanostructures	207
	7.7.5	Selected Methods for Production of Ferrite Nanoparticles	208
7.8	Conclusion		209
References			209

Subject Index   **217**

About the Authors

Dr Vivek Polshettiwar obtained his Ph.D. from DRDE, Gwalior, and he then investigated nanostructured silica-catalysis, with Prof. J. J. E. Moreau, during his postdoctoral research at ENSCM, Montpellier (France). He then moved to the U. S. Environmental Protection Agency to research nanocatalysis and MW-assisted new synthetic methods for green chemistry. Currently, he is working as senior research scientist at KAUST catalysis center directed by Prof. J. M. Basset. His research interests are in the area of advanced nanomaterials for perfect catalysis and green chemistry. He has over 60 publications including various review articles and book chapters.

Dr. Rajender S. Varma, a former chemistry professor, is senior scientist at US EPA with over 35 years of research experience in management of multi-disciplinary technical programs ranging from natural products chemistry and nanomaterials, to development of environmentally benign synthetic methods using microwaves, ultrasound and alternative solvents. He has to his credit 300 scientific peer-reviewed publications and has been awarded 6 US Patents.

Dr. Victorio Cadierno studied chemistry at the University of Oviedo (Spain) and obtained his Ph.D. degree in 1996 working under the supervision of Prof. J. Gimeno. He then joined the group of Prof. J. P. Majoral at the LCC-CNRS (Toulouse, France) for a two-year postdoctoral stay. Thereafter, he returned to the University of Oviedo, where he is currently Associated Professor of Inorganic Chemistry. His research interests cover the chemistry of ruthenium complexes and their catalytic applications, and he is coauthor of more than 90 publications in the field of organometallic chemistry.

Dr. Aziz Fihri received his Ph.D. from the Burgundy University (France) under the guidance of Prof. J.-C. Hierso. He focused his efforts on the synthesis of ferrocenylphosphines and their application in catalyzed cross-coupling reactions. He then worked as a post-doctoral fellow in the group of Prof. M. Fontecave (Atomic Energy Center, France). Currently he is working as a principal research scientist at the ESCOM, France. His research interests are in the area of coupling reactions and green catalysis and he has published several articles and reviews in prestigious international journals including Science.

About the Authors

Dr. Hua Zhao completed his Ph.D. degree at New Jersey Institute of Technology (USA), and a post-doc training at Rutgers University (New Brunswick). He is currently an Associate Professor of Chemistry at Savannah State University. His research interests include microwave-assisted enzymatic reactions, biocatalysis in ionic liquids, and the production of biofuels (cellulosic ethanol and biodiesel) from biomass using ionic liquids.

Dr. Catherine Marestin earned her Ph.D. in Macromolecular Chemistry, under the direction of Dr Jérôme Claverie at the University of Lyon in 1998. After one year in Milan (CNR) as postdoctoral fellow with Professor Tritto, she joined the LMOPS laboratory as CNRS permanent researcher. Her research in polymer chemistry is presently focused on aromatic, heterocyclic and ionic conducting polymers synthesis with emphasis on the microwave assisted polymerisation reactions.

CHAPTER 1
Fundamentals of Aqueous Microwave Chemistry

VIVEK POLSHETTIWAR* AND RAJENDER S. VARMA

Sustainable Technology Division, National Risk Management Research Laboratory, U.S. Environmental Protection Agency, 26 W. Martin Luther King Dr., MS 443, Cincinnati, Ohio 45268, USA

1.1 Introduction

The first chemical revolution changed modern life with excellent amenities and services, but also created the serious problem of environmental pollution. Now, 150 years later, we need to develop the concept of 'green chemistry' to help safeguard human life. The core principle of this concept is to protect the environment, not by cleaning it up, but by discovery of new chemistry and chemical processes that avert pollution.[1,2] The concept of green chemistry prompts the chemical and pharmaceutical industries to consider the impact on human life when new chemicals are produced and introduced into our society. Thus, we can develop innovative pathways by rethinking chemical design from the ground up, to create products that stimulate our economy and lifestyles, without damaging our environment and surroundings. Green chemistry has emerged as a discipline that permeates all aspects of chemistry.

A myriad of drugs are required by society in short periods of time. To achieve this goal, medicinal chemists have been under increased demands to produce

---

*Current address: KAUST Catalysis Center (KCC), King Abdullah University of Science and Technology, Thuwal 23955, Kingdom of Saudi Arabia.

RSC Green Chemistry No. 7
Aqueous Microwave Assisted Chemistry: Synthesis and Catalysis
Edited by Vivek Polshettiwar and Rajender S. Varma
© Royal Society of Chemistry 2010
Published by the Royal Society of Chemistry, www.rsc.org

bio-active drug molecules, because of the high molecular complexity in drug discovery processes accompanied by time constraints. The central focus of pharmaceutical green chemistry is the development of efficient and environmentally benign synthetic protocols.[3]

1.2 Green Chemistry Approach

Green chemistry is a vast and multifaceted field. The principles of green chemistry can be used to evaluate the greenness of a particular synthetic protocol.[1] These principles deal with several issues, such as preventing the use of volatile and toxic solvents, the quantity and reusability of catalyst and reagents employed, the use of benign chemicals, atom-economic synthetic methods with a minimum number of chemical steps (which selectively generate the desired product without producing any by-products), energy efficient and mild reaction conditions, and chemical waste produced (Figure 1.1). It is not anticipated that any synthetic protocol will satisfy all green chemistry principles, but the more it satisfies the greener the process will be.

Solvents play a key role in deciding the environmental fate of a chemical process and have a huge impact on its cost, safety and health issues. The volatile and

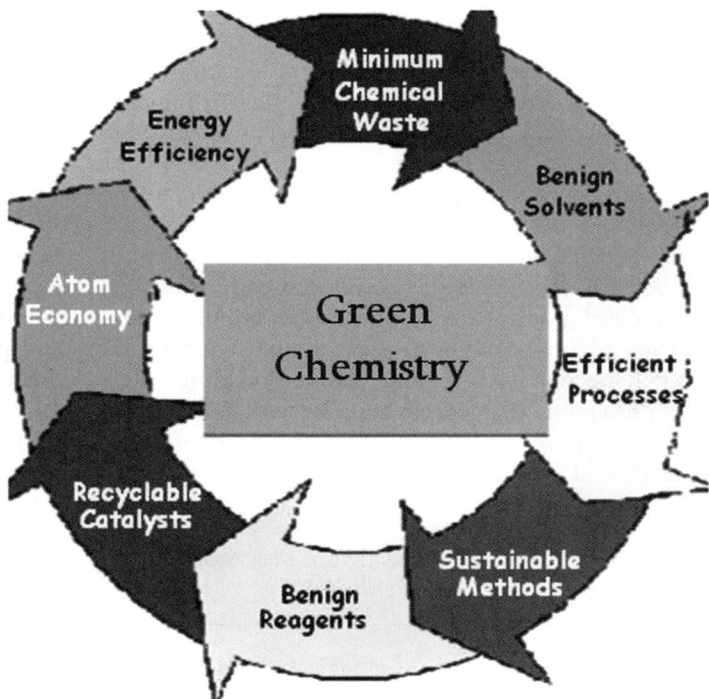

Figure 1.1 Principles of green chemistry.

highly flammable solvents that are commonly used are the foremost source of ecological pollution and are rapidly rising on the green chemistry agenda.[4] The use of reaction solvents, however, cannot be avoided as they are necessary for various steps such as the mixing of reactants, harmonized supply of heat and energy, and also in some cases to control the regio- and chemoselectivity of reactions.

There are assorted approaches for the development of environmentally friendly methods. The replacement of toxic reaction solvents with benign media is one of the best ways to make a protocol greener. While solvent replacement is a successful approach, it alone may not be enough. The entire development process must be well thought out, considering factors such as atom economy, energy efficiency and use of benign and naturally available renewable resources, and the solvent should be only one part of this. In addition, toxic solvents should not only be avoided during reaction but also after completion of the reaction; their use for isolation and purification of products (which involves the use of excess amounts per mass of final products) should be prevented or minimized.

Toxic solvents can be replaced by various non-conventional alternatives with superior ecological, health and safety properties, such as:

- bio-solvents: solvents produced from renewable resources, *e.g.* ethanol produced by fermentation of sugar-containing feeds and starchy feed materials;
- supercritical fluids, *e.g.* CO_2;
- benign ionic liquids that have low vapor pressure and restrain release into the environment;
- fluorous and re-generable biphasic media.

1.3 Water as Green Solvent

To reduce the environmental impact resulting from the use of toxic solvents in chemical production, the identification and use of "green" solvents is a top priority. A solvent-free process is another answer, as one of the green chemistry principles states that the use of no toxic solvent makes the protocol green; however, this is not true in every case, and in fact this is how we misinterpret these principles. This particular green chemistry principle is only valuable if the developed solvent-free protocol works at the industry level (or at least the pilot plant level). Simply carrying out the reactions at small scale in the laboratory has no value as, at the bench-scale, small amounts of reactants can be mixed without solvent, but this is not feasible at the kilogram level, where a lack of reaction medium may lead to overheating of the reaction mixture because of the inadequate heat- and mass-transfer.[5]

Biphasic technologies, using fluorous and ionic liquids[6] along with aqueous systems[7] and supercritical carbon dioxide, have formed the main thrust of this green solvent movement. However, the cost and toxicity of ionic liquids are prohibitive.[8] Water appears to be a better option because of its abundant, non-toxic, non-corrosive, and non-flammable nature.[9] In addition, water can be contained because of its relatively higher vapor pressure as compared to organic solvents, making it a green and sustainable alternative.[9,10]

The major difficulty with using water as a solvent is the insolubility of most of organic reactants, making reaction mixtures heterogeneous. One way to overcome this is by using phase-transfer catalysts, but their expensive nature means that the resulting method is not economical. Product isolation from aqueous reaction mixture is another critical issue. Usually, evaporation of water is an option, but this is not an energy-efficient technique. Interestingly, these challenges can be tackled successfully by using the microwave (MW) heating technique for reactions in aqueous medium.

1.4 Why Microwaves?

Conventional processes of chemical synthesis are orders of magnitude too slow to satisfy the current demand for the generation of new compounds. Although the fields of combinatorial and automated chemistry have emerged to meet this burgeoning demand, most of these techniques generate considerable quantities of chemical waste. Chemists have been under growing pressure to develop new methods that are rapid and environmentally benign. One of the alternatives is the use of nano-catalysis in conjunction with non-conventional MW heating technology. The efficiency of MW flash-heating has resulted in dramatic reductions in reaction times – from days to minutes – that are potentially important in process chemistry for the expedient generation of fine chemicals.[11] In the last few years, MW-assisted chemistry has blossomed into a mature and useful technique for various applications. Although MW-assisted reactions in conventional solvents have developed rapidly, the center of attention has now shifted to environmentally benign processes,[10,12] which use nano-catalysts and greener solvents such as water.

MW-enhanced chemistry is based on the efficiency of the interaction of molecules in a reaction mixture (substrates, catalyst and solvents) with electromagnetic waves generated. This process mainly depends on the specific polarity of molecules. Since water is polar it has good potential to absorb microwaves and convert them into heat energy, consequently accelerating the reactions in an aqueous medium as compared to results obtained using conventional heating.[10] This can be explained by two key mechanisms: dipolar polarization and ionic conduction of water molecules (Figure 1.2 below). Irradiation of a reaction mixture in an aqueous medium by MW results in the dipole orientation of water molecules and reactants in the electric field.

1.4.1 Thermal Effect

Dielectric heating ensues from the tendency of dipoles (mostly water molecules in addition to reactants) to follow the inversion of alternating electric fields and induce energy dissipation in the form of heat through molecular friction and dielectric loss, which allows more regular repartition in reaction temperatures compared to conventional heating (Figure 1.2).

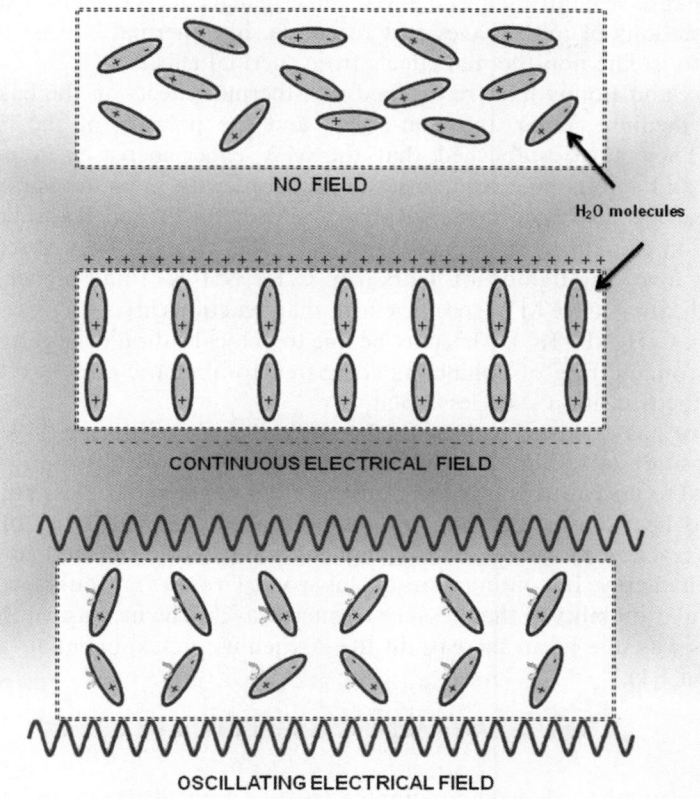

Figure 1.2 Effect of microwaves on the reaction mixture in aqueous medium.

Antonio and Deam recently proposed a new hypothesis based on enhanced diffusion that stated "If the transport of an active species is a rate limiting step in a reaction (such as for diffusion limited reactions), and if microwave enhances the diffusion of that species, then overall reaction rate would change under microwave heating compared with conventional heating".[13] Notably, various organic reactions can be carried out in an aqueous medium using MW irradiation without using any phase-transfer catalyst (PTC). This is because water at higher temperature behaves as a pseudo-organic solvent, as the dielectric constant decreases substantially and an ionic product increases the solvating power towards organic molecules to be similar to that of ethanol or acetone.[14]

1.4.2 Non-thermal Effect

Non-thermal effects have been envisaged to have several origins, including thermodynamic parameters. Molecular shake-up and movement are other factors that have contribute to the MW effect. Such effects can also occur from the interactions of microwaves and reactants, like thermal effects. It is very difficult to isolate non-thermal effects from thermal effects.

Perreux and Loupy have researched non-thermal effects on the basis of the reaction medium (polar and non-polar) and the polarity of the transition state.[15] They have established that the MW effect increases in non-polar solvents and solvent-free reactions and reactions with polar transition states. This effect has also been demonstrated by Polshettiwar and Kaushik using a "tighter ion pair effect" in aminolysis of enolizable esters.[16] They observed that reactions involving tighter ion pairs (*e.g.* $C_6H_{11}NH–K^+$) had higher reaction rates, indicating more MW-specific effect, than reactions involving less tight ion pairs (*e.g.* $C_6H_5NH–K^+$). This may be due to delocalization of negative charge on the aromatic ring of aniline (as compared to aliphatic cyclohexyl amine), making the transition state less polar.

Miklavc has observed a decrease in activation energy during MW-assisted decomposition of sodium bicarbonate.[17] The exposure of substrates (dielectric materials) to microwaves induced rapid rotation of the molecules. This in turn generated heat due to friction and also increased the probability of contact between reactant molecules, thus enhancing the reaction rate by reducing the activation energy. In continuation of this work, Cross *et al.* found an increase in molecular mobility in the presence of microwaves. The increase in the rate of reactions was due to an increase in the Arrhenius pre-exponential factor (A) (Equation 1.1):[18]

$$K = Ae^{-\Delta G/RT}, A = \gamma \lambda^2 \Gamma \qquad (1.1)$$

where, γ = number of neighbor jump sites, λ = jump distance and Γ = jump frequency.

As per Equation (1.1), the Arrhenius pre-exponential factor (A) depends on the frequency of vibration of the atoms at the reaction interface and therefore is affected by MW irradiations.

Kappe and co-workers have performed an in depth study of this MW effect in ring-closing metathesis and Biginelli reactions.[19] They observed no considerable difference between conventional heating and microwave heating. This conclusion is only true for these two reactions and cannot be generalized. The subject of a non-thermal specific MW effect is still divisive and not absolutely known.

In addition to these microwave "thermal effect" and "non-thermal effect",[20] there are some additional advantages of using microwaves for aqueous protocols.

1.4.3 Selectivity towards Water

MW heating depends on the composition and structure of molecules (*i.e.* their dielectric properties) and this property can facilitate selective heating. Microwaves initiate rapid intense heating of polar molecules such as water, while non-polar molecules do not absorb the radiation and in turn do not contribute to heating. Strauss[21] and Hallberg[22] have demonstrated that this selective heating can be exploited to develop a high yield rapid MW protocol using a two-phase (polar–non-polar) solvent system. They determined that the use of water was advantageous in MW chemistry and expedited the protocol with more energy efficiency.

1.4.4 Selectivity towards Catalyst

Selective heating can be exploited in heterogeneous catalysis protocols. This was demonstrated in the MW-assisted rapid molybdenum-catalyzed allylic reaction by Larhed and his co-workers,[23] and in the case of oxidation of alcohol using Magtrieve™ by Bogdal et al.[24] They established that polar catalysts absorbed extra energy and heated at higher temperatures than the overall reaction temperature, making the protocol more energy efficient.

1.4.5 Catalyst as Susceptors

Susceptors are materials that efficiently absorb MW irradiation and transfer the generated thermal energy to molecules in the vicinity that are weak microwave absorbers. Although transmission of the energy occurs through conventional mechanisms, it is more rapid than conventional heating. Kappe[25] and Leadbeater,[26] pioneers in the field of MW chemistry, have used silicon carbide and ionic liquid, respectively, as susceptors, and determined that addition of these materials to the reaction mixture enhanced its overall capacity to absorb MWs and significantly reduced the required MW energy. The addition of these materials as susceptors to the reaction mixture, however, adds to the overall cost of the protocol and makes it expensive. Ideally, if nanomaterials can play a dual role of catalyst and susceptor then all the interesting advantages related to it can be enjoyed without the need for additional material as a susceptor. This can be achieved by using a nano-catalyst, such as a polar ferrite-based material, which can act not only as a catalyst but also as a susceptor.

1.4.6 Stability of Catalyst

Since MW-assisted reactions are rapid, the residence time of nano-catalyst at high temperature is minimal. Catalytic processes with shorter reaction times safeguard the catalyst from deactivation and decomposition, consequently increasing the overall competence of the catalyst, as well as the entire protocol.

1.5 Conclusion

It appears that the approach of fusing the MW technique with water (as a benign reaction medium) can offer an extraordinary synergistic effect with greater potential than these two individual components in isolation. The advantages of aqueous MW chemistry in rapid and greener synthesis of fine chemicals, polymers and nanomaterials as well as in enzymatic and nanocatalysis are illustrated in various chapters of this book.

References

1. P. T. Anastas and J. C. Warner, *Green Chemistry: Theory and Practice*, Oxford University Press, Oxford, 2000.
2. M. Poliakoff and P. Licence, *Nature*, 2007, **450**, 810.
3. (a) V. Polshettiwar and R. S. Varma, *Pure Appl. Chem.*, 2008, **80**, 777; (b) V. Polshettiwar and R. S. Varma, *Acc. Chem. Res.*, 2008, **41**, 629; (c) V. Polshettiwar, M. N. Nadagouda and R. S. Varma, *Aust. J. Chem.*, 2009, **62**, 16.
4. J. H. Clark and S. J. Tavener, *Org. Process Res. Dev.*, 2007, **11**, 149.
5. R. S. Varma, *Green. Chem.*, 1999, **1**, 43.
6. N. V. Plechkova and K. R. Seddon, *Chem. Soc. Rev.*, 2008, **37**, 123.
7. C. J. Li and L. Chen, *Chem. Soc. Rev.*, 2006, **35**, 68.
8. Y. Zhang, B. R. Bakshi and E. Sahledemessie, *Environ. Sci. Tech.*, 2008, **42**, 1724.
9. (a) J. H. Clark, *Green Chem.*, 2006, **8**, 17; (b) C. J. Li and L. Chen, *Chem. Soc. Rev.*, 2006, **35**, 68; (c) V. Polshettiwar and R. S. Varma, *J. Org. Chem.*, 2008, **73**, 7417; (d) V. Polshettiwar and R. S. Varma, *J. Org. Chem.*, 2007, **72**, 7420.
10. (a) V. Polshettiwar and R. S. Varma, *Chem. Soc. Rev.*, 2008, **37**, 1546; (b) D. Dallinger and C. O. Kappe, *Chem. Rev.*, 2007, **107**, 2563; (c) V. Polshettiwar and R. S. Varma, *Alternative Heating for Green Synthesis in Water (Photo, Ultrasound, and Microwave), Handbook of Green Chemistry*, ed. P. T. Anastas and C.-J. Li, Wiley-VCH Verlag GmbH, Weinheim, 2009; (d) R. S. Varma, Clean chemical synthesis in water, *Org. Chem. Highlights* 2007, http://www.organic-chemistry.org/Highlights/2007/01February.shtm.
11. (a) A. Loupy, G. Bram and J. Sansoulet, *New J. Chem.*, 1992, **16**, 233; (b) A. K. Bose, M. J. Manhas, B. K. Banik and E. W. Robb, *Res. Chem. Intermed.*, 1994, **20**, 1; (c) C. R. Strauss and R. W. Trainor, *Aust. J. Chem.*, 1995, **48**, 1665; (d) S. Caddick, *Tetrahedron*, 1995, **38**, 10403; (e) C. Gabriel, S. Gabriel, E. H. Grant, B. S. J. Halstead and D. M. P. Mingos, *Chem. Soc. Rev.*, 1998, **27**, 213; (f) P. Lidstrom, J. Tierney, B. Wathey and J. Westman, *Tetrahedron*, 2001, **57**, 9225; (g) M. Larhed, C. Moberg and A. Hallberg, *Acc. Chem. Res.*, 2002, **35**, 717; (h) C. O. Kappe, *Angew. Chem. Int. Ed.*, 2004, **43**, 6250; (i) A. Loupy, ed., *Microwaves in Organic Synthesis*, 2nd ed.; Wiley-VCH: Weinheim, 2006.
12. (a) P. T. Anastas, *ChemSusChem*, 2009, **2**, 391; (b) V. Polshettiwar and R. S. Varma, *Environmentally benign chemical synthesis via mechanochemical mixing and microwave irradiation, inEco-Friendly Processes and Procedures for the Synthesis of Fine Chemicals*, ed. R. Ballini, RSC Green

Chemistry Series, Royal Society of Chemistry, Cambridge, ch. 8, 2009, pp. 275–292.
13. C. Antonio and R. T. Deam, *Phys. Chem. Chem. Phys.*, 2007, **9**, 2976.
14. (a) C. R. Strauss and R. W. Trainor, *Aust. J. Chem.*, 1995, **48**, 1665; (b) C. R. Strauss and R. S. Varma, *Top. Curr. Chem.*, 2006, **266**, 199.
15. (a) L. Perreux and A. Loupy, *Tetrahedron*, 2001, **57**, 9199; (b) A. Loupy and R. S. Varma, *Chim. Oggi*, 2006, **24**, 36–40.
16. V. Polshettiwar and M. P. Kaushik, *Indian J. Chem. B*, 2005, **44**, 773.
17. A. Miklavc, *ChemPhysChem*, 2001, 552.
18. J. G. P. Binner, N. A. Hassine and T. E. Cross, *J. Mater. Sci.*, 1995, **30**, 5389.
19. (a) S. Garbacia, B. Desai, O. Lavaster and C. O. Kappe, *J. Org. Chem.*, 2003, **68**, 9136; (b) A. Stadler and C. O. Kappe, *J. Chem. Soc., Perkin Trans. 2*, 2000, **2**, 1363.
20. A. Hoz, A. Diaz-Ortiz and A. Moreno, *Chem. Soc. Rev.*, 2005, **34**, 164.
21. K. D. Raner, C. R. Strauss and R. W. Trainor, *J. Org. Chem.*, 1995, **60**, 2456.
22. P. Nilsson, M. Larhed and A. Hallberg, *J. Am. Chem. Soc.*, 2001, **123**, 8217.
23. N. F. K. Kaiser, U. Bremberg, M. Larhed, C. Moberg and A. Hallberg, *Angew. Chem., Int. Ed.*, 2000, **39**, 3595.
24. D. Bogdal, M. Lukasiewicz, J. Pielichowski, A. Miciak and Sz. Bednarz, *Tetrahedron*, 2003, **59**, 649.
25. T. Razzaq, J. M. Kremsner and C. O. Kappe, *J. Org. Chem.*, 2008, **73**, 6321.
26. N. E. Leadbeater and H. M. Torrenius, *J. Org. Chem.*, 2002, **67**, 3145.

CHAPTER 2
Metal-catalyzed Reactions in Water under MW Irradiation

VICTORIO CADIERNO, PASCALE CROCHET AND
SERGIO E. GARCÍA-GARRIDO

Departamento de Química Orgánica e Inorgánica, Instituto Universitario de Química Organometálica "Enrique Moles" (Unidad Asociada al CSIC), Facultad de Química, Universidad de Oviedo, Julián Clavería 8, 33006, Oviedo, Spain

2.1 Introduction

Over the past two decades, increasing environmental concerns have triggered the development of new synthetic protocols that minimize the generation of chemical wastes.[1] In this context, remarkable research endeavors have focused on the replacement of traditional organic solvents, which are generally toxic, flammable and non-renewable, by water.[2] Besides its inherent advantages (harmless, non-flammable, abundant, renewable, inexpensive), fulfilling the requirements of the Green Chemistry principles,[1] the use of water as solvent can also provide a notable difference in reactivity. In this sense, its high polarity combined with the hydrophobic effects enables water to enhance reaction rates and selectivities of some organic processes.[3] Therefore, it is not surprising that in recent years a plethora of studies devoted to the development of metal-catalyzed organic transformations in pure water or aqueous biphasic media have been reported,[4] with some of them disclosing highly efficient eco-friendly synthetic protocols.

RSC Green Chemistry No. 7
Aqueous Microwave Assisted Chemistry: Synthesis and Catalysis
Edited by Vivek Polshettiwar and Rajender S. Varma
© Royal Society of Chemistry 2010
Published by the Royal Society of Chemistry, www.rsc.org

Although generally neglected by chemists until recently, another important aspect of reducing the environmental impact of a synthetic procedure is the optimization of energy consumption.[1] In this context, the use of microwave (MW) irradiations as a source of heat represents a good alternative to conventional heating sources.[5] Effectively, the comparison of energetic costs required for a conventional oil-bath heated reaction and a microwave-assisted process evidences that the latter could provide significant energy savings (up to 85-fold), at least on the laboratory scale.[6] The effective energy transfer to the system results in an extremely rapid heating, thus shortening considerably the reaction times. The use of MW techniques not only enables high-speed syntheses but also provides a homogeneous temperature throughout the sample, limiting therefore the occurrence of side-reactions and decomposition processes. Consequently, higher selectivities and yields, and a much better reproducibility, can be attained with MW technology, especially with the modern equipment, which allows the fine regulation of temperature and pressure. For all these reasons, MW-assisted chemistry has attracted increasing interest, finding application in many organic transformations and metal-catalyzed reactions.[7] In particular, in the search of "greener" synthetic protocols, a series of studies that combine the benefits of MW-technology and the use of water as solvent has appeared in recent years.[8]

The present chapter provides an overview of developments in the field of MW-assisted metal-catalyzed reactions in aqueous media. Palladium-catalyzed carbon–carbon coupling processes are not considered since they are highlighted in Chapter 3. Literature published up to November 2009 is covered.

2.2 Addition Reactions

The hydration of nitriles to access amides is a process of great significance since the latter constitute versatile synthetic intermediates in organic chemistry with a broad range of industrial applications.[9] Although this process has been extensively studied,[10] a series of difficulties remains unaddressed: (i) traditional methodologies, based on the use of a large amount of acid or base, require a final neutralization step that leads to an extensive salt formation with inconvenient product contamination and pollution effects; (ii) the use of acidic or basic media also favors the undesirable over-hydrolysis of the nitriles into the corresponding carboxylic acids, lowering considerably the amide yields; and (iii) although transition metal catalysts could circumvent these problems, only a very limited number of catalytic systems are effective in pure water, with the use of ecologically hazardous organic solvents being usually required. Recently, efforts to make this hydration process greener have led to the development of methodologies operative in water under microwave heating. In this sense, the selective hydration of both aromatic and aliphatic nitriles into the corresponding amides has been achieved, in water under neutral conditions, using catalytic amounts (5 mol.%) of the water-soluble arene-ruthenium(II) catalyst [$RuCl_2$(η^6-C_6Me_6)(PTA-Bn)] (**1**) (Scheme 2.1).[11] Almost quantitative conversions were

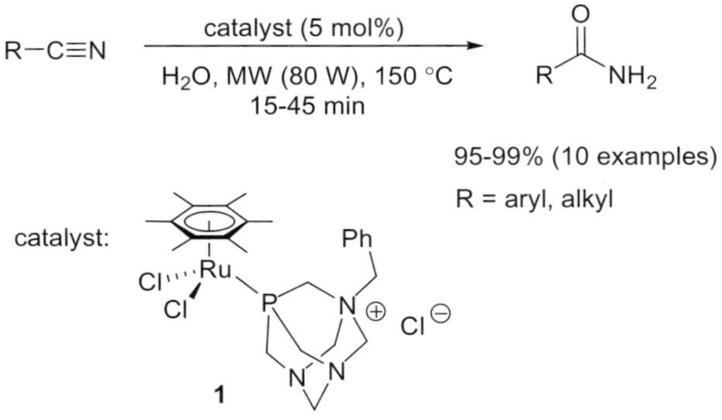

$$R-C\equiv N \xrightarrow[\text{15-45 min}]{\text{catalyst (5 mol\%)}\atop\text{H}_2\text{O, MW (80 W), 150 °C}} R-\underset{\text{O}}{\overset{\|}{C}}-NH_2$$

95-99% (10 examples)
R = aryl, alkyl

catalyst: [Ru complex 1 with Cp*, two Cl ligands, and PTA-CH$_2$Ph⁺ Cl⁻ ligand]

Scheme 2.1

in all cases observed within 45 min of MW heating at 150 °C. Under this protocol both electron-rich and electron-poor benzonitriles could be satisfactorily transformed, with the substitution pattern of the aromatic ring not affecting the reactivity of the substrates. Notably, (i) complex **1** represents one of the most efficient catalysts reported to date for this transformation in aqueous media, and (ii) the activity of **1** under MW heating largely surpasses that observed under conventional heating.

The heterogeneous nanoferrite-supported ruthenium hydroxide **2** has also proved to be a selective catalyst for this transformation, using pure water as solvent and MW heating (Scheme 2.2).[12] In particular, this nano-Ru(OH)$_x$ catalyst showed remarkable activity for the hydration of a range of activated and inactivated aromatic and heteroaromatic nitriles. The rates of the reactions were barely influenced by the nature or position of the substituents on the aromatic rings, and the protocol showed excellent chemoselectivity and tolerance to functional groups, being compatible with the presence of halide, alkoxy, amino, nitro and ether functionalities in the substrates. Only the hydration of the isonicotinonitrile N-oxide did not follow the desired reaction path – isonicotinoamide instead of isonicotinoamide N-oxide being formed as the result of the concomitant N→O reduction. Interestingly, after completion of reaction, the magnetic nanoferrite **2** could be separated easily from the reaction mixture with the help of an external magnet and recycled at least three times without any significant change in the catalytic activity. After separation of the catalyst, the aqueous solution was cooled and crystals of the amides with acceptable purity were precipitated. Thus, the use of organic solvents was not required during the reaction or in the work-up steps. It is also notable that, under the same reaction conditions, **2** was also able to promote the hydrolysis of oximes into the corresponding carbonyl compounds (82–85% yield).[12]

Recently, Nolan and co-workers have reported the first example of a gold-based catalytic system for nitrile hydration.[13] The reactions were performed using the cationic gold(I) N-heterocyclic carbene complex [(IPr)Au(NTf$_2$)] as catalyst in a water–THF mixture under microwave irradiation (Scheme 2.3). A

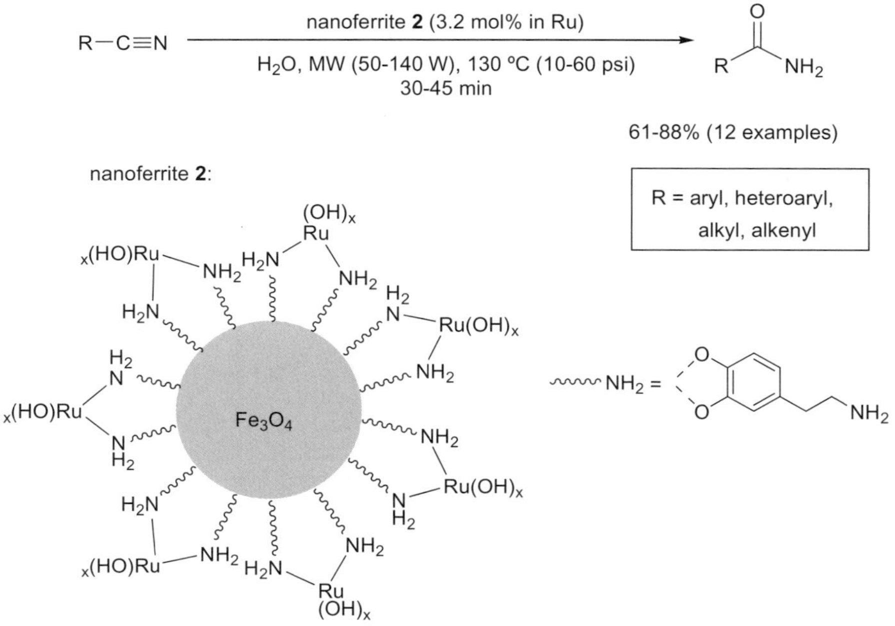

Scheme 2.2

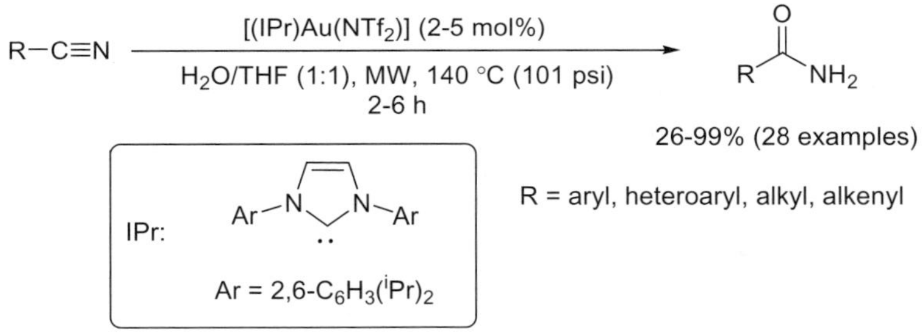

Reaction: R–C≡N → R-C(=O)-NH$_2$
Conditions: nanoferrite **2** (3.2 mol% in Ru), H$_2$O, MW (50–140 W), 130 °C (10–60 psi), 30–45 min
Yield: 61–88% (12 examples)
R = aryl, heteroaryl, alkyl, alkenyl

nanoferrite **2**: Fe$_3$O$_4$ nanoparticle functionalized with H$_2$N–NH$_2$ linkers bound to Ru(OH)$_x$ centers.

$\sim\!\!\sim\!\!\sim\,$NH$_2$ = (3,4-methylenedioxyphenyl)ethylamine linker

Scheme 2.3

Reaction: R–C≡N → R-C(=O)-NH$_2$
Conditions: [(IPr)Au(NTf$_2$)] (2–5 mol%), H$_2$O/THF (1:1), MW, 140 °C (101 psi), 2–6 h
Yield: 26–99% (28 examples)
R = aryl, heteroaryl, alkyl, alkenyl

IPr: Ar–N⌒N–Ar (imidazol-2-ylidene); Ar = 2,6-C$_6$H$_3$(^{i}Pr)$_2$

broad spectrum of nitriles, including aromatic, heteroaromatic and aliphatic examples, were efficiently converted into the corresponding amides (60–99% yield), confirming the wide scope of the process. For benzonitriles, the hydration was clearly favored by the presence of electron-donating groups on the aromatic ring; electron-withdrawing groups gave rise to a significant decrease in efficiency. Although the process was also operative with sterically hindered benzonitriles bearing substituents in the *ortho*-position, in these cases the yields were only modest (26–55%).

Interestingly, when this methodology was applied to dicyano derivatives, selective mono- or double-hydration processes were observed depending on the

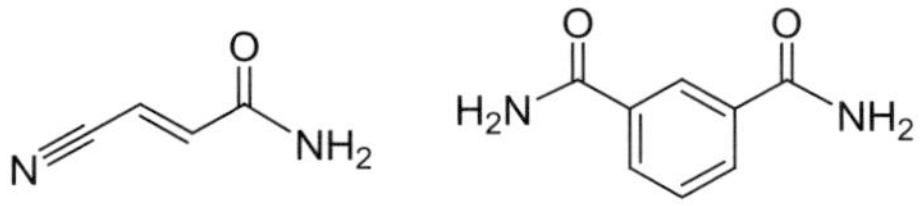

Figure 2.1

Scheme 2.4

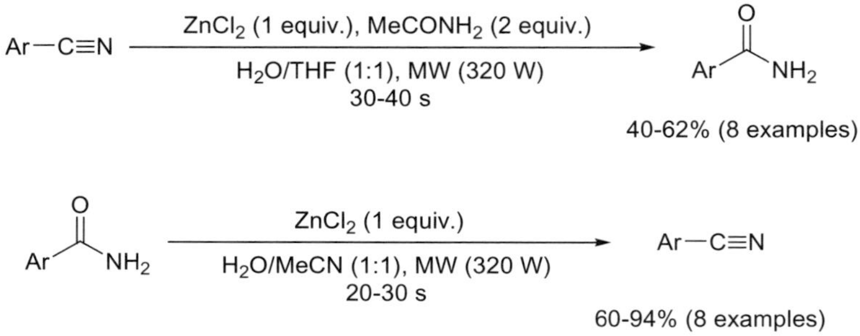

Ar—C≡N → [ZnCl$_2$ (1 equiv.), MeCONH$_2$ (2 equiv.), H$_2$O/THF (1:1), MW (320 W), 30-40 s] → Ar–C(O)–NH$_2$

40-62% (8 examples)

Ar–C(O)–NH$_2$ → [ZnCl$_2$ (1 equiv.), H$_2$O/MeCN (1:1), MW (320 W), 20-30 s] → Ar—C≡N

60-94% (8 examples)

aromatic or aliphatic nature of the substrate. Thus, while the hydration of fumaronitrile only afforded the monohydrated product **3**, the aromatic bis-amide **4** was selectively obtained starting from 1,3-dicyanobenzene (Figure 2.1).[13]

The MW-assisted conversion of different *para-* and *meta-*substituted benzonitriles into the corresponding amides has also been achieved using a stoichiometric amount of ZnCl$_2$ (Scheme 2.4).[14] The reactions were performed with a domestic microwave oven in a 1:1 THF–H$_2$O mixture and in the presence of a two-fold excess of acetamide. Under these conditions, moderate to good yields were reached in extremely short reaction times (30–40 s). Oddly enough the reverse process, *i.e.* the ZnCl$_2$-promoted dehydration of amides into nitriles, could also be developed in aqueous medium (Scheme 2.4).[14] The feasibility of both processes seems to be controlled by the additives employed – acetamide (hydration) or acetonitrile (dehydration) – but unfortunately their role was not specified by the authors.

Hydration of alkynes is a useful method of preparing carbonyl compounds. Such a process, extensively studied in organic media under conventional heating, is usually catalyzed by Brønsted acids[15] or Lewis acid metal-complexes.[16] Studies focused on the use of microwave irradiation in superheated water have evidenced that alkynes can also be hydrated without the aid of acid catalysts.[17] Effectively, water becomes more acidic as the temperature increases to the near-critical conditions (200–300 °C) and can therefore promote the reaction by itself. Nevertheless, in some cases the presence of a metal catalyst is still necessary to reach high conversions. In this context, an AuBr$_3$ – promoted hydration of functionalized phenylacetylenes has been developed in superheated water (Scheme 2.5).[18] Ketones, resulting from the Markovnikov addition of the water molecule to the terminal C≡C bond, were selectively obtained in moderate to good yields. In general, the presence of electron-donating substituents on the phenyl ring led to almost quantitative conversions, while electron-withdrawing ones gave rise to a dramatic decrease in catalytic efficiency. Unfortunately, the scope of the process is limited to aromatic substrates since aliphatic alkynes decompose at such elevated temperatures.

Transition-metal catalyzed addition of $R_2P(O)$–H bonds to alkynes is a powerful synthetic route to generate vinylphosphine oxides.[16a,19] Nevertheless, although efficient and selective protocols have been described for both aromatic and aliphatic acetylenes, the use of propargylic alcohols as substrates remains a challenge since competitive formation of a wide range of by-products (propargyl-, allenyl- or dienyl-phosphine oxides) is usually observed.[20,21] In this context, Stockland and co-workers have described the selective synthesis of some vinylphosphine oxides derived from ethynyl steroids, using aqueous MW conditions and the rhodium(III) complex [RhMe$_3$(PPhMe$_2$)$_3$] as catalyst (Scheme 2.6).[21] The desired hydrophosphinylation products were obtained in moderate to good yields after only 15 min of irradiation. The transformations were not only selective towards the formation of the vinylphosphine oxides but also stereoselective, with the thermodynamically more stable (E)-isomers being

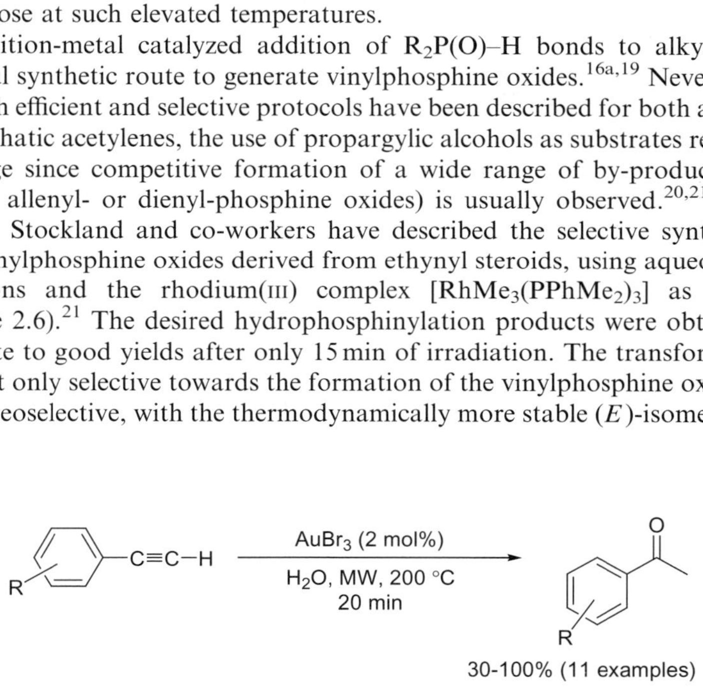

Scheme 2.5

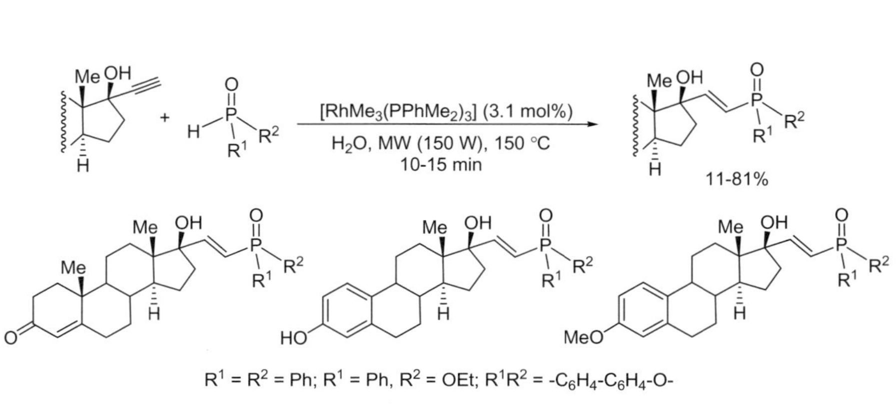

Scheme 2.6

formed exclusively. However, notably, the use of *P*-stereogenic phosphine oxides (*i.e.* $R^1 \neq R^2$) led to non-separable mixtures of diastereoisomers. Remarkably, all these addition reactions were insensitive to air and so the use of inert atmospheres was not required.

Propargylation of carbonyl compounds constitutes an important transformation in organic synthesis since the resulting homopropargylic alcohols are versatile building blocks and relevant structural units in various biologically active compounds.[22] In this context, Barbier-type propargylation processes, consisting of the coupling between aldehydes and propargyl halides in the presence of aluminium-, zinc-, indium- or tin-based promoters, have emerged as useful synthetic tools.[23] However, the long reaction times usually required to achieve full conversions and the contamination of the desired products with the corresponding allenyl isomers, especially in water media, have limited their applications. More recently, rapid and selective microwave-assisted Barbier-type processes have been described by Li and co-workers.[24] The best performances were achieved using a Sn–In–SnCl$_2$ mixture as catalyst in a biphasic THF–water medium, in the presence of the phase-transfer reagent [NMe$_3$Ph][Br] (Scheme 2.7). The reactions, realized in a domestic microwave oven with a power setting of 160 W, afforded selectively the corresponding homopropargylic alcohols in good yields (usually >85%). The combined use of tin and indium metals played a crucial role in the selectivity of the process; transformations mediated exclusively by Sn gave rise to a mixture of the desired homopropargylic alcohols along with their allenyl tautomers. However, sterically demanding benzaldehydes bearing substituents in *ortho*-position, such as 2,6-dimethoxybenzaldehyde or 2-chlorobenzaldehyde, were also found to favor the formation of allenic side-products, reducing considerably the yields of the corresponding homopropargylic alcohols.

One-pot, three-component coupling of aldehydes, alkynes and amines (A^3 coupling) represents a convenient synthetic route to propargylic amines.[25] In this context, an efficient microwave-assisted protocol in aqueous medium has been reported by Tu and co-workers using inexpensive copper(I) iodide as catalyst (Scheme 2.8).[26] The reactions were performed by successive irradiations of 1 min in a domestic microwave oven. This methodology was found to be operative with aromatic aldehydes bearing electron-withdrawing as well as electron-donating groups, albeit conversions of the latter were generally slower. Interestingly, aliphatic aldehydes also displayed a high reactivity; the usual

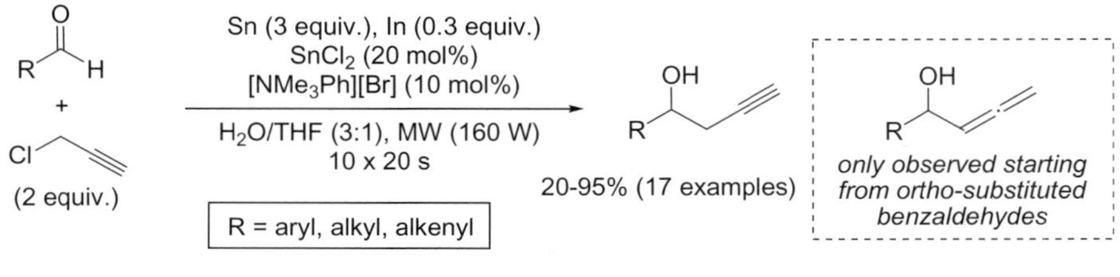

Scheme 2.7

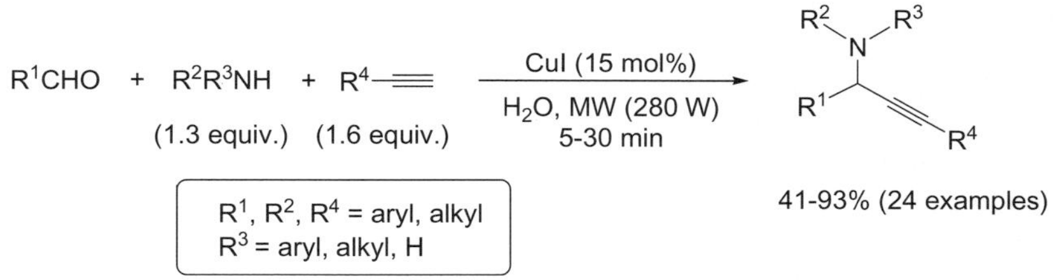

Scheme 2.8

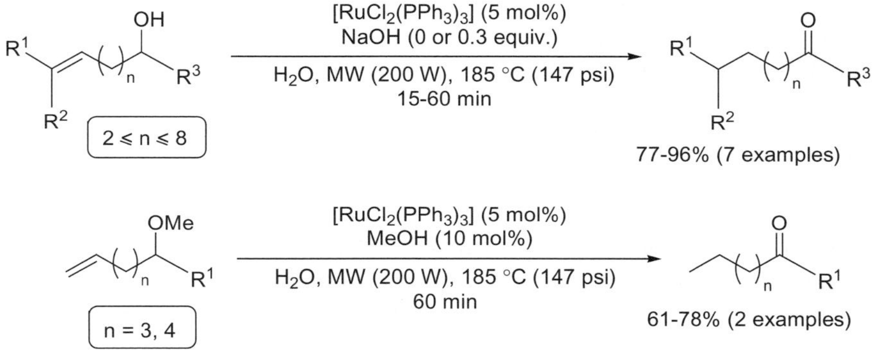

R^1CHO + R^2R^3NH + $R^4{-}{\equiv}{-}$ 　(1.3 equiv.) 　(1.6 equiv.)

CuI (15 mol%) / H_2O, MW (280 W), 5–30 min

→ R^1–C(NR^2R^3)–C≡C–R^4

R^1, R^2, R^4 = aryl, alkyl
R^3 = aryl, alkyl, H

41–93% (24 examples)

Scheme 2.9

Allylic/homoallylic alcohol substrate ($2 \leq n \leq 8$) with OH, R^1, R^2, R^3 substituents

[RuCl$_2$(PPh$_3$)$_3$] (5 mol%), NaOH (0 or 0.3 equiv.)
H_2O, MW (200 W), 185 °C (147 psi), 15–60 min

→ saturated ketone product

77–96% (7 examples)

Methyl ether substrate with OMe, R^1 (n = 3, 4)

[RuCl$_2$(PPh$_3$)$_3$] (5 mol%), MeOH (10 mol%)
H_2O, MW (200 W), 185 °C (147 psi), 60 min

→ saturated ketone product

61–78% (2 examples)

---

competitive trimerization[27] of such substrates did not occur under these aqueous-MW conditions. In addition, both primary and secondary amines could be successfully applied in this A³ coupling, and the use of chiral ones, such as (S)-proline methyl ester, (S)-α-methylbenzylamine or (S)-N-benzyl-1-phenylethylamine, also allowed control to some extent of the configuration of the new stereogenic carbon centre formed, leading to diastereomeric ratios ranging from 67:33 to 95:5.

2.3 Isomerization Reactions

In contrast to allylic alcohols, which can be isomerized efficiently into the corresponding carbonyl compounds by a great variety of metal catalysts under mild conditions,[28] unsaturated alcohols with more than one sp³ carbon between the alkene and hydroxyl groups are much less reactive.[29] The use of microwave heating has shown a beneficial effect in the isomerization of such substrates, accelerating significantly these processes. Thus, in the presence of catalytic amounts of the ruthenium(II) complex [RuCl$_2$(PPh$_3$)$_3$], different cyclic and acyclic alkenols have been converted into the corresponding saturated ketones in good yields through an aqueous MW-assisted protocol (Scheme 2.9).[30,31] The process takes place through successive C=C bond migrations along the hydrocarbon chain and subsequent tautomerization of the resulting final enol. Under the aqueous MW conditions employed, the reaction medium became slightly acidic since the ruthenium catalyst hydrolyses to form hydrochloric acid. Therefore, addition of 0.3 equivalents of NaOH was required to extend the methodology to acid-sensitive substrates such as 1-cyclopropylhex-3-en-1-ol or 1-(cyclohex-3-enyl)-3-phenylpropan-1-ol. Furthermore, a slightly modified protocol using 10 mol.% of methanol as additive allowed also the formation of saturated ketones starting from unsaturated ethers, through a tandem isomerization–hydrolysis process (Scheme 2.9).

In addition to the above-mentioned C=C bond migrations, homoallylic alcohols are also prone to isomerize into allylic alcohols *via* a metal-catalyzed transposition of the hydroxyl unit.[32] As an example, using a mesoporous silica-supported titanium-ruthenium catalyst (abbreviated as Ti-Ru-SBA-15), 1-phenyl-3-buten-1-ol could be predominantly converted into 4-phenyl-3-buten-2-ol in water under MW heating (Scheme 2.10).[33] Several reaction conditions were tested to minimize the formation of the saturated ketone 1-phenylbutan-1-one, with the highest selectivity in the allylic alcohol (97%) being reached at 100 °C with a catalyst charge of 5.6 mol.% in Ru. The use of lower metal loadings gave rise to larger proportions of the ketone. Similarly, reactions performed under solvent-free conditions also resulted in lower selectivities.

Using the same catalytic system a tandem allylation–isomerization process has also been developed.[33] Thus, as shown in Scheme 2.11, 4-phenyl-3-buten-2-ol could be efficiently generated in a one-pot manner starting directly from benzaldehyde and tetraallyltin. Under these conditions, the *in situ* generated homoallylic alcohol 1-phenyl-3-buten-1-ol evolved rapidly into the desired

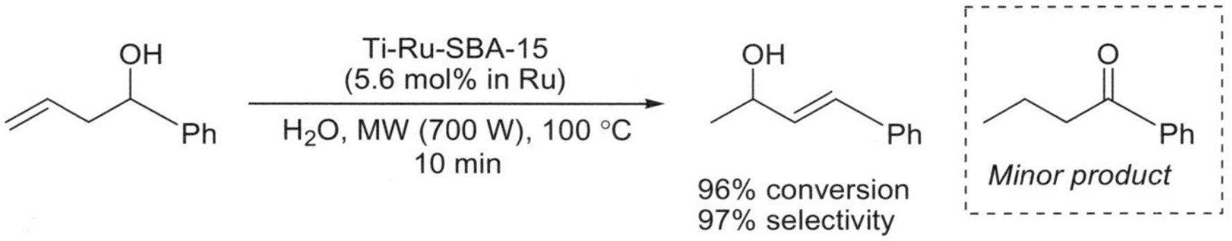

Scheme 2.10

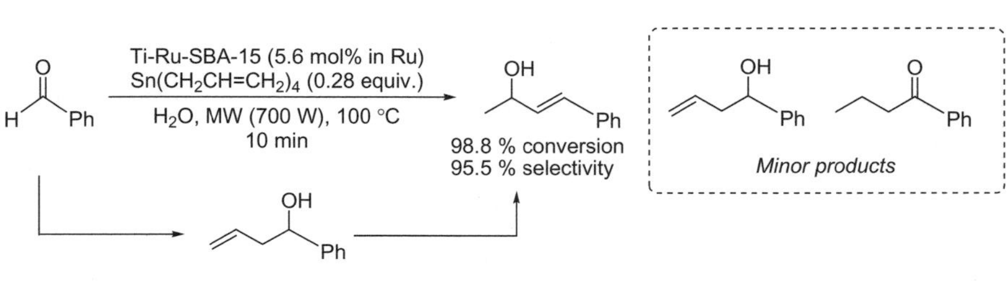

Scheme 2.11

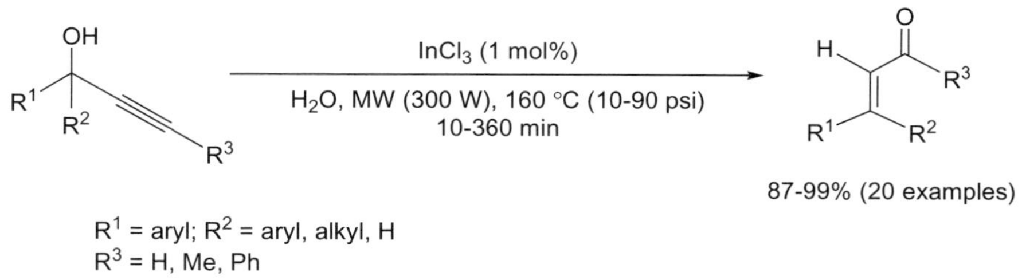

Scheme 2.12

allylic alcohol, with only trace amounts of the by-products 1-phenyl-3-buten-1-ol and 1-phenylbutan-1-one observed in the crude reaction mixtures. Moreover, taking advantage of its insolubility, the catalyst could be recycled by simple filtration and reused at least five consecutive times without any significant changes in activity and selectivity.

The Meyer–Schuster rearrangement of propargylic alcohols is an appealing and atom-economic route for the preparation of synthetically useful α,β-unsaturated carbonyl compounds. Such a process is traditionally promoted by strong Brønsted acids under harsh reactions conditions, which often give rise to non-regioselective transformations.[34] However, in recent years, these limitations have been overcome by the aid of transition-metal complexes and Lewis-acid catalysts.[35] In this context, some of us have recently developed an extremely efficient MW-assisted protocol for the selective Meyer–Schuster isomerization of terminal propargylic aryl carbinols ($R^3 = H$) in a pure aqueous medium using inexpensive $InCl_3$ (1 mol.%) as catalyst (Scheme 2.12).[36] As a general trend, alkynols with electron-withdrawing groups on the aryl substituents were less reactive than those with electron-rich substrates, requiring longer reaction times to reach satisfactory conversions. Remarkably, starting from secondary alcohols ($R^2 = H$) the corresponding enals were formed with complete (E)-stereoselectivity. It is also notable that the scope of the reaction is not restricted to substrates bearing terminal C≡C bonds; internal propargylic alcohols ($R^3 = Me, Ph$) are also efficiently transformed into the corresponding enones.

2.4 Substitution Reactions

Deuterium-labeled compounds have found widespread applications in mechanistic investigations as well as internal standards for mass spectrometry analyses.[37] One of the most employed strategies to access such compounds is based on H/D-exchange reactions using D_2O as the deuterium source. These processes, traditionally performed under harsh conditions (*i.e.* requiring excess of acid or base, very high temperatures and pressures and long reaction times), have been greatly improved in recent years by the aid of MW technology, which leads to faster and cleaner protocols, especially for functionalized substrates.[38] For example, primary alcohols were readily (30 min) and selectively deuterated

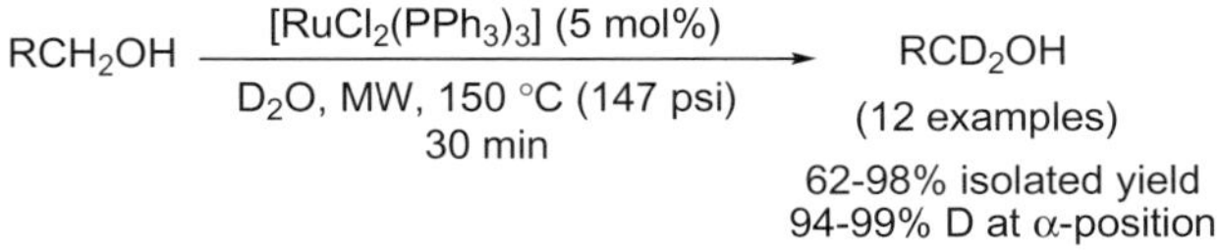

$$\text{RCH}_2\text{OH} \xrightarrow[\substack{\text{D}_2\text{O, MW, 150 °C (147 psi)} \\ \text{30 min}}]{[\text{RuCl}_2(\text{PPh}_3)_3]\ (5\ \text{mol\%})} \text{RCD}_2\text{OH}$$

(12 examples)
62-98% isolated yield
94-99% D at α-position

Scheme 2.13

at the α-position when D_2O solutions were irradiated at 150 °C in the presence of catalytic amounts of [RuCl$_2$(PPh$_3$)$_3$] (Scheme 2.13).[39] High degrees of deuteration (>94% D) were reached with both aromatic and aliphatic substrates. As proposed by the authors, the process proceeds through the initial dehydrogenation of the alcohol and subsequent deuteration of the resulting aldehyde. Interestingly, this protocol was successfully applied to optically active alcohols bearing a stereogenic centre at the β-position, but a lower temperature (100 °C, 22 psi) and a longer reaction time (60 min) were necessary in these cases to avoid racemization. Acid-sensitive monoprotected-diols could also be deuterated efficiently at 100 °C, with the reactions now requiring of the presence of one equivalent of NaOD. Finally, notably, similar H/D-exchanges have also been described starting from primary and secondary amines, with the initial formation of an imine intermediate being postulated.[39]

Following an analogous protocol, the perdeuterated cyclohexene-d_{10} could be prepared (80–96% yield) by heating unlabeled cyclohexene in D_2O at 140 °C under microwave irradiation.[40] Here, the observed H/D-exchanges were explained on the basis of successive Ru-catalyzed C=C bond migrations along the carbocycle, via a repetition of hydroruthenation and β-elimination processes. The best catalytic performances were reached when the catalyst [RuCl$_2$(PPh$_3$)$_3$] was associated to 10 mol.% of the surfactant sodium dodecyl sulfate (SDS) and 20 mol.% of EtOH as co-solvent. Under these conditions, almost quantitative incorporation of deuterium (96% D) was observed. The use of a higher reaction temperature (185 °C) and/or the absence of the SDS and EtOH additives gave the desired deuterated product along with small amounts of benzene (5–15%), generated through a competitive dehydrogenation pathway. Larger cycloalkenes could also be perdeuterated using this protocol (deuterium contents ranging from 79 to 94% depending on the position; see Scheme 2.14).

Isotopic exchanges in carbohydrates have been described using a digested and pre-sonicated Raney nickel alloy catalyst in a THF–D_2O mixture (10 : 1).[41] Reactions were performed through successive short irradiations (15 s intervals) in a domestic microwave oven. As shown in Scheme 2.15, under these conditions the monosaccharide 1-*O*-methyl-β-D-galactopyranoside incorporated deuterium at C_2, C_3 and C_4, and the disaccharide sucrose at C_2, C_3, C_4, $C_{3'}$ and $C_{4'}$. Cooling in an ice bath after each irradiation was required to avoid epimerization of the products.

The H/D-exchange processes are not restricted to aliphatic compounds as several metallic precursors can also promote the isotopic labeling of aromatic

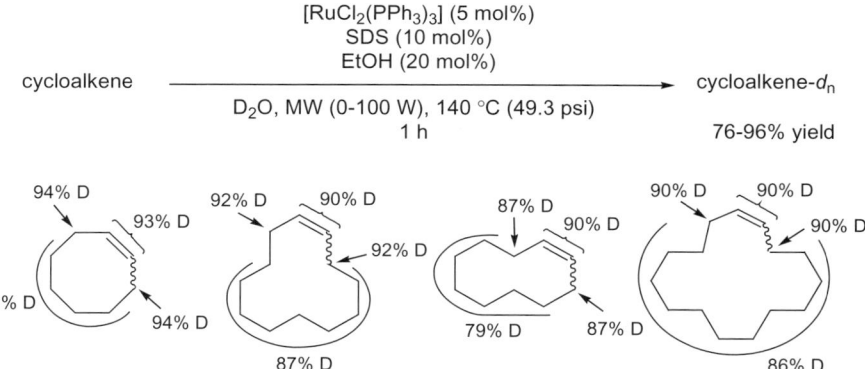

Scheme 2.14

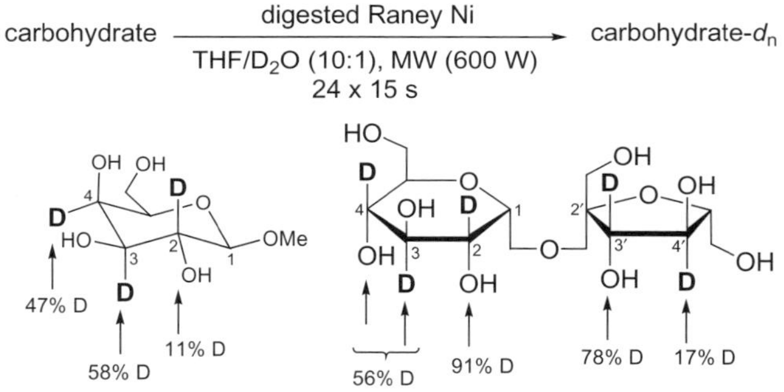

Scheme 2.15

derivatives. In this context, deuterated arylsilanes, which represent important intermediates for the preparation of more elaborated deuterium-containing aromatic compounds, have been synthesized through MW-assisted metal-catalyzed H/D-exchange procedures in deuterium oxide. Thus, MW-heating of Ph_3SiOH and $PhMe_2SiOSiMe_2Ph$ with 5 mol.% of PtO_2 in D_2O gave rise to the corresponding D-labeled compounds in good yields (Scheme 2.16).[42] Full incorporation of deuterium took place at the *meta-* and *para-*positions, while the more crowded *ortho-*positions remained untouched or only slightly deuterated (15% of D).

Platinum-promoted isotopic exchanges have also been described in several other substituted-benzenes (Scheme 2.17).[43] Both homogeneous Na_2PtCl_4 and heterogeneous pre-reduced PtO_2 catalysts were active in these processes, the former usually requiring lower metal loadings (0.3–1.4 *vs* 0.7–27 mol.%). The deuterium distribution at the different positions of the phenyl ring was found to

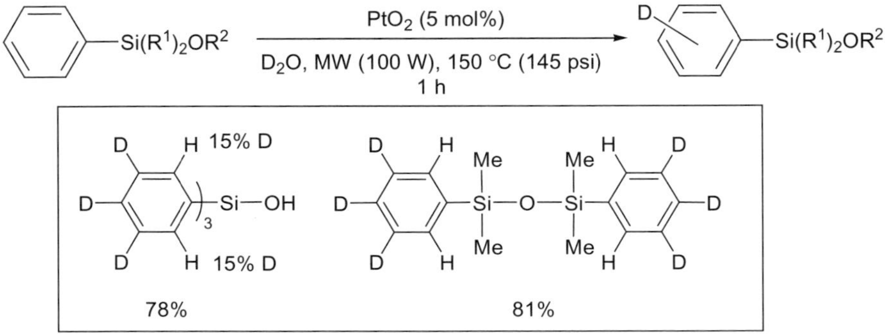

Scheme 2.16

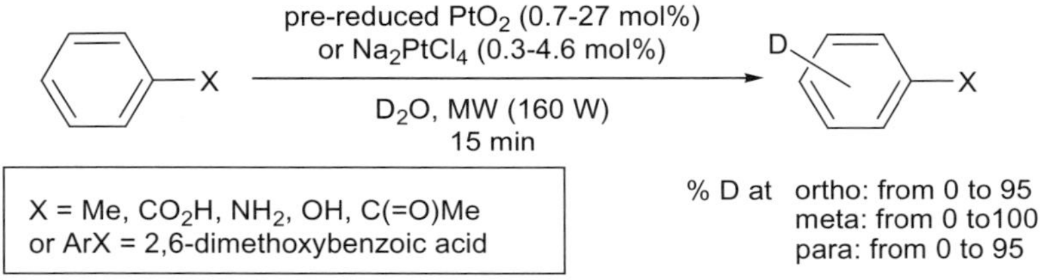

Scheme 2.17

be strongly dependent on the nature of the substituent and the metallic precursor involved. As an example, 2,6-dimethoxybenzoic acid underwent the H/D-exchange exclusively at the *meta*-carbons using Na_2PtCl_4, while deuteration occurred predominantly at the *para*-position with PtO_2 (95% D in *para* and 5% D in *meta*). In contrast, isotopic exchange in aniline (X = NH_2) took place mainly at the *ortho*-carbons regardless of the catalyst used. As a general trend, except for benzoic acids, the selectivity was in all cases moderate to low.

Aromatic heterocyclic compounds, such as indole and *N*-methylindole, have also been subject to H/D-exchange processes (Scheme 2.18).[44] Reactions were performed in a domestic microwave oven through intermittent irradiations using deuterated Raney nickel prepared, before use, from classical Raney Ni by treatment with NaOD and D_2O. Remarkably, these H/D-exchange reactions were 500-fold faster than those performed under conventional heating (at 40 °C); the deuterium distribution also differed significantly,[45] a discrepancy ascribed to the lack of temperature control associated with the use of a domestic oven.

Derdau and co-workers have reported a more recent and extensive study on the isotopic labeling of a wide range of aromatic and heteroaromatic derivatives.[46] Reactions were performed at 150 °C in D_2O with $RhCl_3$ or Pd/C catalysts activated by $NaBD_4$ (Scheme 2.19). Remarkably, MW-heating led to

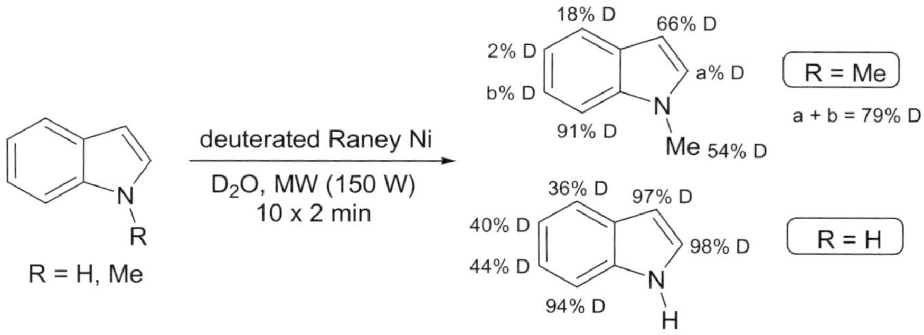

Scheme 2.18

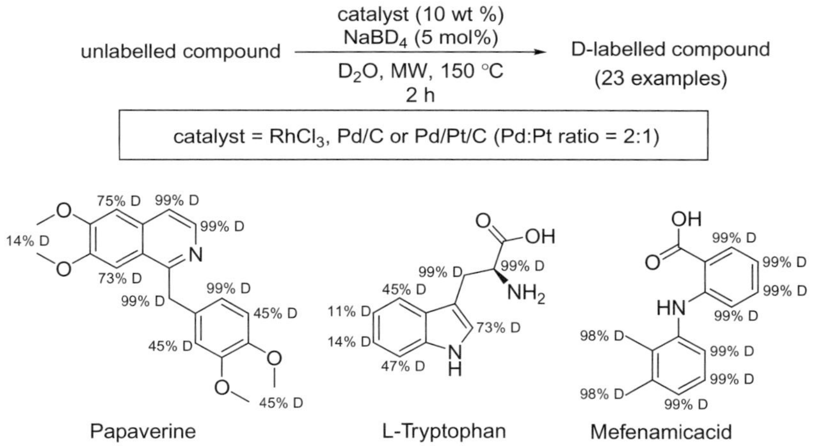

Scheme 2.19

significantly higher deuterium uptake than conventional thermal conditions. In general, the palladium precursor was more effective and versatile than the rhodium one, promoting the H/D-exchanges not only in aromatic but also in aliphatic carbons of the substrates. Involvement of different *para*-substituted anilines evidenced that the amino group oriented the substitution preferably towards *ortho*-positions. For pyridines, all the positions of the ring were affected and high incorporation of deuterium was generally observed. In contrast, electron-poor substrates presented lower reactivity and satisfactory results were only obtained when mixing Pd/C with Pt/C (ratio 2:1) as a result of their well-known synergic effect.[47] This protocol has been successfully applied in the D-labeling of pharmaceutically active compounds such as papaverine, L-tryptophan and mefenamic acid (Scheme 2.19).

The development of effective hydrodechlorination processes of highly toxic substances is an important research field due to their practical interest in the remediation of soil and water purification.[48] In this context, microwave-assisted reductive degradation of mono-, di- and penta-chlorobenzene, chosen as models for polychlorinated aromatic pollutants, has been explored in alkaline aqueous media using sodium hypophosphite as reductant and Pd/C as catalyst (temperatures between 70 and 180 °C).[49] Under these conditions, the haloaromatic compounds underwent Cl/H-exchanges leading to benzene formation in short times (15 min).

Copper-mediated arylation of amines (Ullmann-type reactions) usually require extremely high temperatures (≈ 200 °C), long reaction times and stoichiometric amounts of copper reagents to generate the desired arylated amines in only moderate yields. All these facts limit the synthetic applicability of these reactions to activated aryl halides containing electron-withdrawing groups. However, recent advances have considerably extended the scope of these processes.[50] In particular, addition of ligands able to increase the catalyst solubility and stability, preventing the aggregation of the metal, strongly accelerates the reaction, allowing the use of milder conditions. Therefore, higher functional group tolerance and enhanced activities and selectivities could be achieved with only catalytic amounts of copper. In this context, the use of microwave irradiation, which provides a rapid and homogeneous heating, has also proven to improve the efficiency of these Ullmann-type processes. Thus, Wan and co-workers have reported a rapid and effective aqueous MW protocol for the N-arylation of amines using catalytic amounts of a copper source (Scheme 2.20).[51] Optimal performances were achieved using 4 equiv. of the amine with respect to the aryl halide, and a catalytic system based on CuO (25 mol.%) associated with the bis(cyclohexanone)oxalyl dihydrazone ligand (BCO), the phase-transfer agent tetrabutylammonium bromide (TBAB) and KOH. Under these conditions, a large variety of N-aryl amines could be obtained in moderate to good yields after only 5 min of irradiation at 130 °C. This Ullmann-type process was compatible with both electron-rich and electron-poor aryl iodides and bromides, as well as aromatic or aliphatic primary amines and dialkyl amines. As expected, crowded aryl halides bearing *ortho*-substituents were less reactive and required the use of a higher temperature (140 °C) and metal loading (50 mol.%) to be transformed into the desired amines.

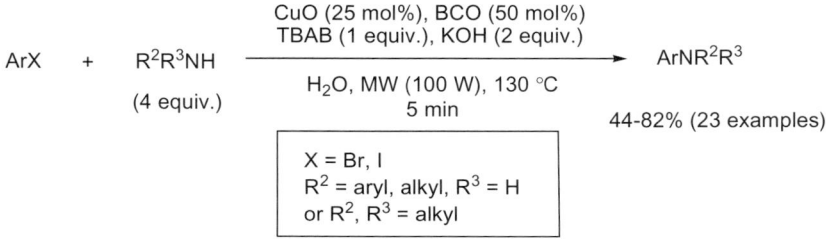

Scheme 2.20

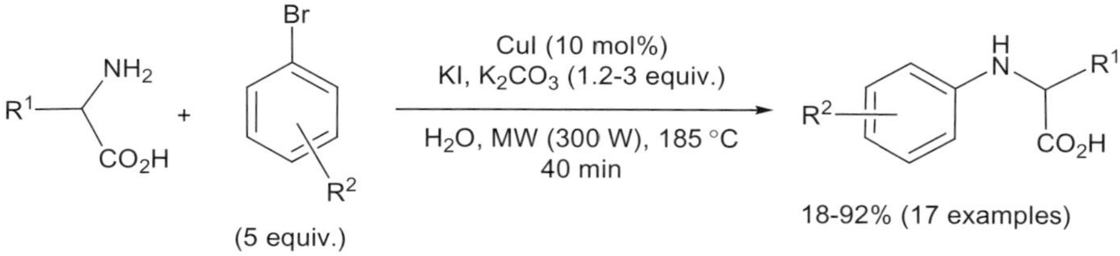

Scheme 2.21

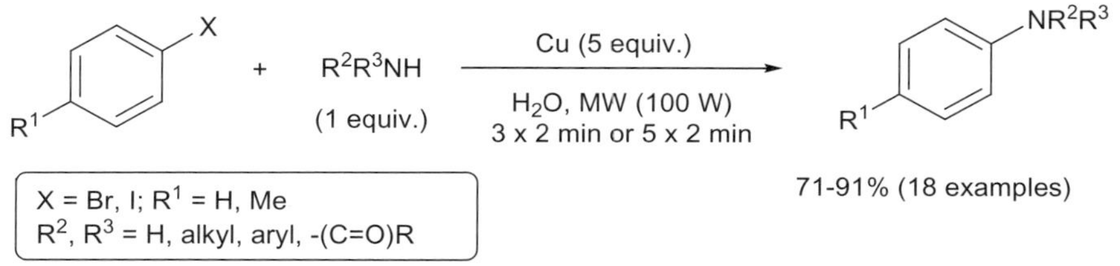

Scheme 2.22

A fast racemization-free protocol for the *N*-arylation of amino acids and amino acid esters with aryl bromides has also been developed in water by the aid of MW heating.[52] Reactions took place at 185 °C using copper(I) iodide as catalyst in the presence of KI and K_2CO_3 (Scheme 2.21). The amount of base proved to be critical to achieve high conversions, with the optimum quantity (from 1.2 to 3 equiv.) being dependent on the nature of amino acid. After optimization, the selected conditions turned out to be suitable for different amino acids (L- and D-Phe, L-Trp, Gly, L-Val, L-Leu and L-Pro) and several aryl bromides substituted at the *ortho*-, *meta*- or *para*-position. Once again, the use of crowded *ortho*-substituted substrates led to the lowest conversions. Interestingly, reactions could also be efficiently performed starting from *N*-Boc-protected amino acids, *in situ* deprotection and subsequent *N*-arylation taking place in these cases. Moreover, hydrogen chloride salts of amino acid esters (methyl, ethyl, *t*-butyl, benzyl and allyl esters) were also converted in good yields into the corresponding aryl amino acids through a deprotonation, ester hydrolysis and *N*-arylation sequence.

Base-free Ullmann-type couplings between aryl halides and amines in an aqueous medium have been described by Yadav and co-workers.[53] The reactions, promoted by a five-fold excess of copper, were realized in a domestic microwave oven through intermittent irradiations (Scheme 2.22). Both aromatic and aliphatic primary and secondary amines proved to be suitable substrates, leading to satisfactory conversions within 10 min. Amides and imides can also participate in these reactions. In particular, the intramolecular *N*-arylation of base-sensitive β-lactams afforded penem and cephem analogues in good yields (Scheme 2.23), where most of the previously existing methods proved inoperative. It is also notable that the use of conventional oil-bath heating, or solvent-free conditions, resulted generally in lower yields.

The synthesis of several N-phenylanthranilic acids, important anti-inflammatory agents, has also been developed in aqueous medium through MW-assisted Ullmann condensation processes.[54] Thus, as shown in Scheme 2.24, the coupling of activated *ortho*-chlorobenzoic acids with different anilines efficiently took place in the presence of copper(II) sulfate and potassium carbonate when heated in a domestic microwave oven (500 W) for 5–10 min.

A cooperative ligand-free iron/copper co-catalyzed N-arylation of amines has been described recently (Scheme 2.25).[55] The cross-coupling reactions were performed in a H_2O–DMSO mixture using a catalytic system composed of

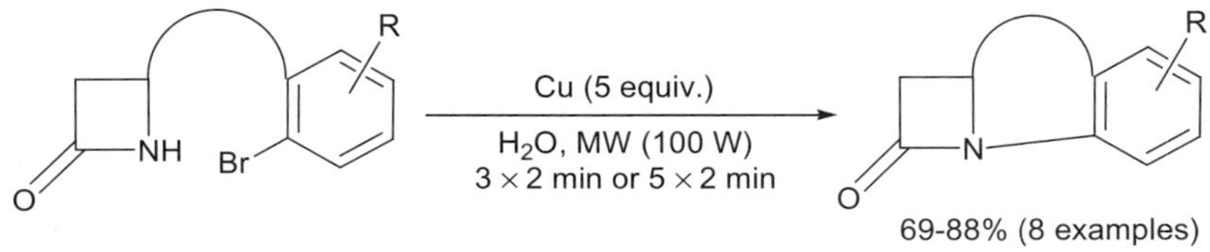

Scheme 2.23

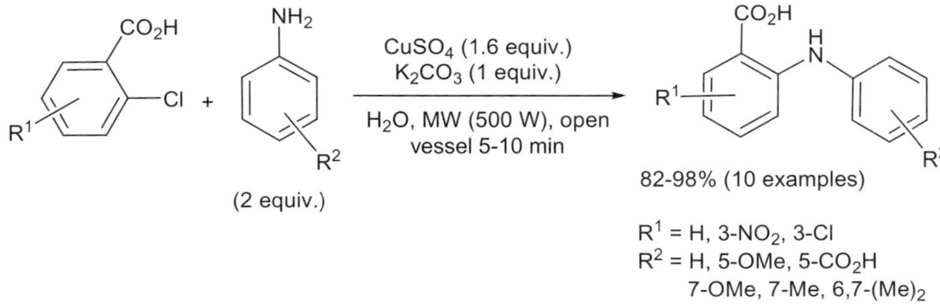

Scheme 2.24

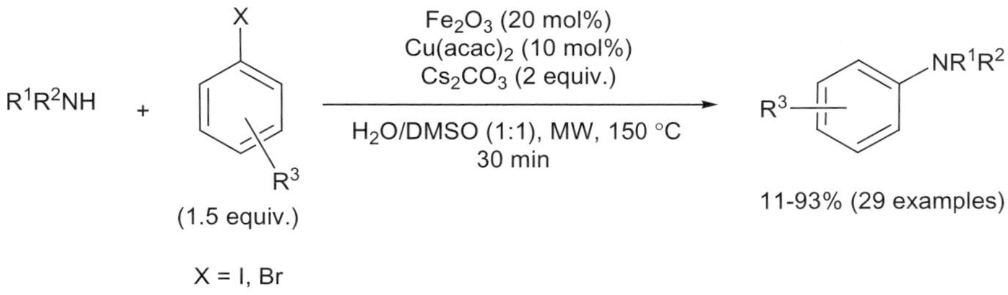

Scheme 2.25

Fe$_2$O$_3$, Cu(acac)$_2$ (acac = acetylacetonate) and caesium carbonate. Starting from aryl bromides, high conversions were usually obtained after 30 min of MW heating at 150 °C. Remarkably, in the absence of the iron source, Cu(acac)$_2$ by itself displayed only moderate activity, while Fe$_2$O$_3$ alone was completely inactive. This protocol was applicable to a great variety of aryl halides and amines, observing the usual trends of reactivity, *i.e.* (i) higher conversions starting from aryl iodides compared to aryl bromides, (ii) slower rates when secondary instead of primary amines are involved, and (iii) lower yields starting from sterically hindered *ortho*-substituted aryl halides.

In contrast to the extensively studied *N*-arylations, related copper-mediated C–S bond-forming reactions have received much less attention.[50a] Thus, as far as we are aware, the only example proceeding under aqueous MW conditions has been reported recently by Bagley and co-workers.[56] As shown in Scheme 2.26, in the presence of copper(I) iodide, a diamine or diol ligand and potassium carbonate, benzenethiol reacted with aryl iodides to afford the corresponding diaryl sulfides in only moderate yields (31–69%) after 2–3 h of MW heating in water. Notably, better results (up to 95% yield) were in all cases achieved using 2-propanol instead of water as solvent.

It has also been demonstrated that Cu-catalyzed conversion of aryl halides into the corresponding phenols, a usually time-demanding process, can be considerably accelerated by the aid of microwaves.[57] In this context, MW-heating of aryl iodides in superheated water (200–300 °C), using a catalytic system based on CuI (10 mol.%) associated with the L-proline ligand and NaOH, has been reported to yield phenols in moderate to good yields within only 30 min (Scheme 2.27).[58] As expected, aryl bromides and chlorides, especially the latter, were less reactive than their iodide counterparts. The yield of the process was also strongly affected by the electronic properties of the aryl substituents, with quantitative conversions being only observed with electron-poor substrates. Sterically hindered aryl halides, bearing substituents at the *ortho* position, where also examined in this process but met with only limited success.

Copper-promoted conversion of aryl halides into aryl azides, in water under MW-heating, has also been described. However, in this particular reaction the

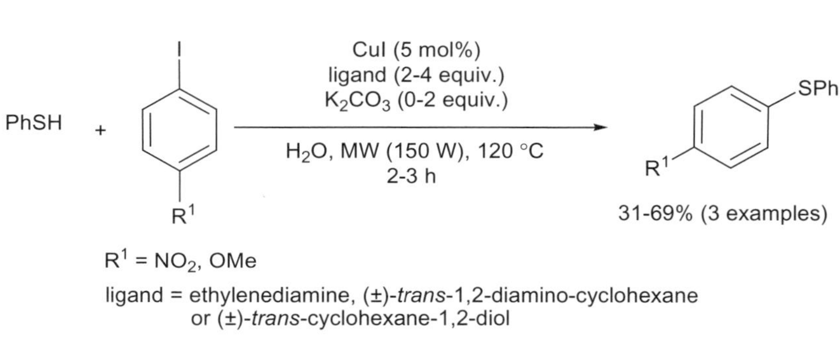

PhSH + [aryl iodide with R^1] → [aryl-SPh with R^1]

CuI (5 mol%)
ligand (2-4 equiv.)
K$_2$CO$_3$ (0-2 equiv.)
H$_2$O, MW (150 W), 120 °C
2-3 h

31-69% (3 examples)

R^1 = NO$_2$, OMe
ligand = ethylenediamine, (±)-*trans*-1,2-diamino-cyclohexane or (±)-*trans*-cyclohexane-1,2-diol

Scheme 2.26

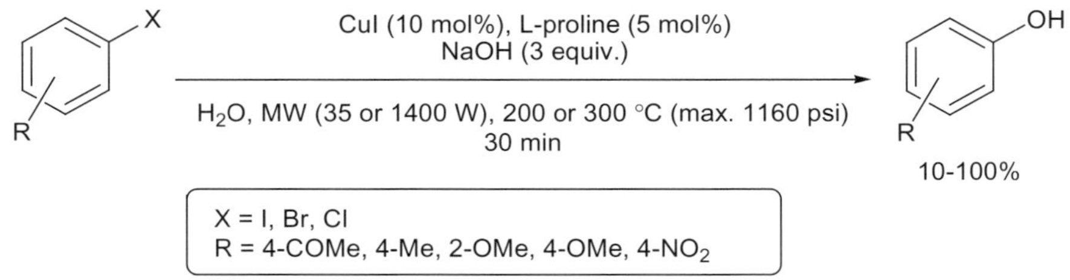

Scheme 2.27

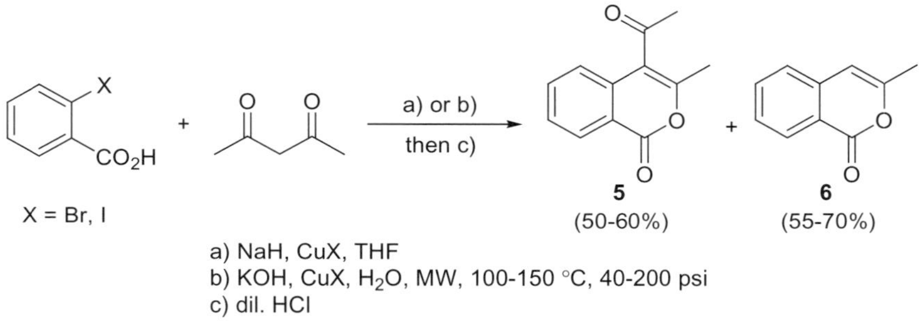

Scheme 2.28

use of MW-techniques was counterproductive as much better results were achieved under conventional thermal conditions.[59]

2.5 Cyclization Reactions

The first approach to MW-induced metal-catalyzed cycloaddition reactions in water was developed by Bryson and co-workers in 2003.[60] In particular, they studied the aromatic substitution of *ortho*-halobenzoic acids with pentane-2,4-dione, in the presence of base and stoichiometric amounts of copper(I) salts (CuX), obtaining isochromenones **5** or **6** depending on the reaction conditions employed (Scheme 2.28). Thus, while isochromenone **5** (50–60% yield) was obtained as the major product under conventional heating conditions (NaH, CuX, THF, reflux), the use of microwave irradiation (KOH, CuX) in water at 100–150 °C (40–200 psi) led to the predominant formation of **6** (55–70% yield), which results from the cleavage of the acyl group of **5** due to the high temperatures employed. Other 1,3-diketones as well as β-keto esters were also employed in this process, furnishing different isochromenone derivatives depending on the temperature, pressure and nature of the activating methylene groups employed.[60]

More recently, Turner and co-workers have developed an efficient and quite general route to *N*-substituted oxindoles through a two-step process based on (i) an initial MW-assisted amide bond formation between 2-haloarylacetic acids and various alkylamines and anilines, and (ii) a subsequent palladium-catalyzed intramolecular amidation under aqueous-MW conditions (Scheme 2.29).[61]

Thus, using optimized reaction parameters [*i.e.*, 3 mol.% of Pd(OAc)$_2$, 6 mol.% of the ligand L, 2 equiv. of NaOH, H$_2$O–toluene (1:1), 100 °C], several *N*-substituted oxindoles could be obtained in moderate to excellent yields (30–96%) after only 30 min of MW irradiation. Remarkably, in the case of alkylamines, the palladium-catalyzed intramolecular amidation step could be performed without prior isolation of the intermediate amide.[61]

The transition-metal mediated conversion of alkynes, alkenes and CO in a formal [2+2+1] cycloaddition process, commonly known as the Pauson–Khand reaction, is one of the most elegant methods presently available for the construction of cyclopentenone scaffolds.[62] In this context, Kwong, Chan and co-workers have described an aqueous-MW protocol for this reaction using formate esters as the CO source and [{RhCl(COD)}$_2$]/dppp (dppp = 1,3-bis(diphenylphosphino)propane) as catalyst (Scheme 2.30).[63] By this way, various oxygen-, nitrogen- and carbon-tethered 1,6-enynes (**7**)

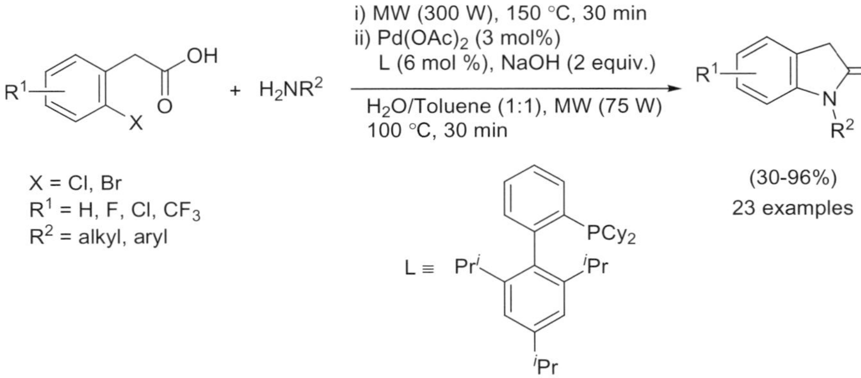

Scheme 2.29

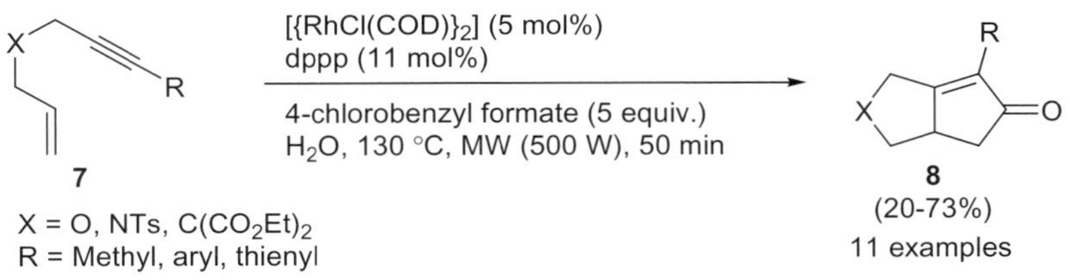

Scheme 2.30

were transformed into their corresponding cyclopentenones (**8**) in moderate to good yields (20–73%) after only 50 min employing previously optimized reaction conditions (5 mol.% of [{RhCl(COD)}$_2$], 11 mol.% of dppp, 5 equiv. of 4-chlorobenzyl formate, 130 °C, MW power 500 W).[63] Interestingly, under conventional oil-bath heating, only trace amounts of the desired products are formed.

In addition, the asymmetric version of this Pauson–Khand reaction was studied (R = Ph, Scheme 2.30) using a battery of chiral diphosphine ligands instead of dppp. Although low yields of **8** were generally obtained (5–72%), in some cases high enantioselectivities could be achieved (up to 98% e.e.).[63]

Metal-catalyzed intramolecular hydroamination of alkynes has emerged in recent years as a useful synthetic tool for the preparation of various *N*-heterocycles, such as indoles, pyrroles, quinolines and isoquinolines.[64] Very recently, an Au(I)/Ag(I)-catalyzed intramolecular hydroamination reaction for the preparation of 2,3-unsubstituted indole-1-carboxamides starting from *N*-(2-alkynylphenyl)ureas, in water under MW irradiation, has been described by Liu and co-workers (Scheme 2.31).[65] *N*-aliphatic, *N*-aromatic and *N*-heterocyclic substituted ureas provided the desired products, in moderate to good yields (40–91%), after only 10 min of irradiation at 150 °C. However, notably, this methodology is only operative with terminal alkynes, the internal ones remaining completely unaffected. It is also interesting that, although both Lewis acids ([AuCl(PPh$_3$)] and Ag$_2$CO$_3$) are catalytically active by themselves in this cyclo-isomerization process, the synergistic effect of the two metals working together improved the yields significantly.

Almost simultaneously, the same authors also described the first report on the synthesis of *N*-heterocyclic compounds by an iron-catalyzed C–N coupling in aqueous media.[66] In particular, an efficient method was developed for the synthesis of various quinazolinone derivatives (**9**, 41–81% yield) starting from substituted 2-halobenzoic acids and amidines, using a mixture of FeCl$_3$ (10 mol.%) and L-proline (20 mol.%) as catalyst, 2 equiv. of Cs$_2$CO$_3$ as base and MW-irradiation as the heating source (Scheme 2.32). Remarkably, the reactions could be also conducted in the absence of any ligand while maintaining acceptable product yields (39–78%).

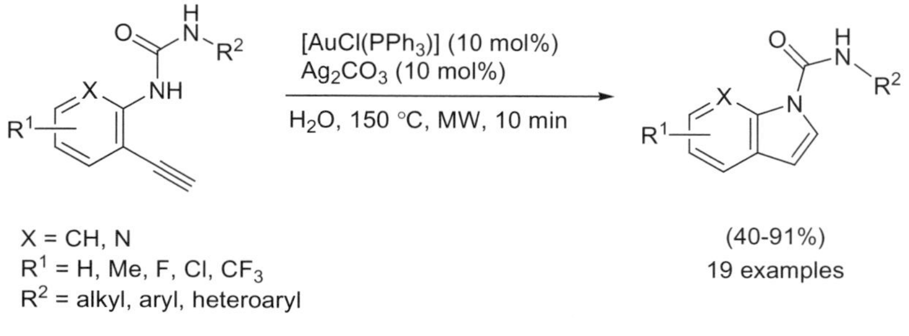

X = CH, N
R^1 = H, Me, F, Cl, CF$_3$
R^2 = alkyl, aryl, heteroaryl

(40-91%)
19 examples

Scheme 2.31

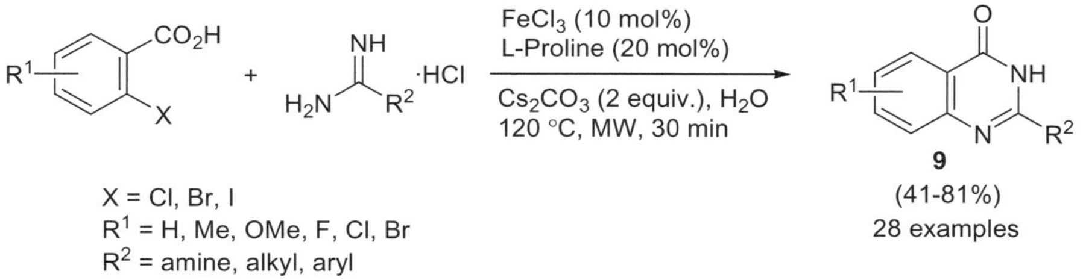

Scheme 2.32

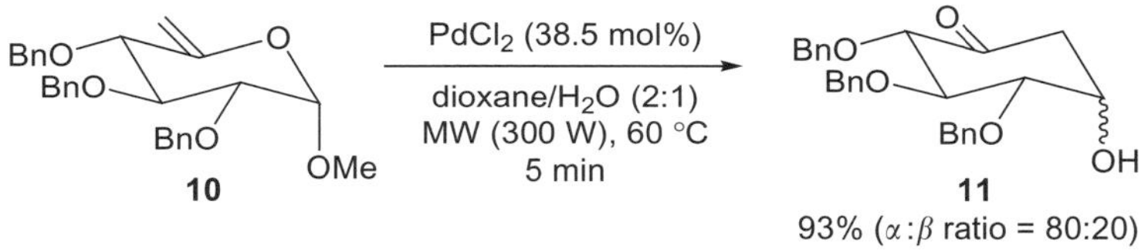

Reagents/conditions: R¹-substituted 2-halobenzoic acid (X = Cl, Br, I; R¹ = H, Me, OMe, F, Cl, Br) + amidine·HCl (H$_2$N-C(=NH)-R²; R² = amine, alkyl, aryl), FeCl$_3$ (10 mol%), L-Proline (20 mol%), Cs$_2$CO$_3$ (2 equiv.), H$_2$O, 120 °C, MW, 30 min → quinazolin-4(3H)-one **9** (41–81%), 28 examples.

Scheme 2.33

Tri-O-benzyl hex-5-enopyranoside **10** (OMe anomeric), PdCl$_2$ (38.5 mol%), dioxane/H$_2$O (2:1), MW (300 W), 60 °C, 5 min → cyclohexanone **11**, 93% ($\alpha:\beta$ ratio = 80:20).

The Ferrier carbocyclization, also known as the Ferrier II reaction, is a useful synthetic tool in carbohydrate chemistry, consisting of the transformation of hex-5-enopyranosides into cyclohexanones.[67] A microwave-assisted Ferrier carbocyclization of **10** into **11** (Scheme 2.33) has been described as a key step in the total synthesis of a carbasugar-1-phosphate.[68] The reaction, catalyzed by 38.5 mol.% of PdCl$_2$, was carried out in a dioxane–water mixture to afford the desired Ferrier product **11** in good yield. Notably, a lower yield and a significantly higher reaction time were observed when the process was performed at the same temperature under conventional heating (81% after 3 h vs 93% after 5 min).

Diels–Alder-type cycloaddition of dienes **12** with dienophiles **13** using the water-soluble organotungsten Lewis acid catalyst [W{κ^3-N,N,N-O=P(C$_5$H$_4$N)$_3$}(CO)(NO$_2$)][BF$_4$]$_2$ (**14**), water as solvent and microwave heating has been described by Yu and co-workers (Scheme 2.34).[69] Reactions were completed in less than 1 min, affording the six-membered ring systems **15** in excellent yields (92–99%) and with good stereoselectivity (endo/exo ratio from 78:22 to 100:0). However, we must note that, under the same reaction conditions, the use of the ionic liquid 1-butyl-3-methylimidazolium hexafluorophosphate (bmimPF$_6$) as solvent instead water resulted in even faster reactions (30 s).

Probably the most important contribution of microwaves to cycloaddition chemistry has been in the field of 1,3-dipolar cycloadditions of azides with alkynes and alkenes. Now known by many as "click reactions", these

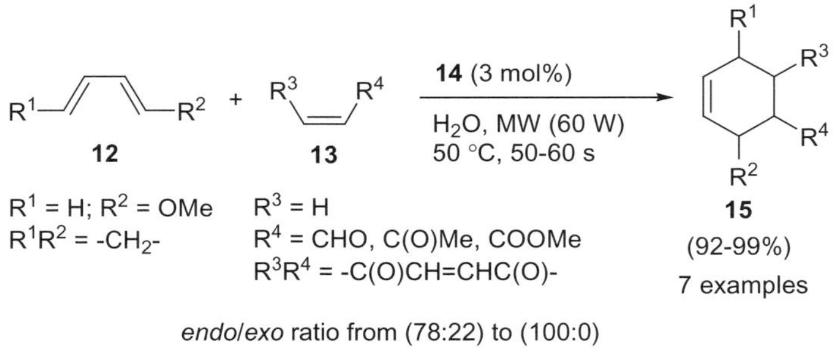

Scheme 2.34

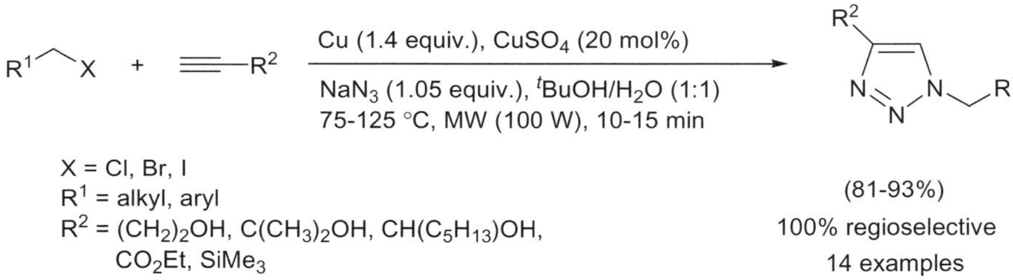

R¹ = H; R² = OMe; R¹R² = -CH₂-
R³ = H; R⁴ = CHO, C(O)Me, COOMe; R³R⁴ = -C(O)CH=CHC(O)-

Reagents: **14** (3 mol%), H₂O, MW (60 W), 50 °C, 50–60 s

Product **15** (92–99%), 7 examples

endo/exo ratio from (78:22) to (100:0)

Scheme 2.35

X = Cl, Br, I
R¹ = alkyl, aryl
R² = (CH₂)₂OH, C(CH₃)₂OH, CH(C₅H₁₃)OH, CO₂Et, SiMe₃

Reagents: Cu (1.4 equiv.), CuSO₄ (20 mol%), NaN₃ (1.05 equiv.), ᵗBuOH/H₂O (1:1), 75–125 °C, MW (100 W), 10–15 min

(81–93%), 100% regioselective, 14 examples

cycloadditions have become extremely useful synthetic tools, especially in the field of biological applications and polymer science.[70] The most useful "click reactions" are, probably, those involving organic azides and alkynes, a reaction known for over 100 years but only studied extensively since the work of Huisgen in the 1960s.[71] The Cu(I)-catalyzed stepwise variant of this process, independently discovered in 2002 by the groups of Fokin and Sharpless[72] and Medal,[73] has become a reliable reaction with an unprecedented level of regioselectivity. In this sense, Fokin, van der Eycken and co-workers were the first to apply microwave techniques to the copper-catalyzed version of the Huisgen reaction. They developed an aqueous process involving the reaction of an alkyl halide with sodium azide, which *in situ* generates the required organic azide, in the presence of different alkynes which led, under MW-irradiation, to a series of 1,4-disubstituted-1,2,3-triazoles in good to excellent yields (81–93%) with complete regioselectivity (Scheme 2.35).[74] In these reactions the required Cu(I) catalyst is generated *in situ* by comproportionation of Cu(0) (Cu turnings, 1.4 equiv.) and Cu(II) (CuSO₄, 20 mol.%).

Application of this methodology to the decoration of 2(1*H*)-pyrazinone scaffolds has also been described (Scheme 2.36).[75] Reactions proceeded smoothly within 3–25 min using only 2 mol.% of CuSO₄, furnishing the corresponding cycloaddition products **16** in moderate to good yields (40–84%).

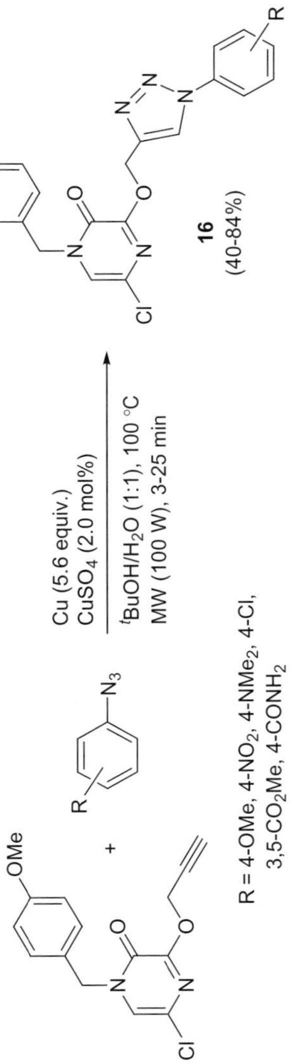
OMe
Cl
N
N
O
O
+
R
N3
Cu (5.6 equiv.)
CuSO4 (2.0 mol%)
tBuOH/H2O (1:1), 100 °C
MW (100 W), 3-25 min
R = 4-OMe, 4-NO2, 4-NMe2, 4-Cl,
3,5-CO2Me, 4-CONH2
16
(40-84%)

Scheme 2.36

Although all these "click reactions" were performed in a ^{t}BuOH–H$_2$O mixture, it is notable that they are also operative using water as sole solvent. However, in these cases longer reactions (ca. 60 min) are required to attain similar conversions. In some cases the introduction of triazole-based ligands (0.8–1.0 mol.%) in the medium resulted in a considerable increase in reactions rate since they can stabilize the air-sensitive catalytically active Cu(I) species.[75]

Using this microwave-assisted protocol, van der Eycken and co-workers also described the coupling of various 3-ethynyl-2(1H)-pyrazinones with different glycosyl β-azides en route to the synthesis of glycopeptidomimetics **17** and **18** (Figure 2.2).[76] On this occasion, the reactions were performed using 2 equiv. of Cu turnings in combination with 5 mol.% of CuSO$_4$, as the Cu(I) source, and 5 mol.% of the stabilizing ligand N,N,N-tris[(1-benzyl-1H-1,2,3-triazol-4-yl)methyl]amine (TBTA) in a THF–H$_2$O (1:1) mixture (85 °C under MW irradiation).

Following a similar approach, the same authors described the synthesis of a small library of hitherto unknown nucleoside analogues **21** by coupling of a set of newly generated furo[2,3-b]pyrazines **19** with five-membered carbosugars **20** (Scheme 2.37).[77] Once again, 5 mol.% of TBTA was introduced in the reaction media to stabilize the catalytically active Cu(I) species. We note also that, in this

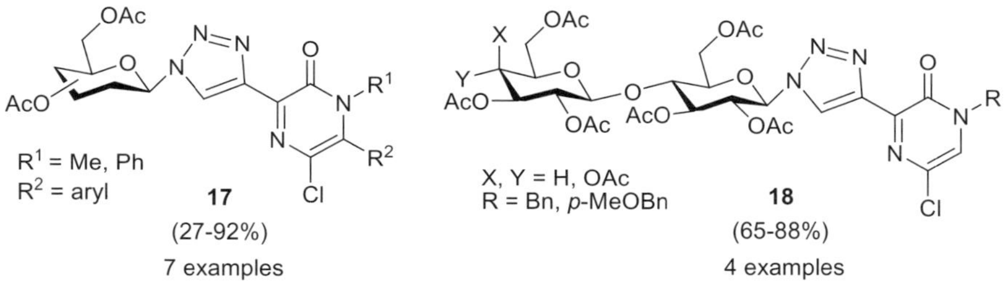

Figure 2.2

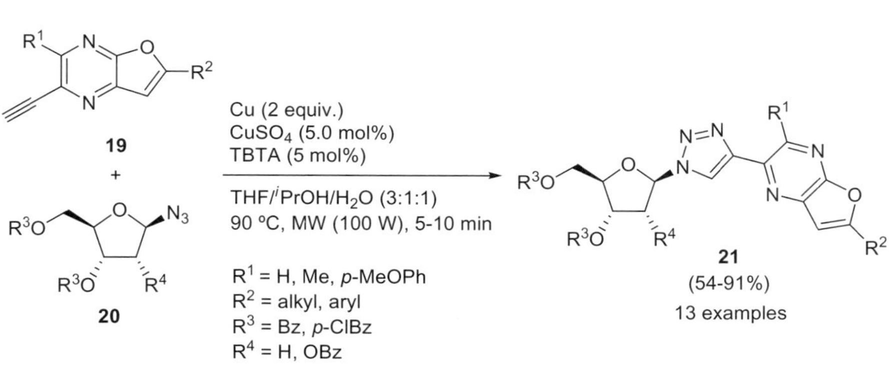

Scheme 2.37

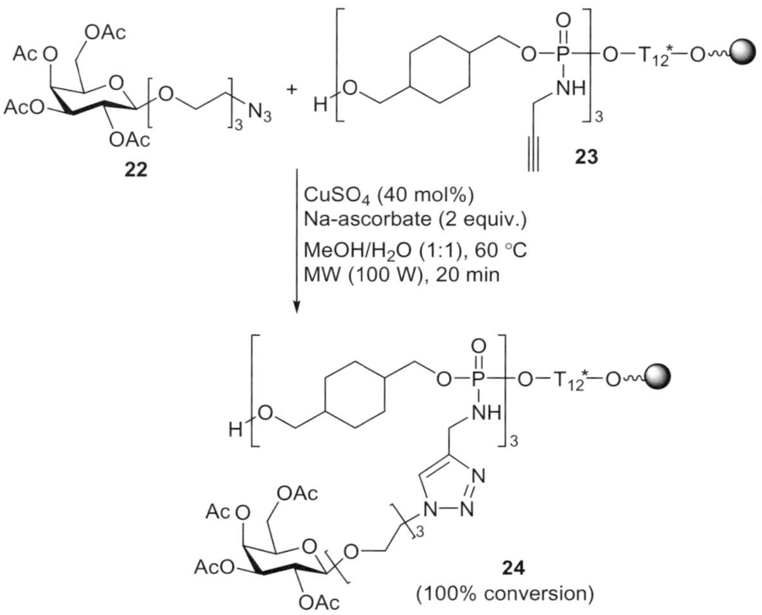

Scheme 2.38

case, the use of a mixture of THF–^{i}PrOH–H$_2$O (3 : 1 : 1) as solvent was required to improve the solubility of the reactants.

A MW-assisted Huisgen 1,3-dipolar cycloaddition was also applied in the anchoring of carbohydrate residues to a solid-supported oligonucleotide.[78] Thus, as shown in Scheme 2.38, the solid-supported trigalactosylated oligonucleotide **24** could be synthesized in quantitative yield by "click" coupling of the alkyne-bearing oligonucleotide **23** (T$_{12}$* = dodecathymidine) with the azide-functionalized galactoside **22**. The reaction, performed in a methanol–water (1 : 1) mixture, proceeded rapidly at 60 °C using CuSO$_4$ (40 mol.%) and sodium ascorbate (2 equiv.) as the Cu(I) source.

A convenient one-pot procedure for generating multivalent triazole-linked structures starting from amines has been reported by Wittmann and co-workers.[79] Thus, in a first step, the corresponding organic azides were generated *in situ* by a Cu(II)-catalyzed diazo transfer reaction from trifluoromethanesulfonyl azide (TfN$_3$) to the amines. These azides subsequently underwent the 1,3-dipolar cycloaddition with the alkynes to generate the corresponding 1,4-disubstituted 1,2,3-triazoles in excellent yields (78–99%). As a representative example, the synthesis of the divalent glucoconjugate **27** is depicted in Scheme 2.39. Starting from diamine **25** and propargyl glucoside **26**, the reaction proceeded cleanly, affording **27** in 86% yield. The essential element of this one-pot procedure is to generate the Cu(I) species required for the Huisgen azide–alkyne cycloaddition by adding a reducing agent (Na-ascorbate) after completion of the Cu(II)-catalyzed diazo transfer reaction.

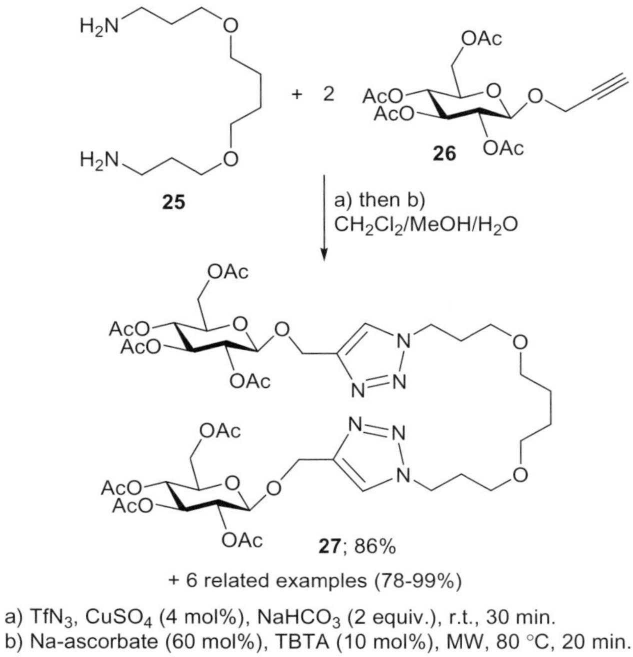

H_2N
H_2N
25
+ 2
OAc
AcO
AcO
26
OAc
a) then b)
$CH_2Cl_2/MeOH/H_2O$
OAc
AcO
AcO
OAc
OAc
AcO
AcO
OAc
27; 86%
+ 6 related examples (78-99%)
a) TfN_3, $CuSO_4$ (4 mol%), $NaHCO_3$ (2 equiv.), r.t., 30 min.
b) Na-ascorbate (60 mol%), TBTA (10 mol%), MW, 80 °C, 20 min.

Scheme 2.39

In the field of dendrimer chemistry, the availability of efficient methodologies for surface decoration with different functional groups is a priority target.[80] In this sense, the application of microwave-enhanced azide cycloaddition chemistry to dendrimer synthesis and functionalization has been reported by several groups. In a pioneering work Liskamp, Pieters and co-workers described the synthesis of di-, tetra-, octa- and hexadecavalent dendrimeric peptides *via* an efficient attachment between (cyclic)peptides and suitable functionalized dendrimers by a 1,3-dipolar cycloaddition reaction, conveniently assisted by microwave irradiation to ensure complete modification of the alkyne end-groups.[81] Thus, after 5–10 min of MW-heating at 100 °C, using a mixture THF–H_2O (1 : 1) or aqueous DMF as solvent, in the presence of $CuSO_4$ (5 mol.%) and sodium ascorbate (50 mol.%) as the Cu(I) source, di- and tetravalent dendrimeric peptides **28** and **29**, as well as octa- and hexadecavalent analogues, were synthesized in moderate to excellent yields (14–96%) (Figure 2.3). When compared with the same reactions performed at ambient temperature, a higher conversion and 20-fold acceleration could be achieved employing microwave irradiation. It is also notable that not only small peptide-based azides but also unprotected biologically relevant larger, and even cyclic, azido peptides could be efficiently converted into the corresponding multiple cycloaddition products.[81]

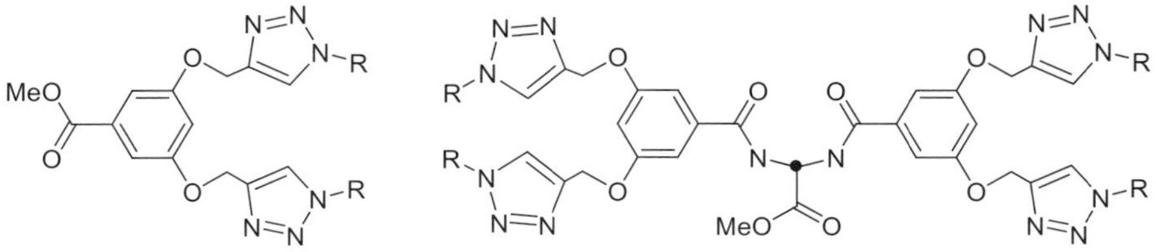

Figure 2.3

28 (48-96%) 9 examples
R = small, large and cyclic peptides
29 (14-91%) 5 examples
+ 2 octavalent examples
+ 1 hexadecavalent example

Scheme 2.40

R = OH, Ac, Bn, Bz

CuSO$_4$ (30 mol%)
Na-ascorbate (60 mol%)
DMF + H$_2$O (several drops)
80 °C, MW, 20 min

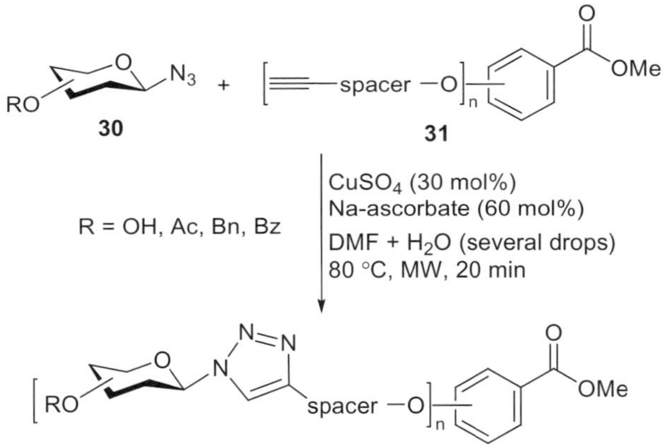

32 (73-98%) 19 examples

Almost simultaneously, the same authors also reported the coupling of several azido carbohydrates with different kinds of alkyne-bearing dendrimers using this MW-assisted Cu(I)-catalyzed 1,3-dipolar cycloaddition process, with the reactions leading to multivalent 1,4-disubstituted 1,2,3-triazole-linked glycodendrimers.[82] Thus, after optimization of the reaction conditions (*i.e.*, 30 mol.% of CuSO$_4$, 60 mol.% of sodium ascorbate, DMF with several drops of water as solvent, 80 °C, MW irradiation) the desired triazole glycodendrimers **32** could be obtained in excellent yields after only 20 min by coupling of the azido carbohydrates **30** with multivalent alkyne-linked dendritic structures **31** (Scheme 2.40).

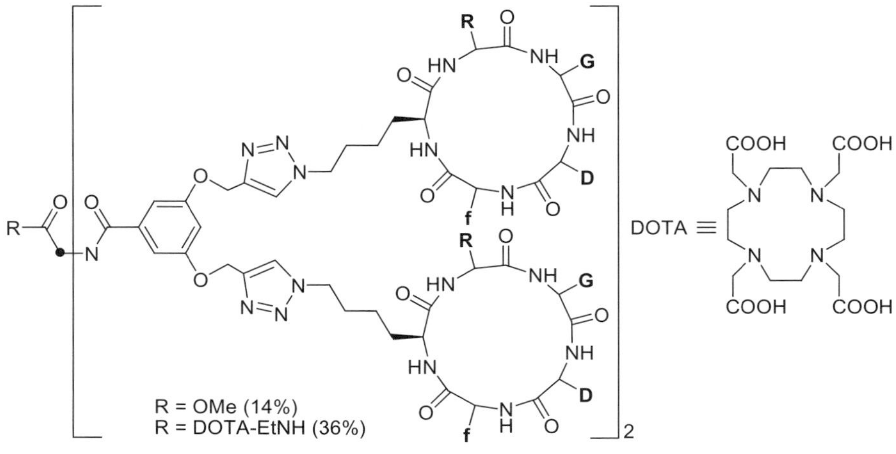

Figure 2.4

More recently, using the same methodology, the same authors also described the preparation of different mono-, di- and tetrameric methoxy and DOTA-conjugated (DOTA = 1,4,7,10-tetraazadodecane-N,N',N'',N'''-tetraacetic acid) cyclo[Arg-Gly-Asp-D-Phe-Lys] (c[RGDfK]) dendrimers (Figure 2.4).[83] Additionally, since multivalence is a well-accepted approach to increase the interaction of weakly interacting individual ligands with their respective receptors, these DOTA dendrimers were radiolabeled with In(III) to evaluate the *in vitro* receptor binding characteristics and *in vivo* tumor targeting properties.[83]

The group of Weck also described the application of MW-assisted Huisgen methodology to the selective and quantitative monofunctionalization of various polyamide based dendrimers.[84] Using 5 mol.% of CuSO$_4$ and 10 mol.% of sodium ascorbate, in a mixture of H$_2$O–^{t}BuOH (1:1) at 100 °C and under MW-irradiation (100 W), a series of previously synthesized dendrimers containing an alkyne or azide moiety reacted with their corresponding complementary couple to give the functionalized derivatives **33** in excellent yields (95–98%) after only 10 min (Figure 2.5).

2.6 Reduction and Oxidation Reactions

Transfer hydrogenation (TH) reactions are a good and safer alternative to the widely used catalytic hydrogenations using flammable and explosive H$_{2(g)}$.[85] In this context, Sinha and co-workers disclosed a microwave-assisted TH-reduction of the olefinic bond of α,β-unsaturated carbonyl compounds.[86] They obtained the corresponding saturated carbonyl derivatives in moderate to excellent yields (8–96%) after only 22–55 min with complete chemoselectivity (Scheme 2.41). The reactions were performed in the presence of silica-supported palladium

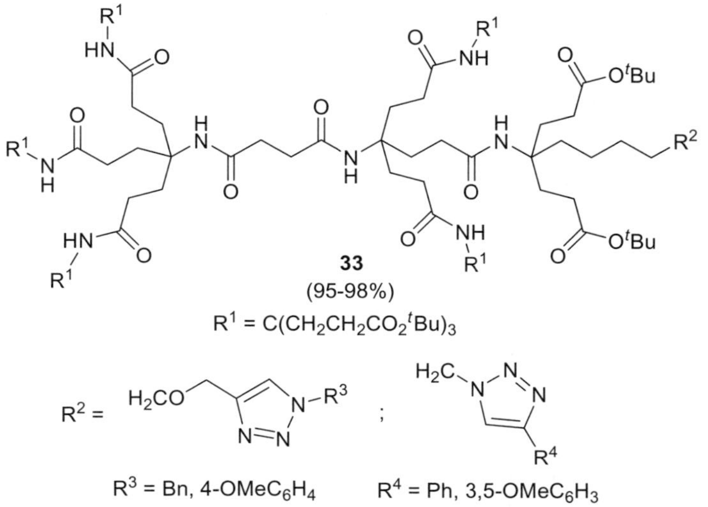

Metal-catalyzed Reactions in Water under MW Irradiation

33
(95-98%)

R¹ = C(CH$_2$CH$_2$CO$_2{}^t$Bu)$_3$

R² = (triazole with H$_2$CO-CH$_2$ and N-R³ substituents) ; (triazole with H$_2$C-N and R⁴ substituents)

R³ = Bn, 4-OMeC$_6$H$_4$ R⁴ = Ph, 3,5-OMeC$_6$H$_3$

Figure 2.5

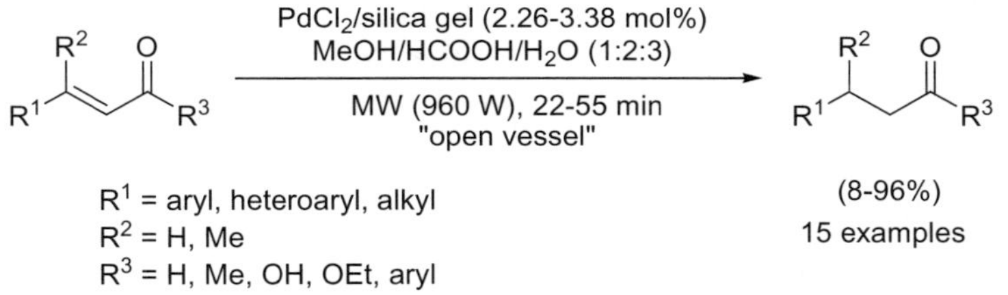

Reaction scheme: R¹-CH=CR²-C(O)R³ → R¹-CH$_2$-CHR²-C(O)R³

Conditions: PdCl$_2$/silica gel (2.26–3.38 mol%), MeOH/HCOOH/H$_2$O (1:2:3), MW (960 W), 22–55 min, "open vessel"

R¹ = aryl, heteroaryl, alkyl
R² = H, Me
R³ = H, Me, OH, OEt, aryl

(8–96%)
15 examples

Scheme 2.41

dichloride as catalyst and a 1:2:3 mixture of MeOH–HCOOH–H$_2$O as the hydrogen source and solvent using a domestic microwave oven (960 W) under open-vessel conditions. The employment of water as a component of the hydrogen source mixture plays a crucial role in these reactions since it causes the rapid release of hydrogen from HCOOH, increasing the yields and reducing the reaction times. Moreover, the partial protonation of the carbonyl group by formic acid prevents palladium interacting with the carbonyl site, thereby avoiding the alternative hydrogenation of the keto group and inducing a complete chemoselectivity over the reduction of the carbon–carbon double bond. Remarkably, attempts to perform this transfer hydrogenation process under conventional thermal heating produced a mixture of the saturated carbonyl compound and its corresponding alcohol. Additionally, the silica-supported palladium catalyst could be reused, remaining active even after five consecutive cycles with a mere 5% loss in activity.

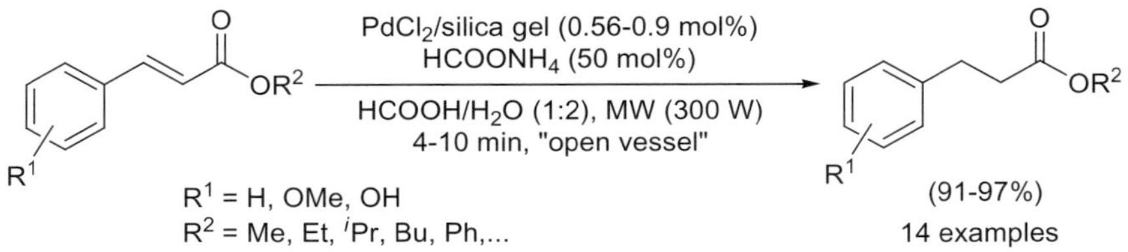

Scheme 2.42

In a subsequent work, the same authors developed an improved TH protocol for the chemoselective synthesis of bioactive dihydrocinnamic esters starting from the corresponding α,β-unsaturated cinnamic esters (Scheme 2.42).[87] The following modifications were introduced: (i) The reactions were performed again under open-vessel conditions but using a focused-microwave irradiator (300 W) instead a domestic microwave oven. (ii) The system HCOOH–H_2O (1:2) was used as the hydrogen source and solvent, the presence of methanol not being required in this case. Further, (iii) the addition of ammonium formate (50 mol.%) to the silica-supported palladium catalyst (0.56–0.9 mol.%) was found to improve both the yields and reaction times. Under these novel conditions, dihydrocinnamic esters were obtained in excellent yields (91–97%) after only 4–10 min of irradiation. Additionally, dimethyl maleate could be reduced to dimethyl succinate in 81% yield after 8 min. This methodology was also successfully applied for the selective hydrogenation of conjugated alkenes other than esters. In all cases, the catalyst was able to discriminate between the olefinic moiety and all other reducible groups present in the molecule, like ketones, carboxylic acids, nitriles and amides.[87]

The catalytic transfer hydrogenation of safflower oil under controlled microwave irradiation conditions has been studied by Prasad and co-workers using aqueous ammonium formate as the hydrogen source and palladium on carbon as catalyst.[88] Thus, using 5 mol.% of Pd/C and 6 equiv. of ammonium formate complete reduction of linoleic acid could be achieved by performing the catalytic reaction directly in water (160 °C; MW power 600 W; 1 h).

Although the selective transformation of amines into ketones is a relatively common biological process, chemical conversions of amines into ketones are rather limited.[89] In this context, the first example of a direct Pd/C-catalyzed transformation of amines **34** into ketones **35**, via a retro-reductive amination pathway, was reported by Miyazawa and co-workers using water as solvent and oxygen source (Scheme 2.43).[90] The reactions were carried out under continuous microwave irradiation (50 W) in a pressure-resistant ampoule, which resulted in a maximum temperature of 170 °C. Conventional heating or microwave heating under reflux conditions resulted in lower yields and selectivities, and were inadequate for practical purposes.

However, notably, the reaction outcome depends strongly on the number of hydrogen atoms on the α-carbon of the amine. Thus, amines containing two hydrogen atoms on the α-carbon (e.g., *n*-butylamine) afforded mainly the

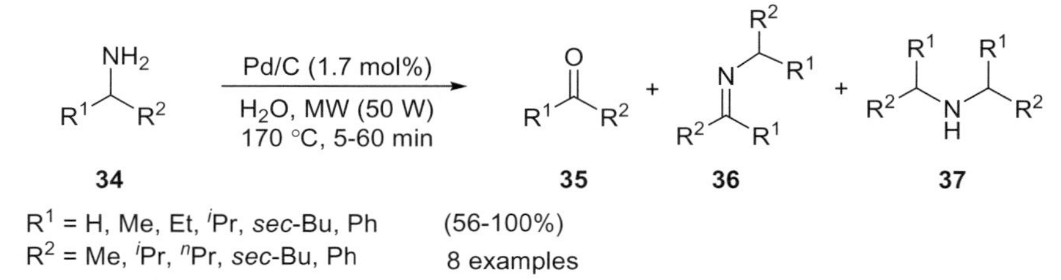

Scheme 2.43

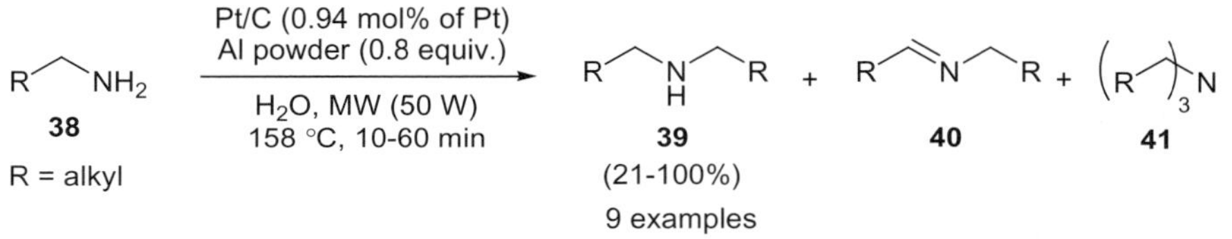

R¹ = H, Me, Et, ^{i}Pr, sec-Bu, Ph
R² = Me, ^{i}Pr, ^{n}Pr, sec-Bu, Ph

(56–100%)
8 examples

Scheme 2.44

R = alkyl

(21–100%)
9 examples

corresponding imines **36** and secondary alkyl-amines **37** instead of the desired ketones **35**. Only those amines possessing one hydrogen (mono- or di-sec-alkylamines) could be selectively converted into the corresponding ketones **35**. Remarkably, this unprecedented reaction proceeds smoothly without the aid of any heavy metal-based oxidant or volatile organic solvents.

Continuing this research line, the same authors developed a new method for the catalytic transformation of primary amines into secondary amines (Scheme 2.44).[91] They discovered that using Pt/C (0.94 mol.% of Pt) as catalyst, in water under continuous microwave irradiation (50 W) and in the presence of aluminium powder (0.8 equiv.), a family of secondary amines **39** could be obtained from the primary ones **38** in moderate to excellent yields (21–100%) after 10–60 min at 158 °C, with tertiary amines **41** being the main side products. Although other catalysts such as Pd/C, Ru/C or Rh/C were tested in this reaction, lower yields were obtained and longer reaction times were required.

Scheme 2.45 depicts the mechanism proposed by the authors for this process. Here, oxidative removal of dihydrogen, which is initially formed *in situ* by the reaction of the starting primary amine **38** with Pt/C, produces the imine **A**. Subsequent hydrolysis of **A** affords ammonia and the aldehyde intermediate **B**, which reacts with a second molecule of the starting primary amine to give the imine **40**. Finally, the Pt/C-dihydrogen species hydrogenates **40** to give the secondary amine **39**, regenerating the Pt/C catalyst. Consequently, the addition of aluminium powder to the reaction mixture reduced the reaction time and prevented the formation of imines **40** since in combination with water additional H_2 for the final hydrogenation step is produced.

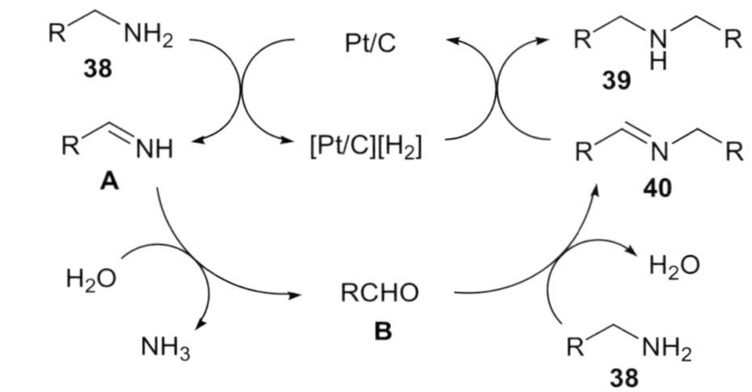

Scheme 2.45

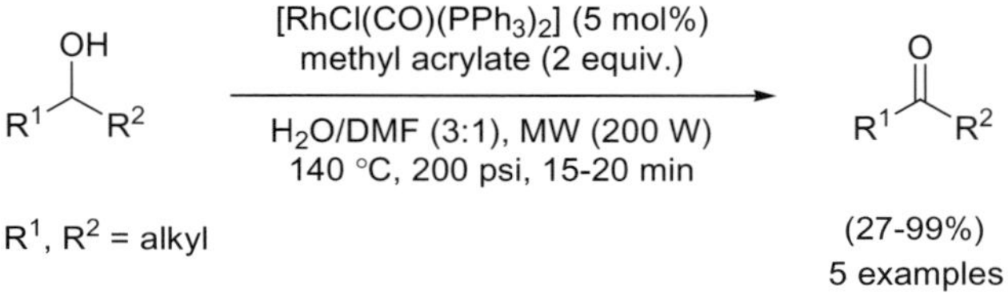

Scheme 2.46

Interestingly, as shown in Scheme 2.45, the first part of the reaction (from primary amine **38** to aldehyde **B**) is a retro-reductive amination and the second part (from aldehyde **B** to secondary amine **39**) is a reductive amination, and these contrasting processes are occurring concurrently in a one-pot manner.

Rhodium-catalyzed oxidation of secondary aliphatic alcohols to ketones under aqueous MW conditions has been described by Matsubara and co-workers (Scheme 2.46).[31] The reactions were performed in the presence of 2 equiv. of methyl acrylate, which acts as the hydrogen acceptor, and 5 mol.% of [RhCl(CO)(PPh$_3$)$_2$] as catalyst. Interestingly, the substitution of microwave irradiation by conventional heating resulted in complete recovery of the starting alcohols. Noticeably, this aqueous MW protocol is only applicable to secondary alcohols; the use of primary ones as substrates results in recovery of the starting materials unchanged even after prolonged reaction times. However, this limitation could be exploited for the selective oxidation of diols containing both primary and secondary alcohol moieties, where the secondary one is selectively transformed into the corresponding ketone while the primary one remains unchanged.

Ketopantolactone[42] is a key intermediate in the manufacture of pantothenic acid, a member of the B complex vitamins and a constituent of coenzyme A. In this context, a process for the oxidation of pantolactone to ketopantolactone with sodium periodate catalyzed by RuCl$_3 \cdot x$H$_2$O, in an aqueous solvent

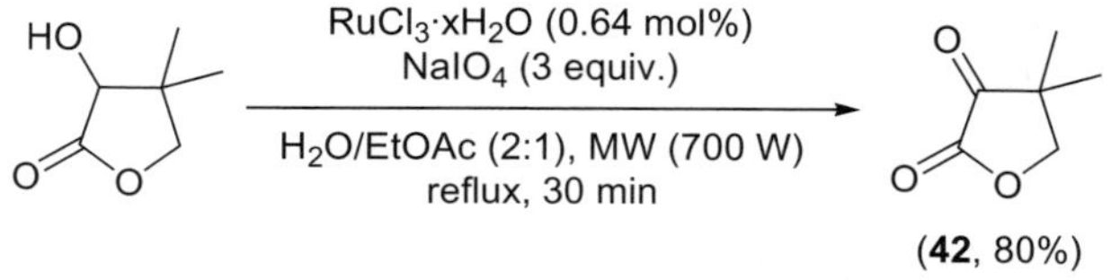

Scheme 2.47

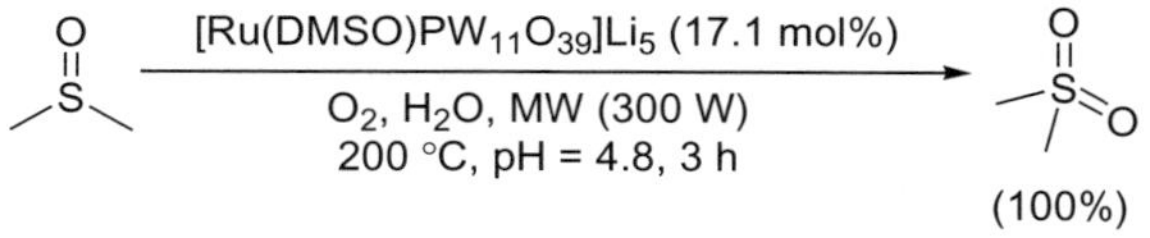

(42, 80%)

Scheme 2.48

(100%)

system under MW irradiation, has been patented by Bonrath and co-workers (Scheme 2.47).[92] Thus, using 0.64 mol.% of $RuCl_3 \cdot xH_2O$ and 3 equiv. of sodium periodate, in a mixture of H_2O–ethyl acetate (2:1) as solvent, pantolactone was oxidized to ketopantolactone in 80% yield after 30 min of heating (reflux temperature; MW power 700 W). Remarkably, the authors found that ketopantolactone is produced in higher conversion and selectivity, and in a shorter reaction time, under MW irradiation than using conventional heating.

Bonchio and co-workers have described the catalytic oxidation of DMSO in water using the polyoxotungstate-ruthenium(II) complex $[Ru(DMSO)PW_{11}O_{39}]Li_5$ (17.1 mol.%) as catalyst (Scheme 2.48).[93] The reaction, which was carried out in an acetate buffer solution (pH = 4.8) under an oxygen atmosphere (1 atm), produced $DMSO_2$ quantitatively after 3 h of MW heating at 200 °C.

Combined with MW-absorbing materials, such as granular activated carbon (GAC), microwave irradiation has been extensively used in the treatment of wastewater. In this field, a series of MW-assisted catalytic oxidations of phenolic pollutants in aqueous solution have been performed by Quan and co-workers.[94–96] In particular, copper[94,95] or platinum[96] particles dispersed onto the surface of GAC can act as reaction centers for the degradation of *para*-nitrophenol (PNP) and pentachlorophenol (PCP). The reactions were carried out by passing the aqueous solutions through a packed bed reactor containing the catalyst (1–12.6 wt% of Cu or Pt), under ambient pressure and continuous flow mode using a microwave field of 160–850 W. Under these conditions, phenolic pollutants present in solution were degraded and mineralized efficiently (86–99% after 6 min to 5 h). In all cases, compared with GAC itself, Cu/GAC and Pt/GAC catalysts showed higher degradation and mineralization efficiencies.

Similarly, the microwave-assisted catalytic oxidation of chlorine dioxide (ClO_2) in wastewater samples was also investigated.[97] In this case, using CuO_n-La_2O_3/γ-Al_2O_3 as catalyst (8.12 and 1.14% of copper and lanthanum, respectively), a 91.66% removal efficiency could be achieved after only 5 min by keeping the microwave power at 50 W.

2.7 Miscellaneous Reactions

An efficient synthesis of 3-aryl and 3-alkyl substituted crotonaldehydes through a microwave-assisted alkyne–enol ether cross-metathesis in aqueous media has been described by Botta and co-workers (Scheme 2.49).[98] The reactions were performed using 10 mol.% of the Grubbs' second generation catalyst [RuCl$_2$(=CHPh)(PCy$_3$)(H$_2$IMes)] (H$_2$IMes = 1,3-dimesityl-4,5-dihydroimidazol-2-ylidene) in a mixture ^{t}BuOH–H$_2$O (1 : 1) under MW-heating at 80 °C. Moreover, the addition of 2 equiv. of CuSO$_4$ proved to be fundamental to deprotect the corresponding enol ethers (RC(=CH$_2$)CH=CHOEt) that were initially formed in this cross-metathesis process. Under these conditions, several aryl and alkyl alkynes smoothly reacted with ethyl vinyl ether to afford the corresponding 3-substituted crotonaldehydes as a 2 : 1 mixture of E/Z stereoisomers. Remarkably, when toluene was used as solvent instead of the ^{t}BuOH–H$_2$O mixture no traces of the aldehydes were detected, with only the intermediate enol ethers RC(=CH$_2$)CH=CHOEt being formed. It is also interesting that quantitative yields of the (E)-isomers could be obtained after treatment of the (E/Z) mixture of crotonaldehydes with I$_2$.

Hosseinzadeh and co-workers have also described a practical deprotection method of N,N-dimethylhydrazones, to generate the corresponding carbonyl compounds, using a mixture of PdCl$_2$ (3 mol.%) and SnCl$_2$ (3 mol.%) as catalyst in water, a process that is accelerated by performing the reactions in a domestic microwave oven (900 W) (Scheme 2.50).[99] Under these conditions, the process proceeded with good to excellent yields (78–98%) in only 90 s. Moreover, a synergetic effect of the two salts employed was observed. Thus, when SnCl$_2$, PdCl$_2$ or both of them were absent, the corresponding carbonyl compounds were obtained in much lower yields (10–32%).

Similarly, Procopio and co-workers proposed a MW-assisted chemoselective deprotecting methodology for the cleavage of isopropylidene acetals in awkward substrates by using pure water as solvent and erbium(III) trifluoromethane sulfonate Er(OTf)$_3$ as Lewis acid catalyst.[100] Thus, using the optimal reaction conditions [i.e., 1–5 mol.% of Er(OTf)$_3$, pure H$_2$O as solvent, 120 °C, MW

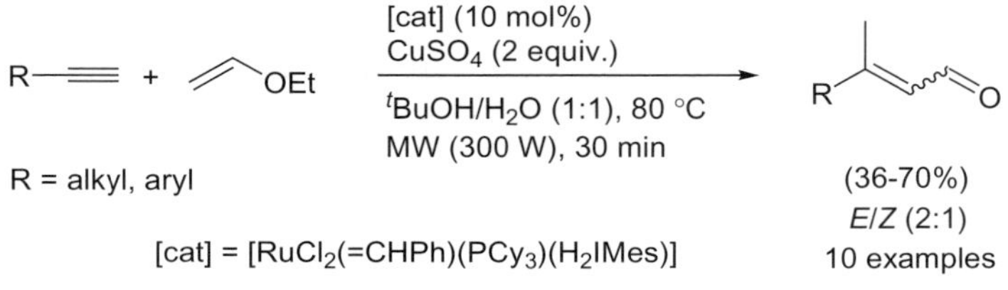

Scheme 2.49

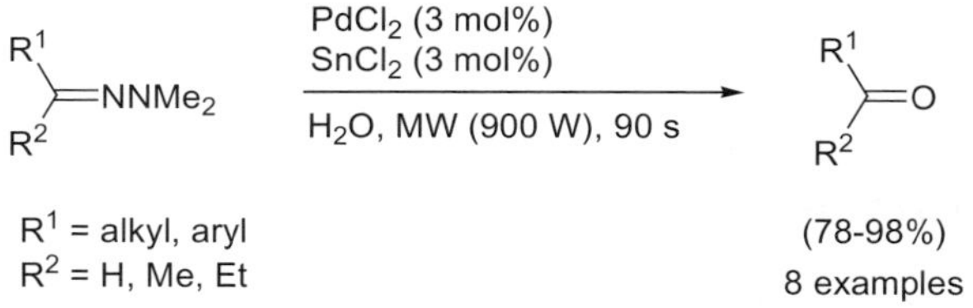

Scheme 2.50

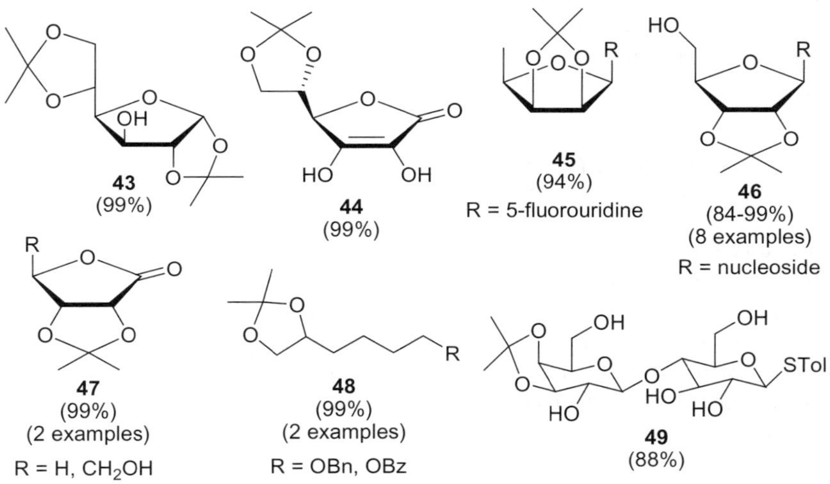

$R^1R^2C{=}NNMe_2$ → $R^1R^2C{=}O$

PdCl$_2$ (3 mol%), SnCl$_2$ (3 mol%), H$_2$O, MW (900 W), 90 s

R^1 = alkyl, aryl
R^2 = H, Me, Et

(78-98%)
8 examples

Figure 2.6

43 (99%)
44 (99%)
45 (94%) R = 5-fluorouridine
46 (84-99%) (8 examples) R = nucleoside
47 (99%) (2 examples) R = H, CH$_2$OH
48 (99%) (2 examples) R = OBn, OBz
49 (88%)

power 1000 W] the deprotection of nucleoside and carbohydrate isopropylidene acetals **43–49** could be achieved selectively and in almost quantitative yield (84–99%) after only 5–30 min (Figure 2.6).[100] Notably, no isopropylidene cleavage was observed upon performing these reactions under conventional heating.

2.8 Representative Experimental Examples

2.8.1 Addition Reactions

2.8.1.1 Synthesis of Amides (Scheme 2.3)[13]

A mixture of the corresponding nitrile (1 mmol), [(IPr)Au(NTf$_2$)] (2–5 mol.%), THF (0.5 mL) and demineralized water (0.5 mL) was introduced in a 2 mL sealed vial and heated in a microwave for 2–6 h at 140 °C (101 psi). After the reaction mixture cooled down to room temperature, volatile compounds were evaporated under reduced pressure and the residue was purified by flash chromatography using pentane–ethyl acetate (from 7:3 to 0:1) to provide the desired amide in 26–99% yield.

2.8.2 Isomerization Reactions

2.8.2.1 Synthesis of Ketones via Alkenols Isomerization (Scheme 2.9)[30]

A mixture of the corresponding alkenol (1 mmol), [RuCl$_2$(PPh$_3$)$_3$] (5 mol.%) and water (3 mL) was introduced into a 10 mL sealed vial and heated by irradiation of microwave (185 °C, 147 psi) for 1 h. After cooling, the resultant mixture was extracted with diethyl ether. Ketones were obtained in 77–96% yield.

2.8.2.2 Synthesis of α,β-Unsaturated Carbonyl Compounds (Scheme 2.12)[36]

A pressure-resistant microwave reactor vial was charged with the corresponding propargyl alcohol (1 mmol), InCl$_3$ (1 mol.%), a magnetic stirring bar and water (1 mL). The vial was irradiated at 160 °C (300 W, 10–90 psi) for 10–360 min. After cooling, the organic product was extracted with diethyl ether and dried with MgSO$_4$. The solvent was evaporated under vacuum and the crude residue purified by flash chromatography over silica gel using EtOAc–hexane (1:10). Products were obtained in 87–99% yield.

2.8.3 Substitution Reactions

2.8.3.1 Synthesis of D-labeled Aryl Silanes (Scheme 2.16)[42]

The substrate (2 mmol), PtO$_2$ (0.1 mmol) and deuterium oxide (1.0 g) were pulverized and mixed completely by a ball mill machine. The mixture was charged in a 10 mL vial with 3 mL of additional deuterium oxide and irradiated, maintaining the temperature at 150 °C (145 psi) for 1 h. The organic product was extracted with diethyl ether, dried over Na$_2$SO$_4$ and concentrated in vacuum. The residue was purified by a short silica-gel column chromatography, to afford the labeled products in 78–81% yield.

2.8.3.2 Synthesis of Phenols (Scheme 2.27)[58]

The aryl halide (5 mmol), copper iodide (0.5 mmol), L-proline (0.25 mmol) and 10 mL of a 1.55 M NaOH aqueous solution were introduced into an 80 mL quartz tube. The vessel was sealed and subjected to a maximum of 1400 W microwave power in a ramp to 300 °C (limited by a maximum pressure of 1160 psi) over a period of 10 min and then held at this temperature for 30 min before being allowed to cool to 50 °C (this takes around 30 min). The reaction mixture was acidified with 2 M hydrochloric acid to pH 5–7. The aqueous layer was extracted with EtOAc (3×15 mL). The organic washings were combined, dried over MgSO$_4$, and then EtOAc was removed under vacuum. Products were obtained in 10–100% yield.

2.8.4 Cyclization Reactions

2.8.4.1 Synthesis of 2,3-Unsubstituted Indole-1-Carboxamides (Scheme 2.31)[65]

A mixture of the N-(2-alkynylphenyl)urea (0.2 mmol), [AuCl(PPh$_3$)] (0.02 mmol) and Ag$_2$CO$_3$ (0.02 mmol) was mixed in water (3–5 mL) under Ar atmosphere. The vial was sealed and the mixture then irradiated for 10 min at 150 °C. After the reaction was cooled to ambient temperature, the crude reaction mixture was extracted three times with ethyl acetate (3×15 mL). The combined organic phase was washed with a saturated NaHCO$_3$ solution, brine, dried with Na$_2$SO$_4$ and concentrated. The residue was purified by column chromatography on CombiFlash® to provide the desired product in 40–91% yield.

2.8.4.2 Synthesis of 1,4-Disubstituted-1,2,3-triazoles (Scheme 2.35)[74]

Attention: copper azides are shock-sensitive when dry. Care should be taken to remove traces of these compounds from the products by washing with basic ammonium citrate or dilute HCl.

Alkyl halide (1.0 mmol), alkyne (1.05 mmol) and sodium azide (0.068 g, 1.05 mmol) were suspended in a 1 : 1 mixture of water–^{t}BuOH (1.5 mL each) in a 10 mL glass vial equipped with a small magnetic stirring bar. Copper turnings (50 mg) and copper sulfate solution (1 M, 200 µL) were added and the vial was then tightly sealed with an aluminium/Teflon crimp top. The mixture was then irradiated for 10–15 min at 75–125 °C, using an irradiation power of 100 W. After completion of the reaction, the vial was cooled to 50 °C by air jet cooling before it was opened. It was then diluted with water (20 mL). The precipitated product was filtered off and washed with cold water (20 mL), followed by 0.25 M HCl (10 mL), and finally with light petroleum (50 mL) to furnish the desired triazole as a white solid (81–93% yield).

2.8.4.3 Synthesis of Triazole Glycodendrimers (**32**, see Scheme 2.40)[82]

The dendrimer **31** (30–100 µmol) and the azido carbohydrate **30** (1.5 equiv. per alkyne group), CuSO$_4$ (30 mol.%), and sodium ascorbate (60 mol.%) were dissolved in DMF (1–1.5 mL) containing several drops of water. The solution was exposed to microwave irradiation at 80 °C for 20 min, then concentrated and purified on a silica column, eluting first with CH_2Cl_2–EtOAc (6:1) to recover the excess of starting azido carbohydrate **30** followed by elution with CH_2Cl_2–MeOH (6:1) to obtain the triazole-linked glycodendrimers **32** in 73–98% yield.

2.8.5 Reduction and Oxidation Reactions

2.8.5.1 Synthesis of Dihydrocinnamic Esters (Scheme 2.42)[87]

$PdCl_2$ (0.05–0.08 g), $HCOONH_4$ (25 mmol) and silica gel (12 g; 60–120 mesh) were thoroughly mixed in a mortar and then transferred to a round bottom flask. The corresponding conjugated alkene (50 mmol) and $HCOOH-H_2O$ (1:2) (15 mL) were added to this powdered mixture. The resultant mixture was then irradiated with a focused monomode microwave (300 W) system for 4–10 min. After completion of the reaction, the solid residue was washed with AcOEt (3×20 mL) and the combined organic layer washed with H_2O (3×15 mL) and brine (10 mL), dried (Na_2SO_4), and evaporated to obtain dihydrocinnamic esters in 91–97% yield.

2.8.5.2 Synthesis of Secondary Amines (**39**, see Scheme 2.44)[91]

A mixture of alkyl amine **38** (1.37 mmol), 50 mg of 5 wt% of Pt/C, 30 mg (1.1 mmol), and 2 mL of water in a pressure-resistant glass ampoule equipped with rubber septum and aluminium cap was introduced into the cavity of microwave apparatus (CEM Discover, 2.45 GHz, CEM Corporation, NC, USA). The reaction mixture was irradiated by microwave (50 W) continuously for 10–60 min. The reaction vessel was cooled in an ice-water bath and the organic materials were extracted with ether. The ether fraction was analyzed by GC and GC-MS for identification and quantification of the secondary amines **39** (21–100% yield).

2.8.6 Miscellaneous Reactions

2.8.6.1 Synthesis of (E) and (Z) 3-Substituted Crotonaldehydes (Scheme 2.49)[98]

The corresponding alkyne (1.0 mmol), ethyl vinyl ether (9.0 mmol), $CuSO_4$ (2.0 mmol), and Grubbs' catalyst [$RuCl_2$(=CHPh)(PCy_3)(H_2IMes)] (0.1 mmol) were suspended in a 1:1 mixture of water and ^{t}BuOH (2.0 mL each) in a 10 mL glass vial equipped with a small magnetic stirring bar. The mixture was heated at 80 °C under microwave irradiation for 3×10 min, using an irradiation power of 300 W. The mixture was then poured into a solution of NH_4Cl (20 mL), NH_4OH (0.5 mL), and Et_2O (10 mL), stirred for an additional 10 min, and then extracted with Et_2O (2×10 mL). The combined organic phases were washed with brine, dried over Na_2SO_4, filtered, and concentrated under reduced pressure. The crude product was purified by flash column chromatography (SiO$_2$) using 1:4 Et$_2$O–hexanes as the eluent to obtain the crotonaldehydes as tan oils (36–70% yield).

2.9 Conclusions

The search for environmentally benign substitutes for volatile and toxic organic solvents, as well as the development of lower energetic cost methodologies, has gained much attention in recent years in view of the increasing importance of green chemistry. In this sense, the combined use of water as solvent and microwave heating has become very popular since a great variety of synthetic organic transformations, in particular transition-metal catalyzed processes, can be carried out very efficiently and rapidly under these environmentally benign conditions. Although in most cases comparison studies with standard organic solvents and/or conventional heating were not performed, it is quite clear from the growing number of publications in this field that there are tremendous advantages in terms of time savings and increasing efficiencies derived from the synergism between water and microwaves. Obviously, the area remains open with many opportunities for new discoveries and, therefore, intense research is expected in this field in the coming years.

Acknowledgements

Financial support from the Spanish MICINN [Projects CTQ2006-08485/BQU, CTQ2009-08746/BQU and Consolider Ingenio 2010 (CSD2007-00006)], the Gobierno del Principado de Asturias (FICYT Project IB08-036), and the University of Oviedo (Grant: UNOV-09-MB-4) is acknowledged. S.E.G.-G. also thanks MICINN and the European Social Fund for the award of a Ramón y Cajal contract.

References

1. See, for example: (a) A. S. Matlack, *Introduction to Green Chemistry*, Marcel Dekker, New York, 2001; (b) M. Lancaster, *Green Chemistry: An Introductory Text*, Royal Society of Chemistry, Cambridge, 2002; (c) P. T. Anastas and M. M. Kirchhoff, *Acc. Chem. Res.*, 2002, **35**, 686.
2. See, for example: (a) U. M. Lindström, *Chem. Rev.*, 2002, **102**, 2751; (b) C.-J. Li, *Chem. Rev.*, 2005, **105**, 3095; (c) *Organic Reactions in Water: Principles, Strategies and Applications*, ed. U. M. Lindström, Blackwell, Oxford, 2007; (d) C. I. Herrerías, X. Yao, Z. Li and C. J. Li, *Chem. Rev.*, 2007, **107**, 2546; (e) F. M. Kerton, *Alternative Solvents for Green Chemistry*, RSC Publishing, Cambridge, 2009.
3. See, for example: (a) S. Narayan, J. Muldoon, M. G. Finn, V. V. Fokin, H. C. Kolb and K. B. Sharpless, *Angew. Chem. Int. Ed.*, 2005, **44**, 3275; (b) Y. Hayashi, *Angew. Chem. Int. Ed.*, 2006, **45**, 8103.

4. See, for example: (a) *Aqueous-Phase Organometallic Catalysis: Concepts and Applications*, eds B. Cornils and W. A. Hermann, Wiley-VCH Verlag GmbH, Weinheim, 1998; (b) F. Joó, *Aqueous Organometallic Catalysis*, Kluwer, Dordrecht, 2001; (c) N. Pinault and D. W. Bruce, *Coord. Chem. Rev.*, 2003, **241**, 1; (d) K. H. Shaughnessy, *Chem. Rev.*, 2009, **109**, 643.
5. See, for example: (a) C. O. Kappe, *Angew. Chem. Int. Ed.*, 2004, **43**, 6250; (b) C. O. Kappe, *Chem. Soc. Rev.*, 2008, **37**, 1127; (c) C. R. Strauss, *Aust. J. Chem.*, 2009, **62**, 3; (d) V. Polshettiwar, M. N. Nadagouda and R. S. Varma, *Aust. J. Chem.*, 2009, **62**, 16.
6. M. J. Gronnow, R. J. White, J. H. Clark and D. J. Macquarrie, *Org. Process Res. Dev.*, 2005, **9**, 516.
7. For recent reviews, see: (a) P. Appukkuttan and E. Van der Eycken, *Eur. J. Org. Chem.*, 2008, 1133; (b) S. Caddick and R. Fitzmaurice, *Tetrahedron*, 2009, **65**, 3355; (c) F. Nicks, Y. Borguet, S. Delfosse, D. Bicchielli, L. Delaude, X. Sauvage and A. Demonceau, *Aust. J. Chem.*, 2009, **62**, 184.
8. For reviews on this topic, see: (a) D. Dallinger and C. O. Kappe, *Chem. Rev.*, 2007, **107**, 2563; (b) V. Polshettiwar and R. S. Varma, *Acc. Chem. Res.*, 2008, **41**, 629; (c) V. Polshettiwar and R. S. Varma, *Chem. Soc. Rev.*, 2008, **37**, 1546.
9. (a) *The Amide Linkage: Structural Significance in Chemistry, Biochemistry and Material Science*, ed. A. Greenberg, C. M. Breneman and J. F. Liebman, John Wiley & Sons, Inc., New York, 2000; (b) I. Johansson, in *Kirk-Othmer Encyclopedia of Chemical Technology*, John Wiley & Sons, Hoboken NJ, 5th edn., 2004, Vol. **2**, p. 442.
10. (a) N. A. Bokach and V. Y. Kukushkin, *Russ. Chem. Rev.*, 2005, **74**, 153; (b) V. Y. Kukushkin and A. J. L. Pombeiro, *Inorg. Chim. Acta*, 2005, **358**, 1.
11. V. Cadierno, J. Francos and J. Gimeno, *Chem. Eur. J.*, 2008, **14**, 6601.
12. V. Polshettiwar and R. S. Varma, *Chem. Eur. J.*, 2009, **15**, 1582.
13. R. S. Ramón, N. Marion and S. P. Nolan, *Chem. Eur. J.*, 2009, **15**, 8695.
14. K. Manjula and M. A. Pasha, *Synth. Commun.*, 2007, **37**, 1545.
15. (a) A. D. Allen, Y. Chiang, A. J. Kresge and T. T. Tidwell, *J. Org. Chem.*, 1982, **47**, 775; (b) T. Tsuchimoto, T. Joya, E. Shirakawa and Y. Kawakami, *Synlett*, 2000, 1777.
16. (a) F. Alonso, I. P. Beletskaya and M. Yus, *Chem. Rev.*, 2004, **104**, 3079; (b) L. Chen and C.-J. Li, *Adv. Synth. Catal.*, 2006, **348**, 1459.
17. J. M. Kremsner and C. O. Kappe, *Eur. J. Org. Chem.*, 2005, 3672.
18. A. Vasudevan and M. K. Verzal, *Synlett*, 2004, 631.
19. L. Coudray and J.-L. Montchamp, *Eur. J. Org. Chem.*, 2008, 3601.
20. M. D. Milton, G. Onodera, Y. Nishibayashi and S. Uemura, *Org. Lett.*, 2004, **6**, 3993.
21. R. A. Stockland Jr., A. J. Lipman, J. A. Bawiec III, P. E. Morrison, I. A. Guzei, P. M. Findeis and J. F. Tamblin, *J. Organomet. Chem.*, 2006, **691**, 4042.

22. H. Yamamoto, in: *Comprehensive Organic Synthesis*, ed. B. M. Trost, I. Fleming and C. H. Heathcock, Pergamon Press, Exeter, 1991, Vol. **2**, ch. 1.3, p. 81.
23. See, for example: (a) L. W. Bieber, M. F. da Silva, R. C. da Costa and L. O. S. Silva, *Tetrahedron Lett.*, 1998, **39**, 3655; (b) M.-J. Lin and T.-P. Loh, *J. Am. Chem. Soc.*, 2003, **125**, 13042.
24. (a) C. Z. Gu, Q. R. Li and H. Yin, *Chin. Chem. Lett.*, 2005, **16**, 1573; (b) Q.-R. Li, C.-Z. Gu and H. Yin, *Chin. J. Chem.*, 2006, **24**, 72.
25. See for example: (a) L. C. Akullian, M. L. Snapper and A. H. Hoveyda, *Angew. Chem. Int. Ed.*, 2003, **42**, 4244; (b) L. Zani and C. Bolm, *Chem. Commun.*, 2006, **20**, 4263.
26. L. Shi, Y.-Q. Tu, M. Wang, F.-M. Zhang and C.-A. Fan, *Org. Lett.*, 2004, **6**, 1001.
27. See, for example: C. Wei and C.-J. Li, *J. Am. Chem. Soc.*, 2003, **125**, 9584.
28. (a) R. C. van der Drift, E. Bouwman and E. Drent, *J. Organomet. Chem.*, 2002, **650**, 1; (b) R. Uma, C. Crévisy and R. Grée, *Chem. Rev.*, 2003, **103**, 27; (c) V. Cadierno, P. Crochet and J. Gimeno, *Synlett*, 2008, 1105.
29. See, for example: (a) S. H. Bergens and B. Bosnich, *J. Am. Chem. Soc.*, 1991, **113**, 958; (b) D. V. McGrath and R. H. Grubbs, *Organometallics*, 1994, **13**, 224; (c) P. Crochet, J. Díez, M. A. Fernández-Zúmel and J. Gimeno, *Adv. Synth. Catal.*, 2006, **348**, 93.
30. K. Ishibashi, M. Takahashi, Y. Yokota, K. Oshima and S. Matsubara, *Chem. Lett.*, 2005, **34**, 664.
31. M. Takahashi, K. Oshima and S. Matsubara, *Tetrahedron Lett.*, 2003, **44**, 9201.
32. See, for example, D. Wang, D. Chen, J. X. Haberman and C.-J. Li, *Tetrahedron*, 1998, **54**, 5129 and references cited therein.
33. G. Liu, Y. Sun, J. Wang, C. Sun, F. Zhang and H. Li, *Green Chem.*, 2009, **11**, 1477.
34. S. Swaminathan and K. V. Narayanan, *Chem. Rev.*, 1971, **71**, 429.
35. (a) D. A. Engel and G. B. Dudley, *Org. Biomol. Chem.*, 2009, **7**, 4149; (b) V. Cadierno, P. Crochet, S. E. García-Garrido and J. Gimeno, *Dalton Trans.*, (DOI: 10.1039/b923602c).
36. V. Cadierno, J. Francos and J. Gimeno, *Tetrahedron Lett.*, 2009, **50**, 4773.
37. *Synthesis and Applications of Isotopically Labelled Compounds*, ed. D. Dean, C. N. Filer and K. E. McCarthy, John Wiley and Sons, Ltd., London, 2004.
38. J. Atzrodt, V. Derdau, T. Fey and J. Zimmermann, *Angew. Chem. Int. Ed.*, 2007, **46**, 7744.
39. M. Takahashi, K. Oshima and S. Matsubara, *Chem. Lett.*, 2005, **34**, 192.
40. K. Ishibashi and S. Matsubara, *Chem. Lett.*, 2007, **36**, 724.
41. (a) E. A. Cioffi, R. H. Bell and B. Le, *Tetrahedron: Asymmetry*, 2005, **16**, 471; (b) S. S. Bokatzian-Johnson, M. L. Maier, R. H. Bell, K. E. Alston, B. Y. Le and E. A. Cioffi, *J. Labelled Compd. Radiopharm.*, 2007, **50**, 380.
42. M. Yamamoto, K. Oshima and S. Matsubara, *Org. Lett.*, 2004, **6**, 5015.

43. J. M. Barthez, A. V. Filikov, L. B. Frederiksen, M.-L. Huguet, J. R. Jones and S.-Y. Lu, *Can. J. Chem.*, 1998, **76**, 726.
44. S. Anto, G. S. Getvoldsen, J. R. Harding, J. R. Jones, S.-Y. Lu and J. C. Russell, *J. Chem. Soc., Perkin Trans. 2*, 2000, **2**, 2208.
45. W.-M. Yau and K. Gawrisch, *J. Labelled Compd. Radiopharm.*, 1999, **42**, 702.
46. V. Derdau, J. Atzrodt, J. Zimmermann, C. Kroll and F. Brückner, *Chem. Eur. J.*, 2009, **15**, 10397.
47. (a) N. Ito, T. Watahiki, T. Maesawa, T. Maegawa and H. Sajiki, *Adv. Synth. Catal.*, 2006, **348**, 1025; (b) N. Ito, T. Watahiki, T. Maesawa, T. Maegawa and H. Sajiki, *Synthesis*, 2008, 1467.
48. See, for example: (a) V. Birke, *Terratech.*, 1998, **7**, 52; (b) F. Kastanek and M. Kuras, *Environ. Studies*, 2004, **10**, 441; (c) F. Kastanek, Y. Maletorova, P. Katanek, J. Rott, V. Jiricny and K. Jiratova, *Desalination*, 2007, **21**, 261.
49. H. Hidaka, A. Saitou, H. Honjou, K. Hosoda, M. Moriya and N. Serpone, *J. Hazard. Mat.*, 2007, **148**, 22.
50. (a) K. Kunz, U. Scholz and D. Ganzer, *Synlett*, 2003, 2428; (b) F. Monnier and M. Taillefer, *Angew. Chem. Int. Ed.*, 2008, **47**, 3096; (c) F. Monnier and M. Taillefer, *Angew. Chem. Int. Ed.*, 2009, **48**, 6954.
51. X. Zhu, Y. Ma, L. Su, H. Song, G. Chen, D. Liang and Y. Wan, *Synthesis*, 2006, 3955.
52. S. Röttger, P. J. R. Sjöberg and M. Larhed, *J. Comb. Chem.*, 2007, **9**, 204.
53. L. D. S. Yadav, B. S. Yadav and V. K. Rai, *Synthesis*, 2006, 1868.
54. A. Martín, R. F. Pellón, M. Mesa, M. L. Docampo and V. Gómez, *J. Chem. Res.*, 2005, 561.
55. D. Guo, H. Huang, Y. Zhou, J. Xu, H. Jiang, K. Chen and H. Liu, *Green Chem.*, 2010, **12**, 276.
56. M. C. Bagley, M. C. Dix and V. Fusillo, *Tetrahedron Lett.*, 2009, **50**, 3661.
57. See, for example: D. D. Weller, E. P. Stirchak and A. Yokoyama, *J. Org. Chem.*, 1984, **49**, 2061.
58. C. M. Kormos and N. E. Leadbeater, *Tetrahedron*, 2006, **62**, 4728.
59. J. Andersen, U. Madsen, F. Björkling and X. Liang, *Synlett*, 2005, 2209.
60. T. A. Bryson, J. J. Stewart, J. M. Gibson, P. S. Thomas and J. K. Berch, *Green Chem.*, 2003, **5**, 174.
61. R. R. Poondra and N. J. Turner, *Org. Lett.*, 2005, **7**, 863.
62. See, for example: (a) T. Shibata, *Adv. Synth. Catal.*, 2006, **348**, 2328; (b) D. Strübing and M. Beller, *Top. Organomet. Chem.*, 2006, **18**, 165.
63. H. W. Lee, F. Y. Kwong and A. S. C. Chan, *Synlett*, 2008, 1553.
64. See, for example: (a) T. E. Müller and M. Beller, *Chem. Rev.*, 1998, **98**, 675; (b) F. Pohlki and S. Doye, *Chem. Soc. Rev.*, 2003, **32**, 104.
65. D. Ye, J. Wang, X. Zhang, Y. Zhou, X. Ding, E. Feng, H. Sun, G. Liu, H. Jiang and H. Liu, *Green Chem.*, 2009, **11**, 1201.
66. X. Zhang, D. Ye, H. Sun, D. Guo, J. Wang, H. Huang, X. Zhang, H. Jiang and H. Liu, *Green Chem.*, 2009, **11**, 1881.

67. O. Arjona, A. M. Gómez, J. C. López and J. Plumet, *Chem. Rev.*, 2007, **107**, 1919.
68. K.-S. Ko, C. J. Zea and N. L. Pohl, *J. Am. Chem. Soc.*, 2004, **126**, 13188.
69. I.-H. Chen, J.-N. Young and S. J. Yu, *Tetrahedron*, 2004, **60**, 11903.
70. See, for example: W. H. Binder and R. Sachsenhofer, *Macromol. Rapid Commun.*, 2007, **28**, 15 and references cited therein.
71. R. Huisgen, *Angew. Chem. Int. Ed. Engl.*, 1963, **2**, 565.
72. V. V. Rostovtsev, L. G. Green, V. V. Fokin and K. B. Sharpless, *Angew. Chem. Int. Ed.*, 2002, **41**, 2596.
73. C. W. Tornøe, C. Christensen and M. Medal, *J. Org. Chem.*, 2002, **67**, 3057.
74. P. Appukkuttan, W. Dehaen, V. V. Fokin and E. van der Eycken, *Org. Lett.*, 2004, **6**, 4223.
75. N. Kaval, D. Ermolat'ev, P. Appukkuttan, W. Dehaen, C. O. Kappe and E. Van der Eycken, *J. Comb. Chem.*, 2005, **7**, 490.
76. D. Ermolat'ev, W. Dehaen and E. van der Eycken, *QSAR Comb. Sci.*, 2004, **23**, 915.
77. D. S. Ermolat'ev, V. P. Mehta and E. van der Eycken, *QSAR Comb. Sci.*, 2007, **26**, 1266.
78. C. Bouillon, A. Meyer, S. Vidal, A. Jochum, Y. Chevolot, J.-P. Cloarec, J.-P. Praly, J.-J. Vasseur and F. Morvan, *J. Org. Chem.*, 2006, **71**, 4700.
79. H. S. G. Beckmann and V. Wittmann, *Org. Lett.*, 2007, **9**, 1.
80. See, for example: G. R. Newkome, C. N. Moorefield and F. Vögtle, in *Dendrimers and Dendrons: Concepts, Syntheses, Applications*, John Wiley & Sons, Inc., New York, 2001.
81. D. T. S. Rijkers, G. W. van Esse, R. Merkx, A. J. Brouwer, H. J. F. Jacobs, R. J. Pieters and R. M. J. Liskamp, *Chem. Commun.*, 2005, 4581.
82. J. A. F. Joosten, N. T. H. Tholen, F. A. E. Maate, A. J. Brouwer, G. W. van Esse, D. T. S. Rijkers, R. M. J. Liskamp and R. J. Pieters, *Eur. J. Org. Chem.*, 2005, 3182.
83. I. Dijkgraaf, A. Y. Rijnders, A. Soede, A. C. Dechesne, G. W. van Esse, A. J. Brouwer, F. H. M. Corstens, O. C. Boerman, D. T. S. Rijkers and R. M. J. Liskamp, *Org. Biomol. Chem.*, 2007, **5**, 935.
84. K. Yoon, P. Goyal and M. Weck, *Org. Lett.*, 2007, **9**, 2051.
85. See, for example: (a) S. E. Clapman, A. Hadzovic and R. H. Morris, *Coord. Chem. Rev.*, 2004, **248**, 2201; b) S. Gladiali and A. Alberico, *Chem. Soc. Rev.*, 2006, **35**, 226; J. S. M. Samec, J.-E. Bäckvall, P. G. Andersson and P. Brandt, *Chem. Soc. Rev.*, 2006, **35**, 237; d) T. Ikariya and A. J. Blacker, *Acc. Chem. Res.*, 2007, **40**, 1300.
86. (a) A. Sharma, B. P. Joshi and A. K. Sinha, *Chem. Lett.*, 2003, **32**, 1186; (b) A. Sharma, V. Kumar and A. K. Sinha, *Adv. Synth. Catal.*, 2006, **348**, 354.
87. V. Kumar, A. Sharma and A. K. Sinha, *Helv. Chim. Acta*, 2006, **89**, 483.
88. B. L. A. P. Devi, M. S. L. Karuna, K. N. Rao, P. S. Saiprasad and R. B. N. Prasad, *J. Am. Oil Chem. Soc.*, 2003, **80**, 1003.

89. See, for example: (a) A. R. Katrizky, O. Meth-Cohn and C. W. Rees, (eds), *Comprehensive Organic Functional Group Transformations*, Pergamon, London, 1995 Vol. **3**; (b) I. T. Harrison and S. Harrison (eds), *Compendium of Organic Synthetic Methods*, John Wiley & Sons, Inc., New York, 1971, Vol. **1**.
90. A. Miyazawa, K. Tanaka, T. Sakakura, M. Tashiro, H. Tashiro, G. K. S. Prakash and G. A. Olah, *Chem. Commun.*, 2005, 2104.
91. A. Miyazawa, K. Saitou, K. Tanaka, T. M. Gädda, M. Tashiro, G. K. S. Prakash and G. A. Olah, *Tetrahedron Lett.*, 2006, **47**, 1437.
92. W. Bonrath, R. Karge, M. Nuechter and B. Ondruschka, *PCT Int. Appl.*, WO 03/091235 A1 (CAN 139:360625), 2003.
93. M. Bonchio, M. Carraro, A. Sartorel, G. Scorrano and U. Kortz, *J. Mol. Catal. A: Chem.*, 2006, **251**, 93.
94. X. Liu, X. Quan, L. Bo, S. Chen, Y. Zhao and M. Chang, *Appl. Catal. A*, 2004, **264**, 53.
95. L. L. Bo, Y. B. Zhang, X. Quan and B. Zhao, *J. Hazard. Mater.*, 2008, **153**, 1201.
96. L. Bo, X. Quan, X. Wang and S. Chen, *J. Hazard. Mater.*, 2008, **157**, 179.
97. X. Bi, P. Wang and H. Jiang, *J. Hazard. Mater.*, 2008, **154**, 543.
98. D. Castagnolo, L. Botta and M. Botta, *J. Org. Chem.*, 2009, **74**, 3172.
99. R. Hosseinzadeh, A. Sharifi, K. Tabar-Heydar and F. Mohsenzadeh, *Monatsh. Chem.*, 2002, **133**, 1413.
100. A. Procopio, M. Gaspari, M. Nardi, M. Oliverio and R. Romeo, *Tetrahedron Lett.*, 2008, **49**, 1961.

CHAPTER 3
Microwave-assisted Coupling Reactions in Aqueous Media

AZIZ FIHRI AND CHRISTOPHE LEN

Transformations Intégrées de la Matière Renouvelable ESCOM-UTC, 1-Allée du Jean-Marie Buckmaster, 60200, Compiègne, France

3.1 Introduction

The emergence of transition metal chemistry in the 1960s has profoundly changed the science of organic synthesis. Used in catalytic or stoichiometric quantities, these metals have provided exceptional tools for the development of countless synthetic protocols, with superb reactivity and selectivity. Palladium has taken centre stage when it comes to the synthesis of complex and functionalized organic molecules involving cross-coupling reactions. Many Pd-catalyzed reactions have been developed over the past three decades and the growing number of scientific publications in this area shows its high demand. Heck, Suzuki, Sonogashira and Stille are among the widely used reactions for the formation of carbon–carbon bonds. These reactions were mostly carried out by homogeneous catalytic systems, because of their high reactivity, high turnover numbers, milder reaction conditions, and notably the possibility of coupling of the widely available and low-cost aryl chlorides.[1] However, the efficient separation and subsequent recycling of homogeneous transition-metal catalysts remains a scientific challenge and an aspect of economical and ecological relevance. Heterogeneous Pd catalyst systems are found to be highly effective in overcoming some of these issues;[2,3] More recently, microwave

(MW)-assisted coupling in aqueous medium has become a dominant choice among chemists.[4,5]

The present chapter describes several sustainable and economical protocols that have been developed in recent decades for cross-coupling reactions using microwave irradiation to identify greener alternatives for the cross-coupling reactions. The main impetus for focusing on this technique is its increasing availability to a wider chemical community, including non-specialists and, in particular, to researchers involved in catalyzed carbon–carbon bond formation or in the domain of total synthesis.

3.2 Suzuki–Miyaura Reaction

The Suzuki–Miyaura coupling reaction is one of the most important synthetic transformations of the twentieth century.[6–8] Its usage has dramatically changed the science of organic synthesis to the extent that it is difficult to find examples in synthetic organic chemistry that do not feature this process. Several sustainable and economical protocols have been developed in past decades for the synthesis of biaryls. Among these, the microwave heating technique has become increasingly popular in synthetic organic reactions and transformations in the last decade. The first report on the general conditions for the enhancement of the Suzuki cross-coupling reaction was provided by Larhed and Hallberg who demonstrated that phenylboronic acid can be coupled with 4-bromotoluene to give a modest yield of desired product (Scheme 3.1).[9,10] The same coupling can be conducted under conventional thermal conditions although longer reaction times are required (4 h).

Ligand-free methodology has been developed for MW-mediated Suzuki coupling in aqueous medium. Various biaryls can be prepared from phenylboronic acid and aryl halides, including aryl chlorides (Scheme 3.2).[11] This approach has many advantages such as low palladium loadings (0.4 mol.%), rapid reaction times (5–10 min) and ease of reaction. The dual role of tetrabutylammonium bromide (TBAB) is that it facilitates the solvation of the organic substrates in the solvent medium and, secondly, it activates the boronic acid by the formation of $[ArB(OH)_3]^-[R_4N]^+$ species. Applying the same protocol, a ten-fold scale-up was possible under MW-assisted open-vessel reflux conditions for 10 min at 110 °C, with nearly identical yields to those obtained in closed-vessel runs. In addition, comparison of the reactions performed using MW heating with those using conventional thermal heating showed no non-thermal microwave effects.[12]

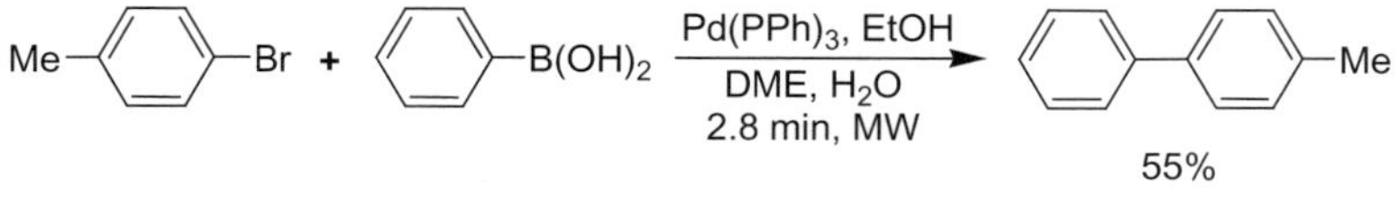

Me—⟨C₆H₄⟩—Br + ⟨C₆H₅⟩—B(OH)₂ →[Pd(PPh₃)₃, EtOH; DME, H₂O; 2.8 min, MW] ⟨C₆H₄⟩—⟨C₆H₄⟩—Me   55%

Scheme 3.1

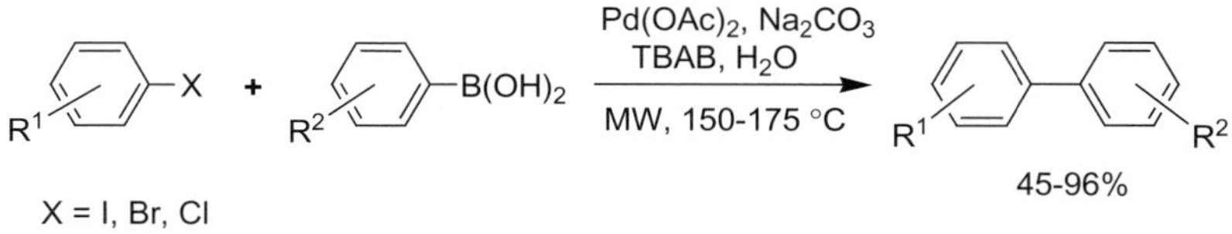

Scheme 3.2

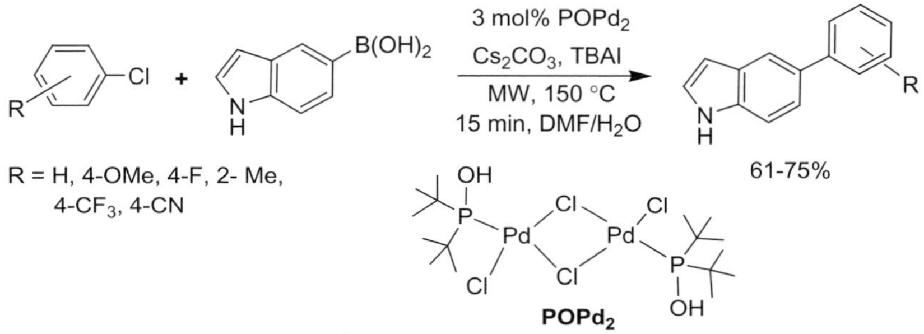

Scheme 3.3

TBAI = tetrabutylammonium iodide

Regarding the organic substrates, remarkable progress has been reported on the activation of the relatively inert aryl bromides and chlorides relative to their iodide analogues; indeed, the low reactivity of chlorides is usually attributed to the strength of the C–Cl bond (bond dissociation energies for Ph–X: Cl = 96, Br = 81 and I = 65 kcal mol^{-1}).[13] As a consequence, the development of a MW-assisted method using the widely available and low-cost aryl chlorides has attracted the attention of several research groups.

In this context a novel and effective method for the Suzuki cross-coupling reaction of aryl chlorides in aqueous media has been studied using the dihydrogen di-μ-chloro-dichlorobis(di-*tert*-butylphosphinito-kP)dipalladate (POPd$_2$).[14] This catalyst system, under microwave conditions (150 °C, 15 min), provided desired products with yields ranging from 61% to 75% (Scheme 3.3). As the authors noted, a decisive advantage of POPd$_2$ is its robustness towards air and moisture, which allows it to be used in water.

Arvela and Leadbeater have developed a MW methodology for the Suzuki coupling of aryl chlorides with phenylboronic acid using palladium on carbon as a catalyst and water as a solvent (Scheme 3.4).[15] They observed that simultaneous cooling in conjunction with microwave heating prolongs the lifetime of the aryl chloride substrates during the reaction. As a result, with substrates bearing electron-neutral or electron-donating substituents, yields of the desired products can be increased as well as overall recovery of material.

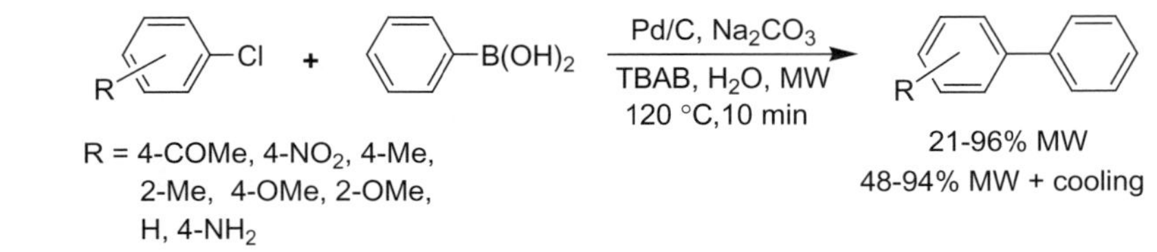

Scheme 3.4

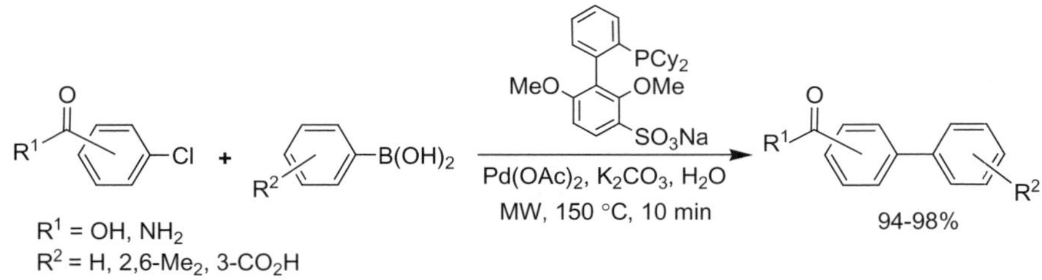

Ar-Cl + Ar'-B(OH)$_2$ → biaryl

R = 4-COMe, 4-NO$_2$, 4-Me, 2-Me, 4-OMe, 2-OMe, H, 4-NH$_2$

Conditions: Pd/C, Na$_2$CO$_3$, TBAB, H$_2$O, MW, 120 °C, 10 min

21-96% MW
48-94% MW + cooling

Scheme 3.5

R^{1}COAr-Cl + Ar'(R^2)-B(OH)$_2$ → coupled product

R^1 = OH, NH$_2$
R^2 = H, 2,6-Me$_2$, 3-CO$_2$H

Ligand: 2-(2',6'-dimethoxybiphenyl)-SO$_3$Na with PCy$_2$

Conditions: Pd(OAc)$_2$, K$_2$CO$_3$, H$_2$O, MW, 150 °C, 10 min

94-98%

Buchwald and Anderson have reported the synthesis and the catalytic activity of a new ligand that incorporates a water-solubilizing sulfonate group, for the Suzuki coupling of aryl chlorides with arylboronic acids in water.[16] Several of both hydrophobic and hydrophilic substrates, including heterocyclic compounds, could be coupled in excellent yields, even at room temperature in some cases. Using microwave irradiation (150 °C) allowed one to both substantially shorten reaction times and reduce the catalyst loading from 2 to 0.1 mol.% without any decrease in activity (Scheme 3.5).

Under similar reaction conditions, an expeditious and highly efficient single-step methodology for the introduction of a phenylalanine moiety into position 8 and 6 of the purine scaffold has been developed based on aqueous-phase Pd-catalysed Suzuki–Miyaura cross-coupling reactions of unprotected 4-boronophenylalanine with 8-bromo- or 6-chloropurines.[17] Classical heating was more efficient for the synthesis of 8-substituted nucleosides and more labile nucleotides, while microwave heating was more efficient for purine bases and 6-substituted nucleosides (Scheme 3.6).

The scope and limitations of this protocol were studied to show that they were fairly general and were applicable to a wide range of 9-unsubstituted 2-, 6-, and 8-halopurine bases with diverse aryl- and alkenylboronic acids that could generate the corresponding arylpurines in a single step.[18] Electron-rich phenylboronic acids tend to be most reactive, whereas hetarylboronic acids showed lower reactivity. The obvious advantage of this system has been that in most cases the products crystallized from the reaction mixtures and could be isolated by simple recrystallization in high yields.

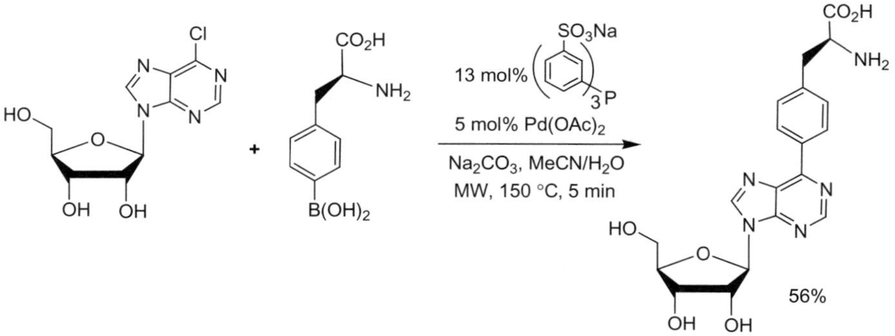

Scheme 3.6

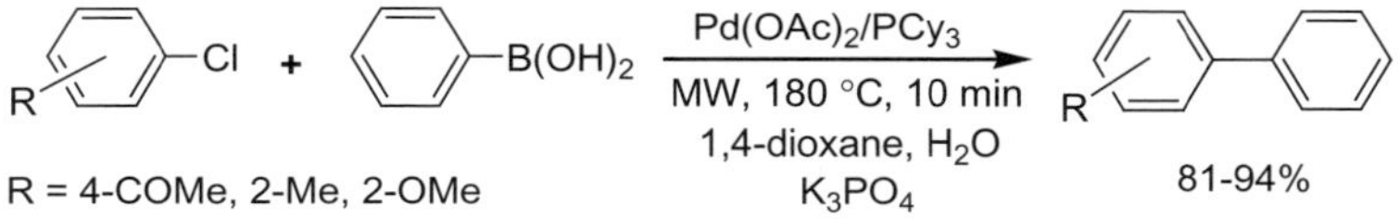

Scheme 3.7

Catalysts formed *in situ* from tricyclohexylphosphine (PCy$_3$) and palladium acetate showed very good activity in the Suzuki coupling of deactivated, non-activated and activated aryl chloride substrates under microwave heating (Scheme 3.7).[19] The reaction conditions enabled the use of several bases but the selection of inexpensive K$_3$PO$_4$ became obvious as it is tolerant of a wide range of substrate functionality. The main drawback of this ligand has been its instability under air, leading to the use of glove box.

An easily accessible and moisture-stable palladium complex containing salicylaldehyde *N*(4)-hexamethyleneiminylthiosemicarbazone (**1**) has been designed by Loupy and co-workers.[20]

This complex was inactive towards Suzuki–Miyaura coupling under aerobic conditions with conventional heating. In contrast, microwave irradiation promoted the effective catalytic activity of the complex for the coupling of aryl bromides with phenylboronic acid in aqueous media, under essentially similar conditions to those used under conventional heating.

These results led the authors to propose the involvement of specific microwave effects rather than thermal effects for the acceleration of this reaction. However, the coupling of the activated 1-chloro-4-nitrobenzene with phenylboronic acid yielded only 25% of coupling product after 1 h at 155 °C (Scheme 3.8).

Despite severe steric hindrance, axially chiral 1,8-bis[3-(3,5-dimethylphenyl)-9-acridyl]naphthalene was obtained *via* microwave-assisted Suzuki coupling of 4-bromo-2-chlorobenzoic acid and 3,5-dimethylphenylboronic acid followed

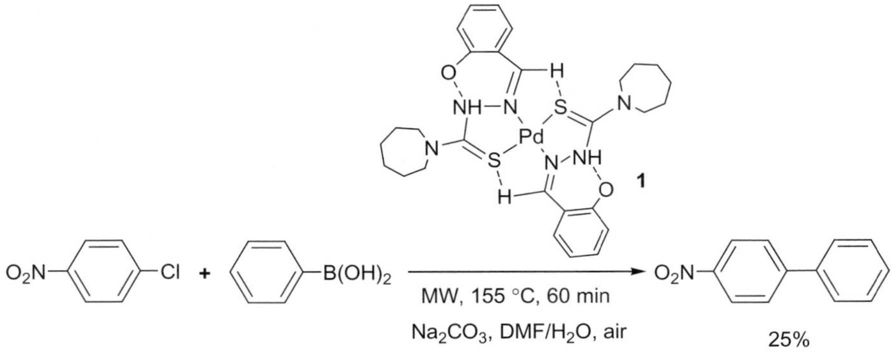

Scheme 3.8

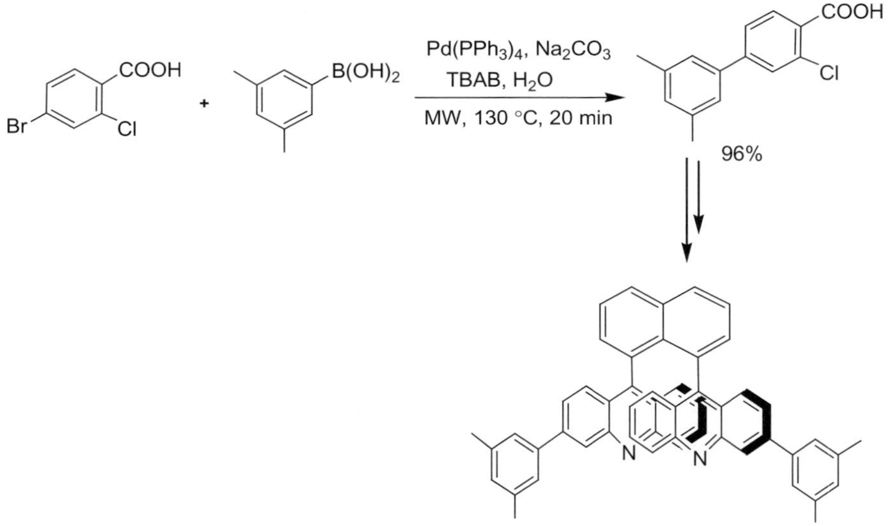

Scheme 3.9

by regioselective amination with aniline and acridine ring construction in phosphorus oxybromide and then two consecutive Stille cross-coupling steps.[21] The product of the Suzuki reaction was obtained in 96% yield by using 1 mol.% of Pd(PPh$_3$)$_4$ and Na$_2$CO$_3$ at 130 °C for 20 min (Scheme 3.9).

Ultrasound has been employed to promote the cross-coupling of 4-chloronitrobenzene with phenylboronic acid in aqueous DME (Scheme 3.10).[22] However, the results obtained remain modest and inferior to conventional thermal activation. The combination of ultrasound and microwaves appear to provide better results, although only three such examples have been demonstrated.

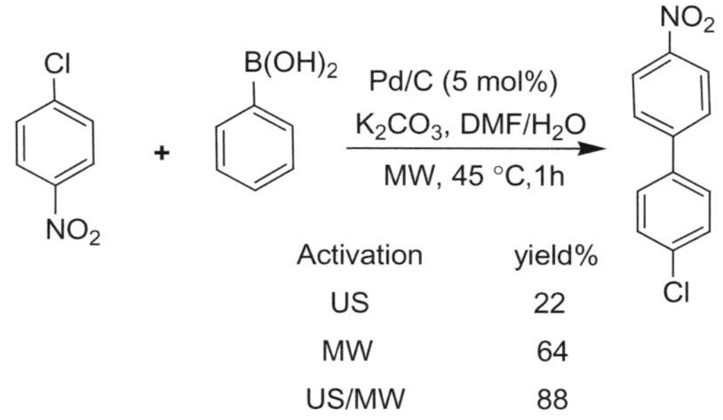

Scheme 3.10

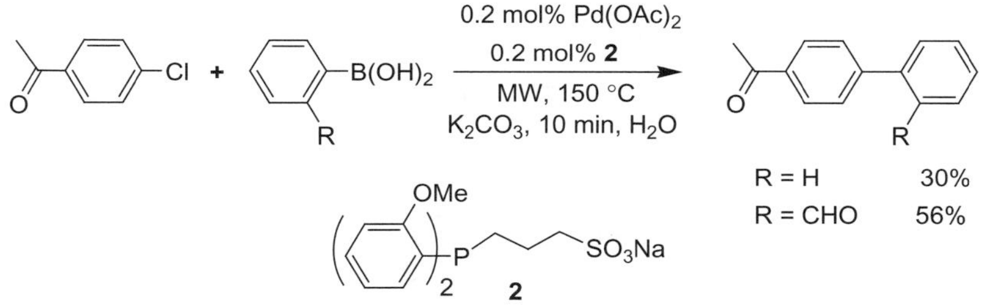

Reaction: 4-chloronitrobenzene + phenylboronic acid, Pd/C (5 mol%), K_2CO_3, DMF/H_2O, MW, 45 °C, 1 h → 4-nitrobiphenyl (with Cl shown)

Activation	yield%
US	22
MW	64
US/MW	88

Scheme 3.11

Reaction: 4′-chloroacetophenone + ArB(OH)$_2$ (R = H or CHO), 0.2 mol% Pd(OAc)$_2$, 0.2 mol% **2**, MW, 150 °C, K_2CO_3, 10 min, H_2O

Ligand **2**: (2-MeO-C$_6$H$_4$)$_2$P-CH$_2$CH$_2$CH$_2$-SO$_3$Na

R = H	30%
R = CHO	56%

Sterically demanding mixed aryl(alkyl)phosphines have also shown promise for aqueous-phase cross-coupling of aryl bromides and chlorides.[23,24] In this context, sterically hindered diarylphosphinopropane sulfonate ligand **2** has been prepared *via* a new sustainable synthetic route.[25] The Pd(OAc)$_2$/number combination (0.05 mol.% Pd, 1:1 L/Pd) provided good yields for Suzuki coupling of aryl bromides at 80 °C, although 4-bromoanisole gave incomplete conversion. This catalyst system could also be applied to activated aryl chlorides to provide modest yields of coupled products using 1–2 mol.% Pd at 150 °C with microwave heating (Scheme 3.11).

Lépine and Zhu have described another nice example of a simple, but potentially valuable, total synthesis of cyclic tripeptide Biphenomycin B, which displayed potent activities against Gram-positive, β-lactam-resistant bacteria such as *Streptococcus aureus*, *Enterococcus faecalis* or *Streptococcus*.[26] The authors performed the macrocyclization to furnish the 15-membered ring system of the title molecule in 50% yield by a microwave-assisted, intramolecular Suzuki–Miyaura cross coupling reaction (Scheme 3.12). The MW-assisted procedure gave a rewarding 33% yield when simple Pd(OAc)$_2$ was used in the absence of any ligands, whereas the use of much more specific catalytic systems met with failure or furnished very low yields of the target biaryl compound when performed under conventional heating conditions.

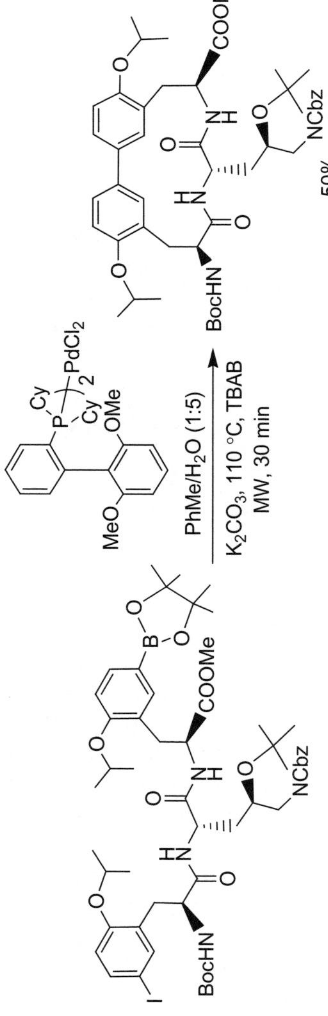

Scheme 3.12

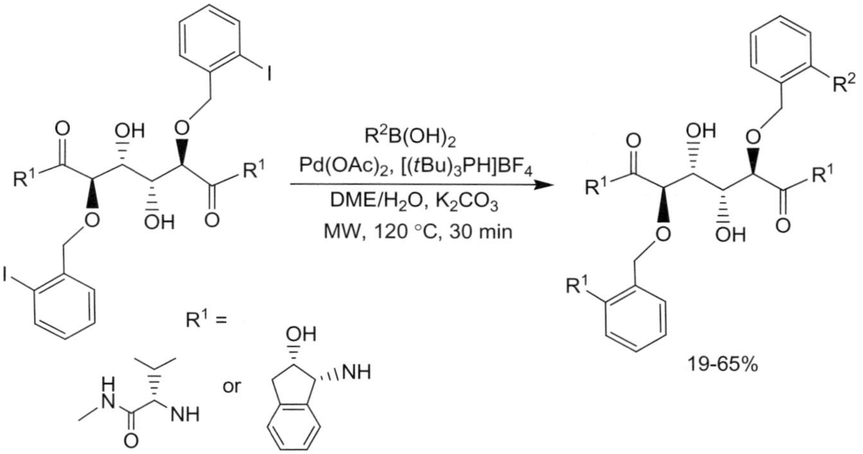

Scheme 3.13

Larhed's group has developed the rapid syntheses of 24 novel C_2-symmetric HIV-1 protease inhibitors from two *ortho*-iodobenzyloxy-containing C-terminal duplicated inhibitors by using various Pd-catalyzed carbon–carbon reactions, including the Suzuki cross-couplings under MW-assisted conditions.[27] The combination of 10 mol.% Pd(OAc)$_2$ and 20 mol.% [(*t*Bu)$_3$PH]BF$_4$ in the presence of K$_2$CO$_3$ as the base proved to be an efficient system for Suzuki reactions in a mixture of water–1,2-dimethoxyethane (1:3) at 120 °C (Scheme 3.13). The future prospects for protein functionalization are rather obvious from this very interesting study.

Many recent examples in the literature have demonstrated the versatility of the Suzuki cross-coupling reactions in aqueous medium using microwave heating. Van der Eycken and co-workers have described a novel microwave-enhanced six-step synthesis for the synthesis of N-shifted buflavine analogues.[28] MW-enhanced Suzuki–Miyaura cross-coupling and ring-closing metathesis reactions were used as the key steps. The reactions were performed with 5 mol.% of Pd(PPh$_3$)$_4$ as catalyst and NaHCO$_3$ as base in a mixture of DMF–H$_2$O (1:1) at 150 °C (Scheme 3.14).

Using similar reaction conditions, the synthesis of ring-expanded buflavine analogues possessing a nine membered medium-sized ring system have been developed (Scheme 3.15).[29] The Suzuki–Miyaura reaction, which is often problematic in the case of cross-coupling between an electron-rich aryl halide and an electron-poor boronic acid, has been accomplished under MW irradiation conditions.

Van der Eycken and co-workers have described the synthesis of a small library of highly electron-rich 2-aryl- and 2-heteroaryl-phenethylamines through a microwave-enhanced Suzuki–Miyaura cross-coupling reaction to develop novel receptor ligands for binding studies.[30] Various boronic acids

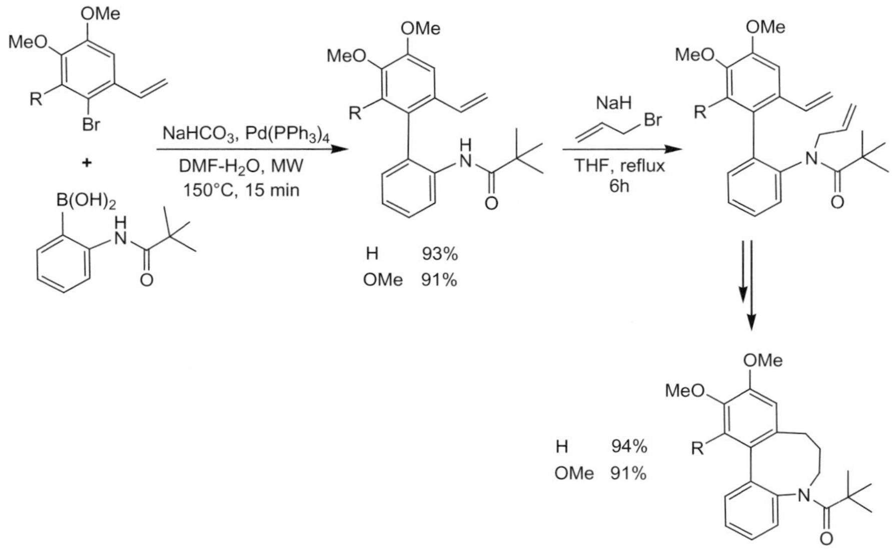

Scheme 3.14

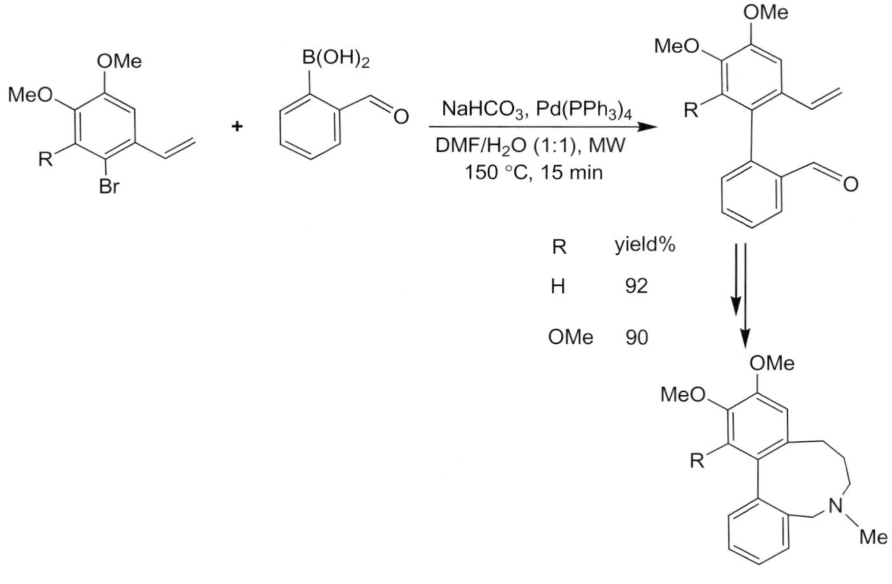

Scheme 3.15

were successfully introduced at the 2-position of the highly electron rich 2-(3,4-dimethoxyphenyl)ethylamine with Pd(PPh$_3$)$_4$ as catalyst and NaHCO$_3$ as base in a mixture of water–dimethylformamide. This strategy has been successfully extended for the coupling of electron-withdrawing and hindered

(2-formylphenyl)boronic acid; however, a change in base was necessary (Scheme 3.16). This reaction could be performed in water as the sole solvent with slightly decreased yield; thus the proto-deboronation could be avoided, clearly demonstrating the benefit of microwave heating.

The proven versatility of the Suzuki reaction has also been demonstrated in the synthesis of natural products and heterocyclic compounds. In this regard, using TBAB and water as solvent, it is possible to couple various arylboronic acids with 2-arylsulfonyl-methyl-6-bromo-3-nitroimidazo[1,2-*a*]pyridines more efficiently with an easier purification of coupled compounds than classical heating methodology.[31] The best results were obtained with 10 mol.% of tetrakis(triphenylphosphine)palladium(0) as catalyst and 5 equiv. of $NaHCO_3$ as base in pure water (Scheme 3.17).

5-Substituted thiazoles with various aromatic groups on the thiazole are commonly found in the chemical structure of natural compounds, such as the coenzyme derived from vitamin B_1, and in various synthetic compounds exhibiting a wide variety of biological activity,[32] such as the utility of gastric antisecretory agents as liquid crystals and cosmetic sunscreens.[33,34] They were synthesized *via* the Suzuki coupling reaction in aqueous medium without organic co-solvent in the presence of tetrabutylammonium bromide (TBAB) and promoted by MW irradiation (Scheme 3.18).[35]

Nucleoside derivatives occupy a pivotal position in the arsenal of drug candidates for combating various viruses.[36,37] 5-Aryltriazole acyclonucleosides with various aromatic groups on the triazole ring belong to an important class of biologically active nucleosides. They were synthesized *via* a simple and

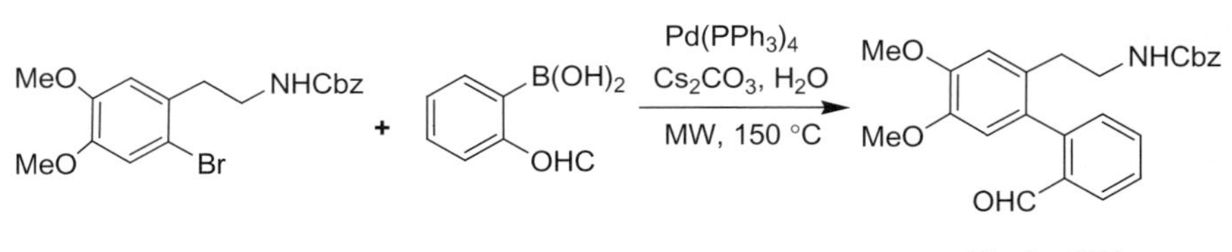

Scheme 3.16

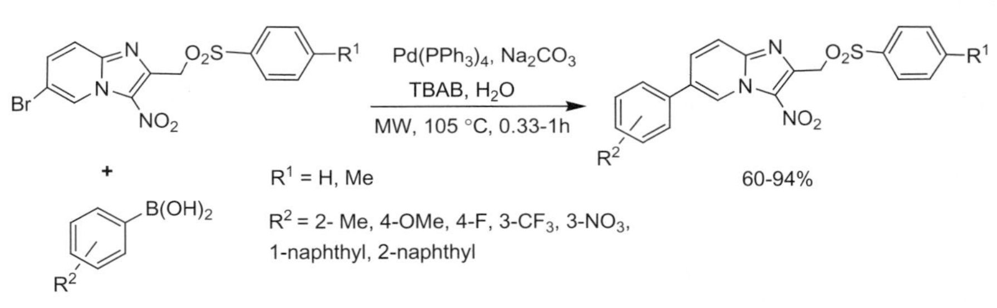

Scheme 3.17

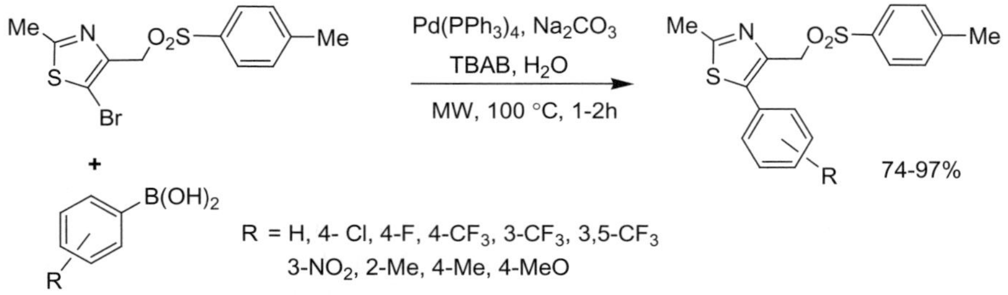

Scheme 3.18

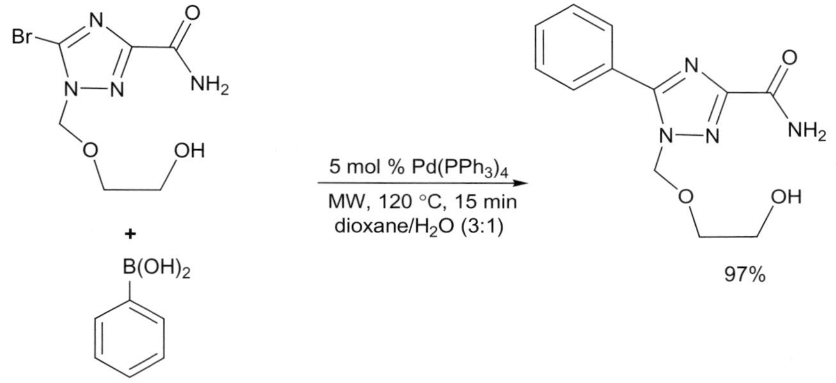

Scheme 3.19

efficient one-step procedure involving the direct Suzuki coupling of the unprotected 5-bromotriazole acyclonucleoside in aqueous solution under microwave irradiation (Scheme 3.19).[38] This coupling method afforded directly the corresponding product in good to excellent yield and involved no protection and deprotection steps.

A novel series of integrin antagonists based on a pyridazinone-substituted phenylalanine scaffold has been synthesized *via* Suzuki palladium cross-coupling reaction.[39] The coupling of 2,6-dichlorobenzoyl-protected 4-borono-L-phenylalanine with diversely substituted 4-halopyridazinones provided a series of pyridazinone-functionalized phenylalanine analogues. The best results were obtained with Na_2CO_3 as the base and 5 mol.% $Pd(PPh_3)_2Cl_2$ in a water–acetonitrile mixture at 150 °C (Scheme 3.20). These compounds are potent antagonists of $\alpha_4\beta_1$/VCAM-1 and/or $\alpha_4\beta_7$/MAdCAM-1 mediated cellular adhesion whether activated by metal ion, lipopolysaccharide or anti-tumour necrosis factor-α.

Generally, Suzuki–Miyaura reactions are carried out utilizing boronic acids or esters; however, recently, alternative boron reagents have been developed for

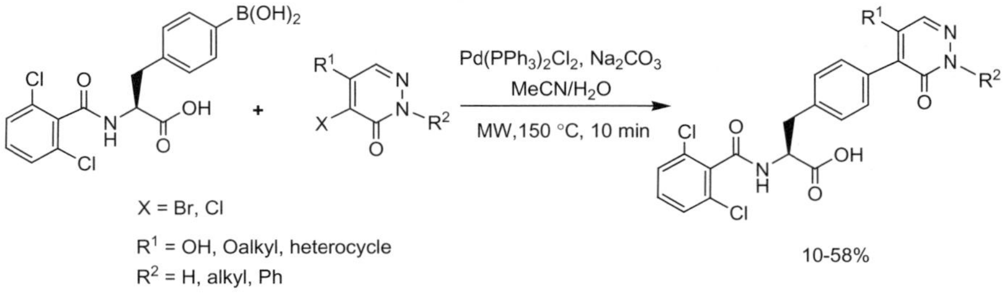

Scheme 3.20

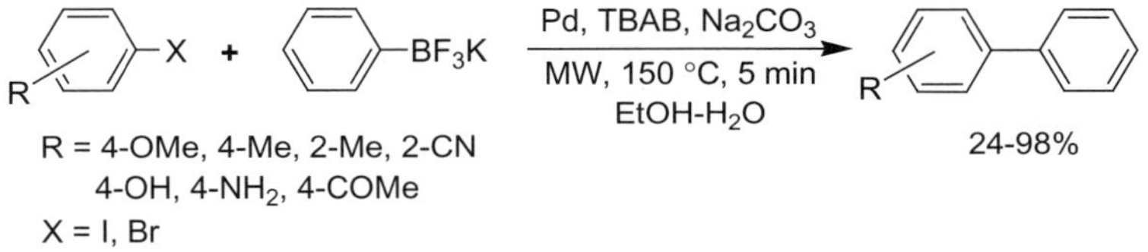

Reagents: B(OH)$_2$-aryl amide + pyridazinone (X = Br, Cl)
Conditions: Pd(PPh$_3$)$_2$Cl$_2$, Na$_2$CO$_3$, MeCN/H$_2$O, MW, 150 °C, 10 min

X = Br, Cl
R^1 = OH, Oalkyl, heterocycle
R^2 = H, alkyl, Ph

10–58%

Scheme 3.21

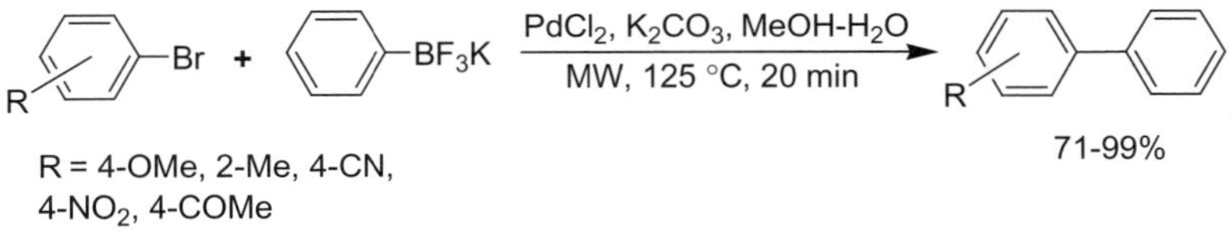

Ar–X + Ph–BF$_3$K → biaryl
Conditions: Pd, TBAB, Na$_2$CO$_3$, MW, 150 °C, 5 min, EtOH–H$_2$O

R = 4-OMe, 4-Me, 2-Me, 2-CN, 4-OH, 4-NH$_2$, 4-COMe
X = I, Br

24–98%

Scheme 3.22

Ar–Br + Ph–BF$_3$K → biaryl
Conditions: PdCl$_2$, K$_2$CO$_3$, MeOH–H$_2$O, MW, 125 °C, 20 min

R = 4-OMe, 2-Me, 4-CN, 4-NO$_2$, 4-COMe

71–99%

Suzuki reactions. Of these, potassium organotrifluoroborate salts are easier to prepare, store and handle than boronic acids or esters.[40] Leadbeater has developed a protocol for coupling of a wide range of aryl iodides and bromides in good yield using low catalyst loading (Scheme 3.21). However, these conditions were not effective for widely available and low-cost aryl chlorides.

MW Suzuki–Miyaura reactions of various aryl halides with potassium organotrifluoroborates have recently described that do not require the use of phosphine ligands or phase-transfer catalysts.[41] The reactions can be carried out using PdCl$_2$ as the catalyst and K$_2$CO$_3$ as the base in aqueous methanol and all reactions are complete in 20 min (Scheme 3.22).

Aryl triflates can be utilized as substrates to substitute for the aryl halide in Suzuki couplings (Scheme 3.23).[42] These couplings can be carried out in aqueous ethanol in the absence of a base or ligand under MW irradiation and the desired products were obtained in good to excellent yields within 15 min at 95 °C.

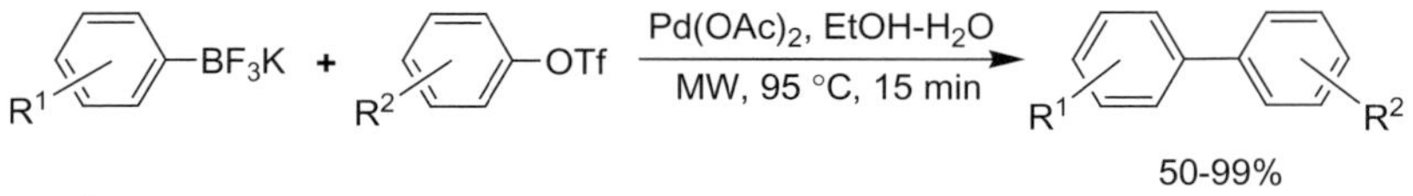

Ar¹–BF₃K + Ar²–OTf →[Pd(OAc)₂, EtOH-H₂O / MW, 95 °C, 15 min] Ar¹–Ar² (50-99%)

R¹ = H, 4-Me, OMe, 4-F, 4-Cl, 4-COMe, 2,5-Me₂
R² = H, 4-NO₂, 4-CN, 4-COMe

Scheme 3.23

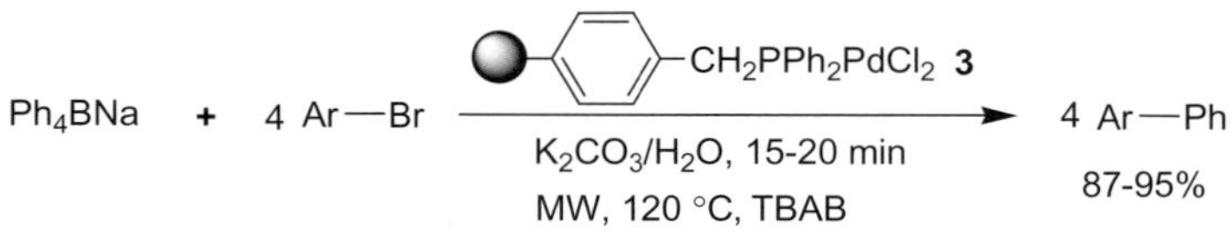

Ph₄BNa + 4 Ar—Br →[PS–CH₂PPh₂PdCl₂ **3** / K₂CO₃/H₂O, 15-20 min / MW, 120 °C, TBAB] 4 Ar—Ph (87-95%)

Scheme 3.24

Phenylboronic acid in Suzuki couplings has been substituted by sodium tetraphenylborate, a very stable and inexpensive reagent. Another advantage is that it can react with 4 equiv. of electrophilic reagents. In this context, a rapid and efficient cross-coupling reaction of sodium tetraphenylborate with aryl bromides was carried out in neat water at 120 °C in the presence of a PS-supported palladium catalyst **3** (1 mol.% palladium) under focused MW irradiation (Scheme 3.24).[43] Water and K_2CO_3 were the best solvent and base, respectively, among those surveyed, and the addition of tetrabutylammonium bromide was beneficial for the reaction rate. All four phenyl groups of sodium tetraphenylborate participated in the reaction and produced polyfunctional biaryls in excellent yields. The PS-supported palladium catalyst was reused ten times without any loss of catalytic activity when another equivalent of the reactants and base were added.

The development of heterogeneous alternatives where the metal is grafted on inorganic or organic supports has attracted great interest because it could combine the advantages of both homogeneous and heterogeneous catalysts in one system.[44,45] Simplification of workup protocols and handling of catalysts is an especially important factor for industrial applications of catalysts. These leach-proof catalysts leave no remnants of metal in the end product, which is important as metal contamination is highly regulated by the pharmaceutical industry.

Kirschning and co-workers have developed a new Pd(II)-precatalyst **4**, which is conveniently prepared by mixing readily available 4-pyridinealdoxime with Pd(II) salts and is insoluble in water as well organic solvents.[46] Under these optimized conditions (water, MW irradiation 120 °C, TBAB, 20 min), fairly good to excellent yields of the coupling products were obtained for the coupling of aryl halides and triflates with various boronic acids (Scheme 3.25). Recycling of this polymeric catalyst has been briefly examined in the coupling of

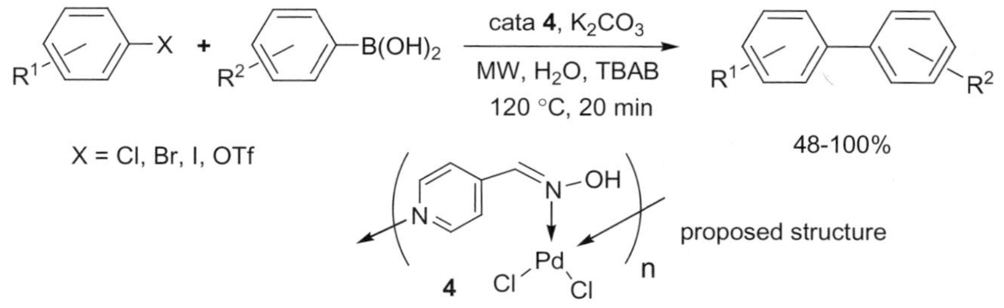

Scheme 3.25

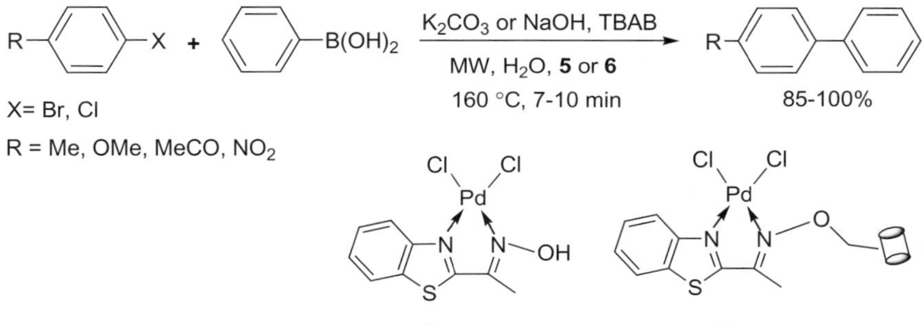

Scheme 3.26

4-bromoacetophenone with phenylboronic acid, wherein recycled catalyst was used in a repeated reaction. Even the 14th run with the reused can give 93% of conversion within 20 min.

Dawood has evaluated the catalytic activity of benzothiazole-based Pd(II)-complexes **5** and **6** in Suzuki–Miyaura cross-coupling reactions of aryl bromides and chlorides with phenylboronic acid both under thermal as well as microwave irradiation conditions in water.[47] These precatalysts were found to be efficient and highly active for activated aryl bromides and heterocyclic bromides, with very high turnover number (TON), under both thermal heating and MW irradiation conditions. Interestingly, deactivated aryl bromides were more effective than their chloride analogues. The immobilized catalyst **6** was highly durable compared with the mobilized catalyst **5**. The high TON associated with the catalytic activity of these catalysts is highly important for mass production on an industrial scale (Scheme 3.26).[47]

A chitosan-supported Pd(0) catalyst, prepared by the adsorption of Pd(OAc)$_2$ on chitosan beads followed by simple reduction and crosslinking, has been used in the Suzuki reaction in water under MW conditions. The use of chitosan beads allowed high loading of palladium because of the high metal ion sorption capacity of chitosan. In the presence of a tetrabutylammonium bromide additive, the catalyst showed excellent catalytic activity in the cross-coupling of various aryl halides with boronic acids. In addition, the catalyst was successfully reused up to five times without significant loss of activity. The recyclability of this catalyst was also examined using the standard procedure of the reaction of 4-bromoacetophenone and phenylboronic acid; the catalyst could be reused for more than five consecutive trials without significant loss of activity.[48]

Bradley and co-workers have reported a crosslinked resin-captured palladium catalyst prepared by diffusing $Pd(OAc)_2$ into swollen resins followed by crosslinking of amino groups on the resin.[49] The resulting supported palladium was highly effective for heterogeneous Suzuki coupling reactions in neat water using microwave irradiation. Unfortunately, a very low yield or no product was obtained in the coupling of aryl chlorides. The applicability of the catalyst was demonstrated in the synthesis of a sulfophthalein dye without any palladium contamination (which could cause serious problems in the optical properties of the product) (Scheme 3.27). This crosslinked, air stable, resin-captured palladium could be simply recovered and reused without significant loss of catalytic activity.

Schotten and co-workers have demonstrated liquid phase "ligandless" palladium acetate-catalyzed Suzuki cross-coupling in water with poly(ethylene glycol-600) (PEG) as soluble support and phase-transfer catalyst without other organic cosolvent.[50] Compared to classical heating, microwave irradiation shortened the reaction time with representative boronic acids from 2 h to 1–4 min (Scheme 3.28). The ester cleavage from the polymer that occurred to an extent of up to 45% under thermal heating was suppressed when microwave heating conditions were employed. The reaction proceeded well on PEG-bound aryl iodides, bromides, triflates and nonaflates and on nonpolymer-bound aryl iodide on various scales.

Fluorous chemistry is an emerging field that capitalizes on the unique physical and solubility properties of perfluorinated organic compounds, and this new strategy could improve the efficiency of Suzuki coupling reactions by

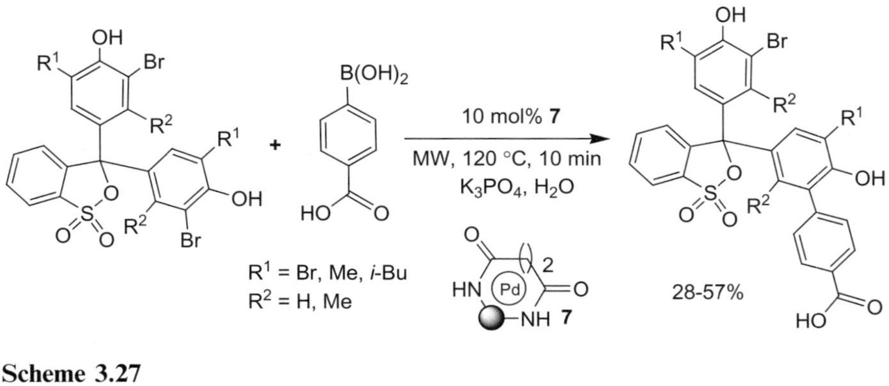

Scheme 3.27

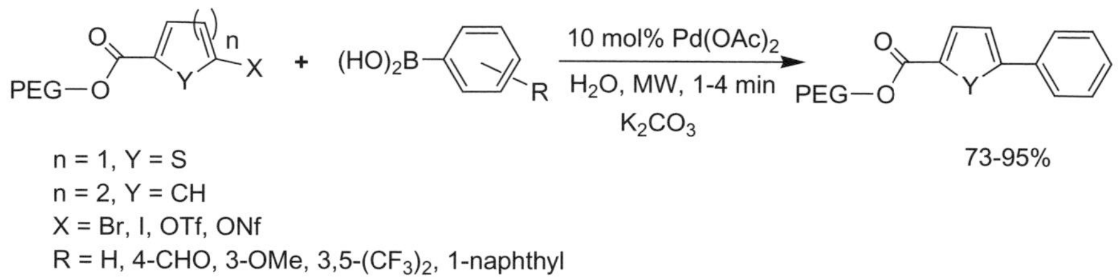

Scheme 3.28

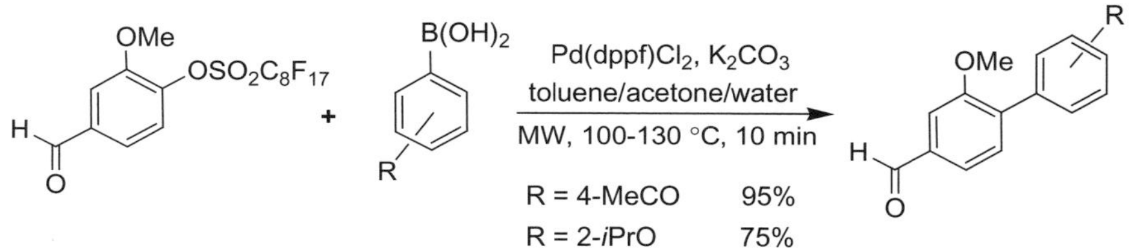

n = 1, Y = S
n = 2, Y = CH
X = Br, I, OTf, ONf
R = H, 4-CHO, 3-OMe, 3,5-$(CF_3)_2$, 1-naphthyl

Scheme 3.29

R = 4-MeCO   95%
R = 2-iPrO   75%

combining a fast microwave reaction with easy fluorous separation.[51] Under MW heating conditions, the aryl-perfluorooctylsulfonates derived from the corresponding phenols were coupled with arylboronic acids to form biaryls in good yields (Scheme 3.29). Application of this tagging strategy was further demonstrated in a multistep synthesis of a biaryl-substituted hydantoin.

3.3 Heck Reaction

The Heck reaction is one of the most noteworthy and comprehensively used reactions for the formation of carbon–carbon bonds, which allows the arylation, alkylation or vinylation of various alkenes *via* their coupling with aryl halides.[3,52] The Heck-type reaction was first developed by Mizoroki and co-workers[53] and Heck and Nolly.[54] Because of the simplicity and high reactivity of this protocol, it has been used for a wide range of functional groups. There are many examples of utilization of this method on an industrial scale for fine chemical synthesis.[55] Academic and industrial interest in this reaction has increased in recent years because of its tolerance for almost any solvent and functional group on the substrates, its high selectivity, and its moderate toxicity.[55,56] The first example of a Heck reaction in aqueous solution was provided by Larhed and Hallberg who observed that the use of MW heating can reduce the reaction times from several hours under conventional reflux conditions to sometimes less than 4 min (Scheme 3.30).[9]

Palladium-catalyzed vinylation of aryl iodides in water without any organic co-solvent *via* microwave irradiation using modified reaction conditions has

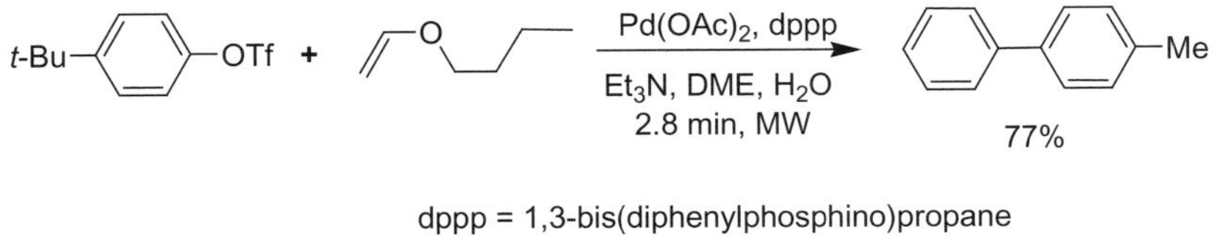

Scheme 3.30

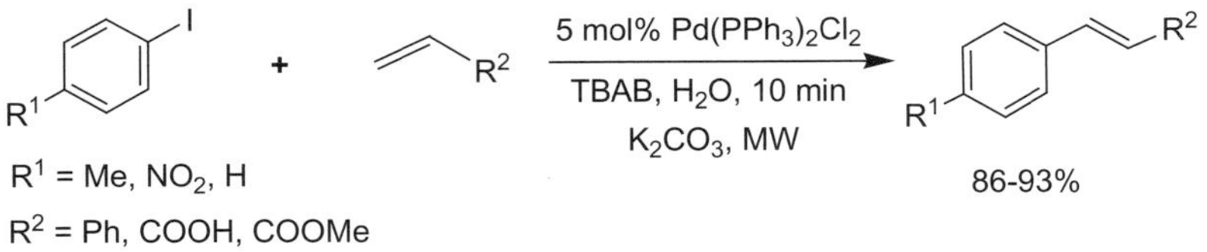

dppp = 1,3-bis(diphenylphosphino)propane

Scheme 3.31

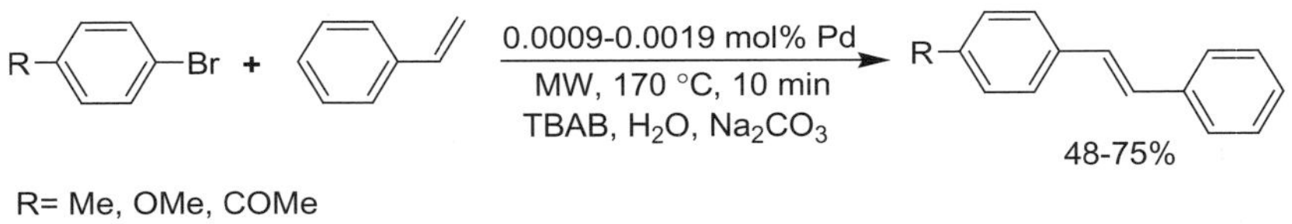

R^1 = Me, NO_2, H
R^2 = Ph, COOH, COOMe

Scheme 3.32

R= Me, OMe, COMe

been reported by Wang and co-workers.[57] K_2CO_3 was the appropriate base and the addition of tetrabutylammonium bromide was found to be beneficial for the reaction rate (Scheme 3.31). Microwave heating provided the products 18–42× faster than under conventional reflux conditions but with comparable yields.

Arvela and Leadbeater have performed the Heck coupling reaction in water using MW irradiation with Pd-catalyst concentrations as low as 500 ppb (Scheme 3.32).[58] The methodology is simple but a study on a more diverse array of substrates would be necessary to determine the scope of these catalysts.

Larhed and co-workers have reported highly regio-selective and fast Pd(0)-catalyzed internal α-arylation of ethylene-glycol vinyl ether with aryl halides in neat water.[59] Aryl bromides were efficiently converted into corresponding acetophenones in good to excellent yields after hydrolysis (Scheme 3.33). The arylation of aryl iodides gave the coupling product in good to moderate yield and MW irradiation was shown to be beneficial in the activation of aryl chlorides towards the internal Heck arylation.

Nájera and co-workers have examined the cross-coupling of 4-bromoacetophenone with styrene in the presence of polymer-supported di(2-pyridyl)methylamine–palladium dichloride complex **8** covalently anchored to a

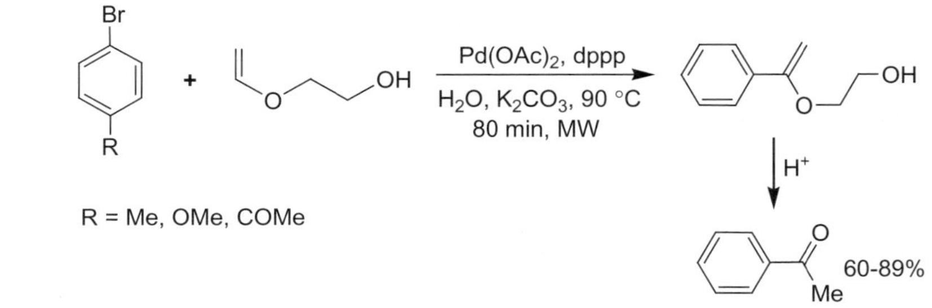

Scheme 3.33

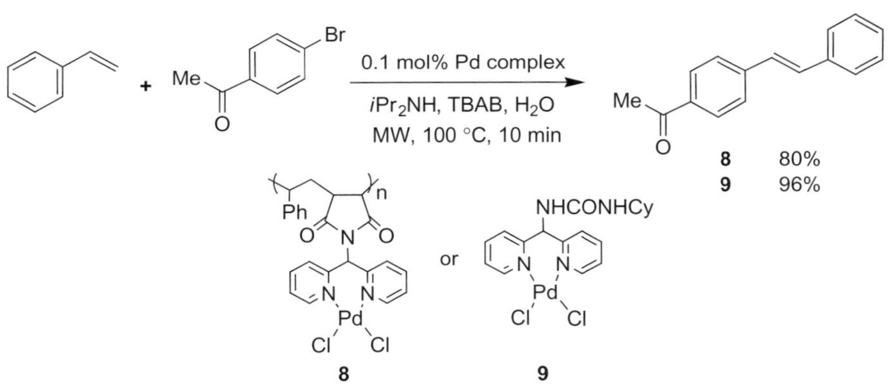

Scheme 3.34

poly(styrene-*alt*-maleic anhydride) copolymer and di(2-pyridyl)methylamine-based palladium dichloride complex **9**.[60] The reactions were performed in neat water with diisopropylamine as the base in presence of TBAB as phase-transfer catalyst. The reaction was faster with the monomeric catalyst **9** (Scheme 3.34). However, the vinylation of aryl chlorides under this reaction conditions failed and the catalyst **8** could not be recycled after microwave heating due to leaching of palladium and degradation of the polymer.

2-Pyridinealdoxime-based Pd(II)-complex covalently anchored *via* the oxime moiety to a glass/polymer composite material **10** has been evaluated under both thermal as well as microwave irradiating conditions.[61] The best results were obtained with *i*Pr$_2$NH as the base and 0.7 mol.% of **10** at 150 °C for 5–20 min in the coupling of aryl and heteroaryl bromides and iodides with *tert*-butyl acrylate under microwave heating – while 5–20 h at 100 °C were necessary under conventional heating – to obtain the coupling products in moderate to excellent yields. Under the thermal approach, *i*PrNH$_2$ was less effective than NaOH in the couplings of aryl and heteroaryl bromides and iodides with styrene, whereas under microwave heating no such differences in yields were observed. Recycling of this precatalyst has been examined in the coupling of 4-bromoacetophenone with *tert*-butyl acrylate; however, the catalytic activity and the recyclability dropped significantly in the first cycle under microwave irradiation and after third cycle under thermal heating (Scheme 3.35).

To date, there are relatively few examples of water-soluble palladium catalysts used in Heck coupling. In this regard, water-soluble oligo(ethylene glycol)-bound SCS palladacycle catalyst **11** was found to be an efficient catalyst for the reaction of water-soluble aryl iodides and bromides with acrylic acid to give cinnamic acid derivatives under aqueous MW irradiation conditions (Scheme 3.36).[62] Recycling of the catalyst using a thermomorphic system was possible even with small oligo(ethylene glycol) groups, and MW irradiation of these thermomorphic mixtures of palladium complexes and substrates could significantly shorten reaction times for simple Heck reactions of aryl iodides.

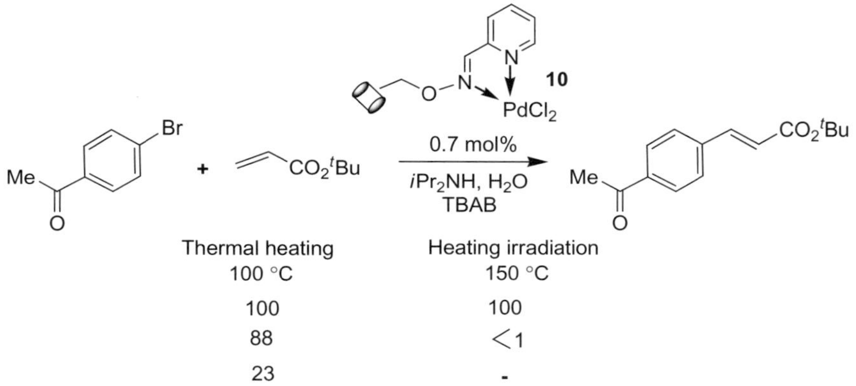

Scheme 3.35

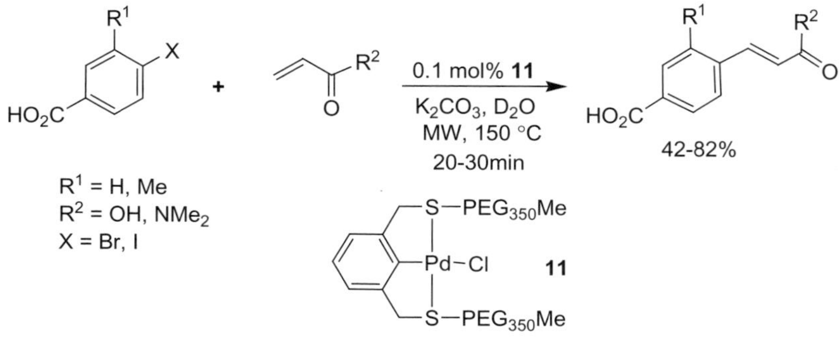

Scheme 3.36

Botella and Nájera have reported that air- and water-stable oxime palladacycles such as **12** represent another class of non-phosphorus-based ligands for Heck arylation of acrolein diethyl acetal.[63] When the reaction was conducted at 120 °C with Cy_2NMe as base and tetrabutylammonium bromide (TBAB) as phase-transfer catalyst, the arylation could be performed in only 10 min (Scheme 3.37); no arylation could be observed with aryl chlorides under these reaction conditions.

Cao and co-workers have reported a fast and convenient base-free Heck reaction of acrylic acid and allyl alcohol with hypervalent iodonium salts under MW irradiation in water.[64] This method has proven its worth in effective and simple, rapid synthesis of *trans*-cinnamic acids and *trans*-cinnamyl alcohols in good yields (Scheme 3.38).

3.4 Sonogashira Reactions

The Sonogashira reaction has been used for the coupling of terminal alkynes and aryl or vinyl halides to provide a powerful method to access aryl alkynes, which are interesting intermediates for the preparation of various target molecules with applications ranging from natural products[65,66] and pharmaceuticals[67] to molecular organic materials.[68,69] The couplings are often carried

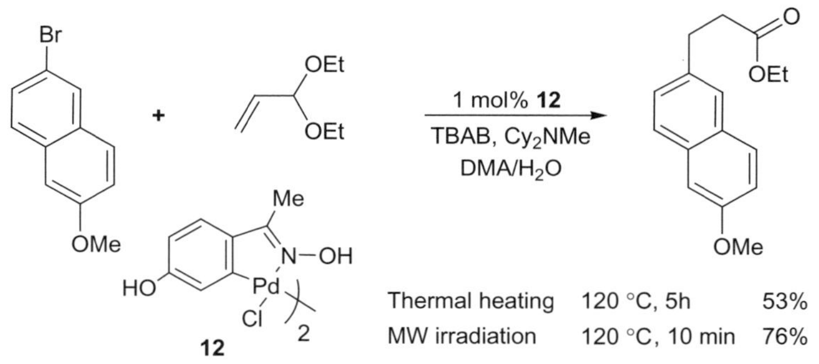

Scheme 3.37

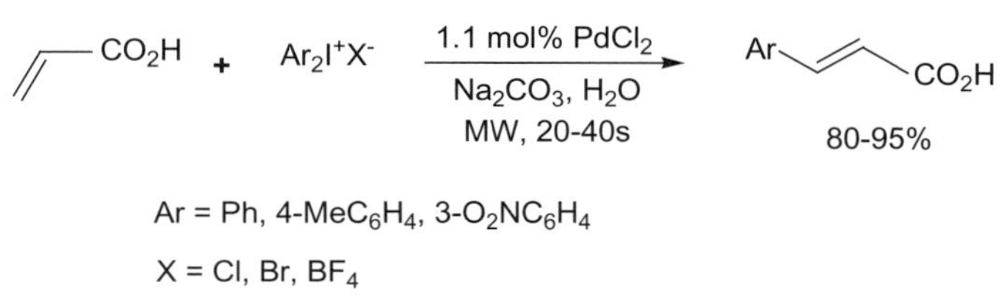

Scheme 3.38

out by palladium catalysts combined with a copper co-catalyst with an amine, which was first reported by Sonogashira and Hagihara in 1975.[70] Several practical and economical modified protocols have been developed more recently, in particular by using microwave heating in conjunction with aqueous media. A Sonogashira cross-coupling reaction under controlled conditions were first reported by Erdélyi and Gogoll.[71] Nájera and co-workers have examined the cross-coupling of 4-chlorobromobenzene with phenylacetylene in the presence of polymer-supported di(2-pyridyl)methylamine–palladium dichloride complex covalently anchored to a poly(styrene-*alt*-maleic anhydride) resin **8** and chelated palladium–dipyridilmethylamine complex **9** with pyrrolidine as a base and TBAB as an additive in refluxing water (Scheme 3.39). Under these conditions, the copper-free reactions gave better yields in shorter reaction times for the polymeric complex than for the monomeric one (66 *versus* 47%).[60]

Heterogeneous palladium(II)-precatalyst **10**, already employed in a Heck reaction, has also been tested in aqueous copper- and phosphine-free Sonogashira couplings under both conventional and MW heating conditions; various aryl and heteroaryl bromides and iodides were successfully alkynylated in good yields. However, under MW heating the reaction times could be reduced from several hours at 100 °C under thermal heating to only 5–20 min (Scheme 3.40).[61]

As with the Suzuki reaction, the research groups of Leadbeater[72] and van der Eycken[73] independently reported that Sonogashira–Heck alkynylations could be performed in the absence of any transition metal compound. Both groups carried out the reactions in water with MW irradiation and verified that transition-metal-free alkynylations did not occur under conventional heating in an oil bath above 170 °C (Schemes 3.41 and 3.42). MW irradiation enabled aryl iodides to be efficiently coupled to phenylacetylene in a water–poly(ethylene

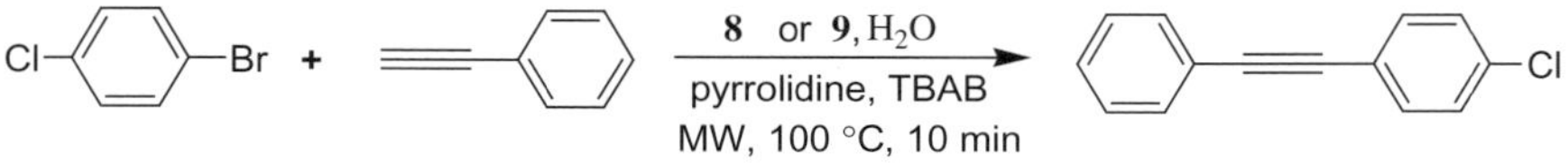

Scheme 3.39

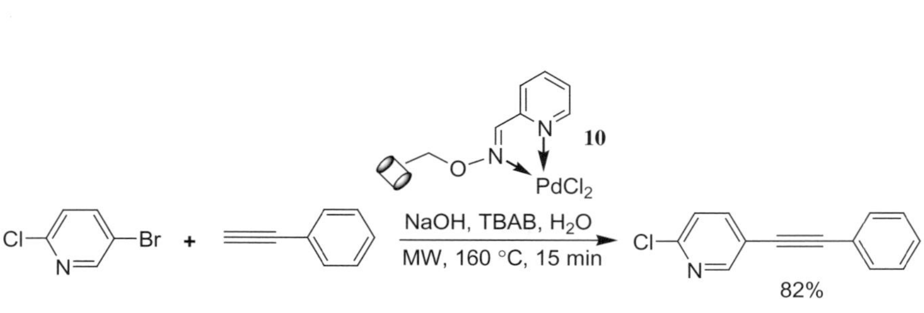

Scheme 3.40

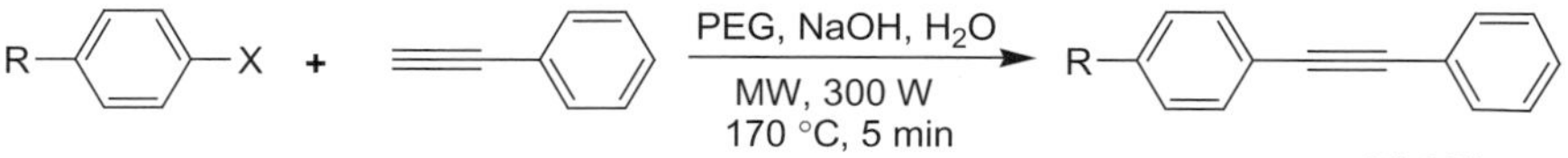

Scheme 3.41

R—C₆H₄—X + HC≡C—C₆H₅ →(PEG, NaOH, H₂O, MW, 300 W, 170 °C, 5 min)→ R—C₆H₄—C≡C—C₆H₅

X = Br, I
R = H, Me, Ac, MeO, CHO

X	R	Yield%
I	Ac	91
I	H	83
I	Me	92
I	OMe	43
Br	Me	9
Br	CHO	0

Scheme 3.42

R—C₆H₄—X + HC≡C—C₆H₅ →(TBAB, Na₂CO₃, H₂O, MW, 150 W, 175 °C, 10-25 min)→ R—C₆H₄—C≡C—C₆H₅ 52–85%

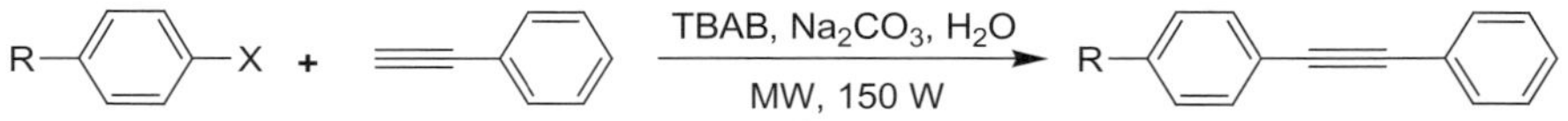

X = I, Br
R = H, OMe, NH$_2$, NO$_2$, Ac

glycol) mixture in the presence of two equivalents of NaOH, while aryl bromides were found to be more difficult substrates.

The experimental conditions proposed by van der Eycken and co-workers were slightly harsher, with higher temperatures and longer reaction times needed, as well as the requirement of addition of a phase-transfer agent (TBAB) and a significant amount of sodium carbonate (Scheme 3.42). Their conditions accommodated the coupling of activated bromides and heteroaryl substrates, while some other iodides and bromides were unexpectedly troublesome. Practical application of these metal-free methods, however, remains rather limited because of the necessary use of a microwave apparatus and the difficulty in anticipating which substrate will react or not, and to date no progress on the coupling of the widely available and low-cost aryl chlorides has been reported. The development of such processes would be advantageous from an economic standpoint as several useful ligands and palladium complexes are considered to be expensive, sensitive or difficult to obtain. However, it was later reported that commercially available Na_2CO_3 containing a palladium contaminant down to a level of 50 ppb could be responsible for the cross-coupling reactions rather than a non-palladium-promoted pathway.[74]

The 4-hydroxycyclohexa-2,5-dien-1-one fragment represents a new pharmacophore in anticancer drug development. Stevens and co-workers have developed[75] a novel procedure, based on a microwave-assisted Sonogashira coupling with concomitant cyclization, leading to the formation of 4-[1-(arylsulfonyl-1H-indol-2-yl)]-4-hydroxycyclohexa-2,5-dien-1-ones in moderate yields (Scheme 3.43).

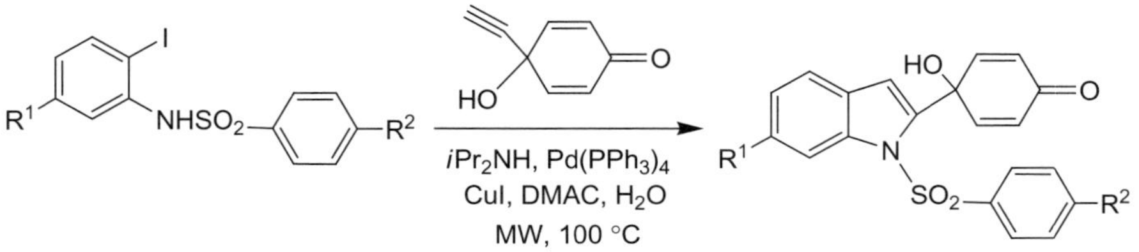

Scheme 3.43

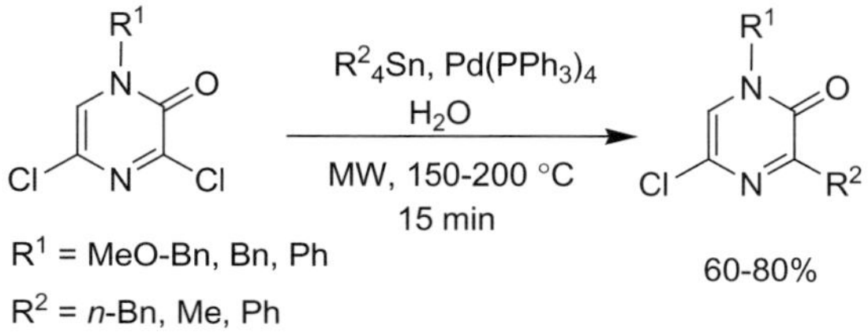

Scheme 3.44

3.5 Stille Reactions

The Stille cross-coupling reaction using organostannane compounds has been used extensively in organic synthesis.[76–78] This is due to the growing availability of the organostannanes, their stability to moisture and air, and their excellent compatibility with a large variety of functional groups. In contrast to Suzuki reactions, there are relatively few examples of microwave-assisted Stille reactions involving organotin reagents as coupling partners. The 2(1H)-pyrazinones have been demonstrated to be versatile building blocks for the synthesis of biologically active compounds. Van der Eycken and co-workers have shown that Stille reactions allow the easy introduction of different substituents at the C3 – and even at the rather unreactive C5 – position of the pyrazinones (Scheme 3.44).[79] For tetraphenyltin, an elevated temperature of 200 °C had to be applied to achieve full conversions. A great acceleration compared to conventional heating in refluxing toluene could be reached, albeit the yields were somewhat lower for the aqueous microwave synthesis.

Although the Stille reaction has emerged as an important reaction in high-throughput chemistry, there are disadvantages such as the toxicity of tin compounds.

3.6 Hiyama Reactions

Palladium-catalyzed cross-coupling of organosilanes with aryl/vinyl halides and triflates, commonly referred to as the Hiyama reaction, has emerged as an

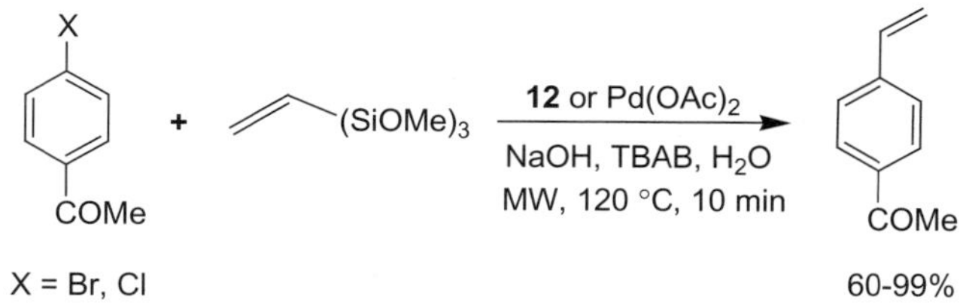

Scheme 3.45

attractive alternative to the Stille coupling because organosilicon compounds are less toxic.[80,81] Several organosilicon reagents, such as alkyl-, fluoro-, chloro-, hydroxy- and alkoxysilanes have been used for this carbon–carbon bond forming reaction.[81,82] Owing to the low reactivity of these reagents, generally they need to be activated by the fluoride anion to generate a more reactive pentacoordinate silicate anion for the transmetallation step of the aryl- or vinylpalladium intermediate. More recently it was found that inorganic bases like KOH, NaOH and K_2CO_3 could form pentacoordinate silicon intermediates.[83,84] To date, few examples of palladium-catalyzed Hiyama couplings in aqueous media using MW irradiation have been reported. The first cross-coupling reaction between vinylalkoxysilanes and aryl bromides or chlorides promoted by aqueous sodium hydroxide under fluoride-free conditions was provided by Nájera and co-workers.[85] The reaction was catalyzed by palladium(II)acetate or a 4-hydroxyacetophenone oxime-derived palladacycle **12** at 120 °C with low catalyst loading in the presence of TBAB as additive in air (Scheme 3.45).

3.7 Cyanation Reactions

Cyanation of aryl halides is a common and useful transformation in organic synthesis. Not only are products containing the nitrile group biologically important[86] but the they are also valuable in the generation of functionalities such as aldehydes, amines, amidines, acids and acid derivatives.[87] Consequently, their preparation has attracted considerable attention, but, for many years, the only method for cyanation of an aryl halide required stoichiometric CuCN and harsh conditions.[88] Metal-catalyzed cyanation of aryl halides has the potential to meet this challenge. Cyanation under MW irradiation has not been explored thoroughly and the first example of palladium-catalyzed cyanation of aryl bromides using MW irradiation was performed in the presence of KCN as the source of cyanide and THF as solvent. Reaction times of between 2 and 2.5 min are reported with product yields comparable to those obtained using longer conventional heating methods.[89] Aryl nitriles can be prepared from aryl iodides using CuCN or NaCN as the cyanide sources, with the latter requiring CuI as an additive. The reactions were performed in water and TBAB as a phase transfer catalyst using MW heating with reaction times of 3–5 min

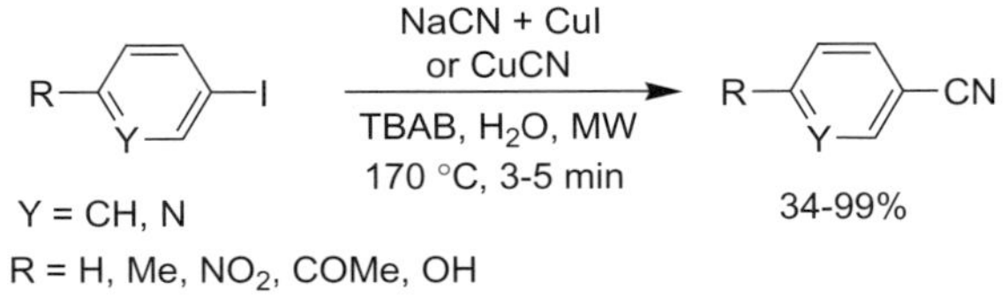

Scheme 3.46

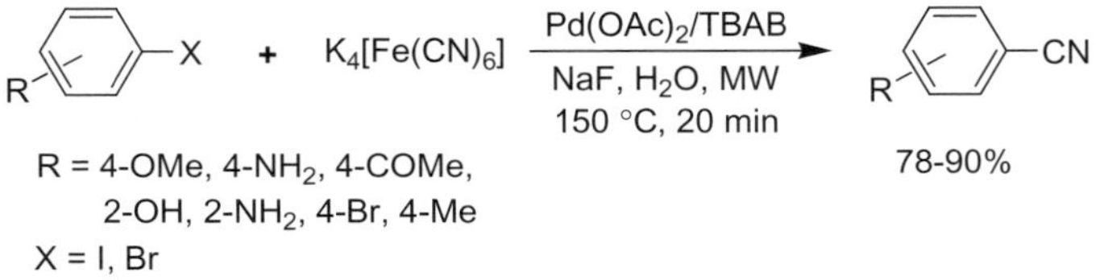

R–[Ar(Y)]–I + NaCN + CuI or CuCN → R–[Ar(Y)]–CN

TBAB, H₂O, MW
170 °C, 3-5 min
34-99%

Y = CH, N
R = H, Me, NO₂, COMe, OH

Scheme 3.47

R–[Ar]–X + K₄[Fe(CN)₆] → R–[Ar]–CN

Pd(OAc)₂/TBAB
NaF, H₂O, MW
150 °C, 20 min
78-90%

R = 4-OMe, 4-NH₂, 4-COMe, 2-OH, 2-NH₂, 4-Br, 4-Me
X = I, Br

(Scheme 3.46). Conventional heating under identical conditions resulted in no product formation; activated aryl bromides did not show any conversion.[90]

More recently, a methodology has been developed for the palladium-catalyzed cyanation of aryl iodides and activated aryl bromides using $K_4[Fe(CN)_6]$ as cyanide source and water as solvent under microwave heating. Sodium fluoride and Pd(OAc)₂ were the appropriate base and palladium complex, respectively, among the systems that were explored (Scheme 3.47).[91]

3.8 Carbonylation Reactions

Palladium-catalyzed carbonylation of aryl halides and triflates provides a powerful method for the synthesis of an array of carbonyl compounds, such as aldehydes, esters and amides, which are versatile building blocks in the synthesis of natural products and pharmaceuticals (Scheme 3.48).[92] However, the methods suffer from some major disadvantages – the difficulty in handling of toxic and gaseous carbon monoxide, including its storage and transport. One of the best ways to overcome theses problems is the use of metal carbonyls.

The earliest examples of carbonylation reactions of aryl halides under MW irradiation were described by the group of Larhed.[93,94] The key feature of these procedures is that the solid Mo(CO)₆ employed as the sole source of carbon monoxide, instead of gaseous CO, could be easily used on a commercially available monomode microwave apparatus with no modifications. During the catalysis, the catalytically active palladium species captured the carbon monoxide released from the metal carbonyls to carbonylate the aromatic halides. Aryl bromides could be rapidly converted into the corresponding secondary and tertiary benzamides under air and in water,[95] by using Mo(CO)₆ as the

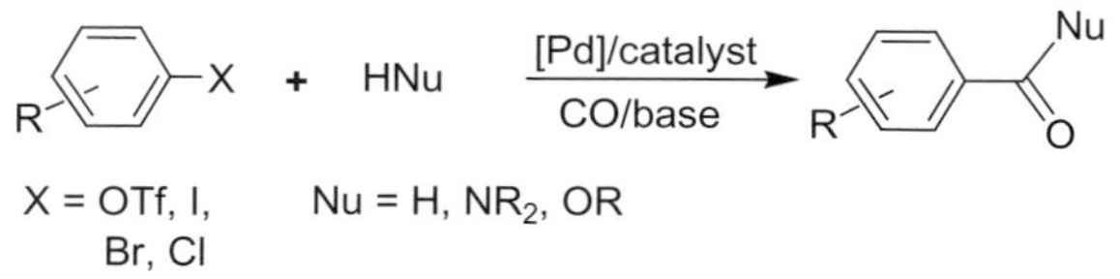

Scheme 3.48

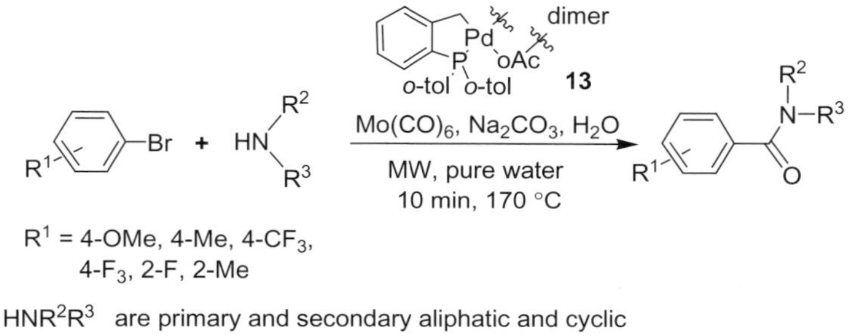

R-Ar-X + HNu →[Pd]/catalyst, CO/base→ R-Ar-C(O)-Nu

X = OTf, I, Br, Cl   Nu = H, NR_2, OR

Scheme 3.49

R^1-Ar-Br + HNR^2R^3 →[Pd dimer with P(o-tol)$_2$, OAc — Herrmann's palladacycle **13**; Mo(CO)$_6$, Na$_2$CO$_3$, H$_2$O; MW, pure water, 10 min, 170 °C]→ R^1-Ar-C(O)-NR^2R^3

R^1 = 4-OMe, 4-Me, 4-CF$_3$, 4-F$_3$, 2-F, 2-Me

HNR^2R^3 are primary and secondary aliphatic and cyclic

source of carbon monoxide, Herrmann's palladacycle **13** as a suitable Pd(0) source and Na$_2$CO$_3$ as a weak base (Scheme 3.49).[96]

Importantly, despite the use of water as solvent, aminocarbonylation strongly dominates over hydroxycarbonylation, providing good yields of both secondary and tertiary benzamides. This air-tolerant aqueous aminocarbonylation protocol provides operational convenience and increased safety for small-scale, high-speed synthesis.

Microwave-assisted aqueous aminocarbonylations have been reported not only for aryl bromides but also for aryl iodides and aryl chlorides using a diverse array of organic amines as nucleophiles.[97] The aryl iodides and bromides gave the corresponding benzamides with acceptable amounts of side products (benzoic acids and arenes). The competing side-reactions could be minimized by adjustment of the reaction temperature and either by using an excess of the nucleophilic amine or by employing the amine as the yield-limiting reactant. The aminocarbonylation of aryl chlorides could be difficult – a problem that was resolved by the use of a thermostable palladacycle and commercially available [(tBu)$_3$PH]BF$_4$ as a ligand source. A small library of benzamides, in high yields and purity, could be formed under the MW-assisted conditions at 110–180 °C in 10–30 min (Scheme 3.50).

This successful protocol was later used in the synthesis of a novel HIV-1 protease inhibitor, albeit the compound was formed in a moderate yield of 34% (Scheme 3.51).

An operationally simple hydroxycarbonylation of diversely substituted aryl triflates to the corresponding carboxylic acids with a palladium-mediated

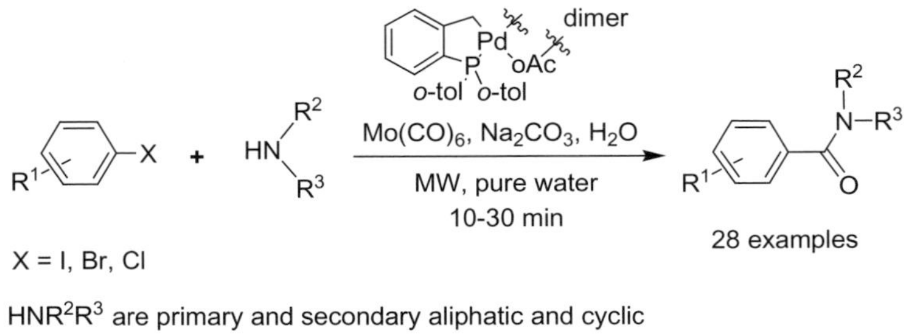

Scheme 3.50

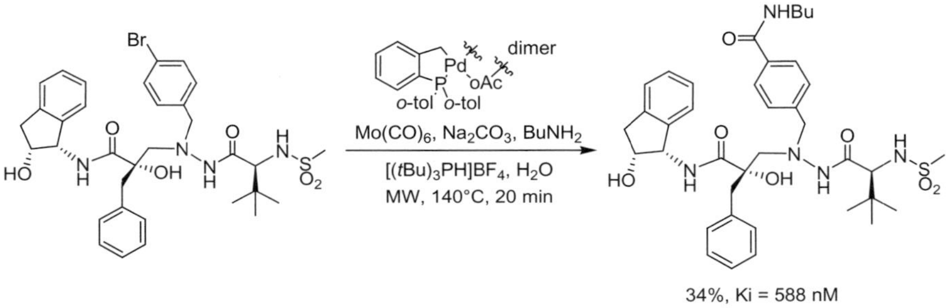

Scheme 3.51

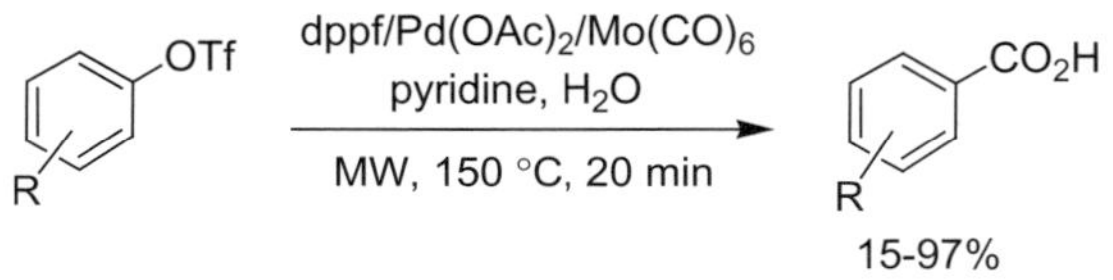

Scheme 3.52

microwave method has been reported by Silvani and co-workers.[98] The best conditions were obtained with the catalyst/ligand system Pd(OAc)$_2$/dppf and Mo(CO)$_6$ as the solid CO-liberating source. Pyridine was found to be the best base. Moderate to excellent yields for aryl carboxylic acids were obtained at 150 °C within 20 min (Scheme 3.52).

Complete chemoselectivity was observed with halogenated compounds, such as trifluoromethanesulfonic acid *p*-chloro- and *p*-bromophenyl esters, which afforded the desired *p*-chloro- and *p*-bromo-benzoic acids in high yields without side product formation.

Another alternative methodology for microwave-promoted hydroxycarbonylation of various aryl iodides in water, employing pre-pressurized

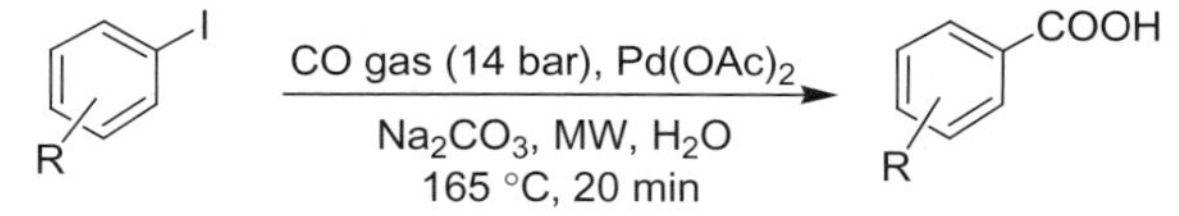

Scheme 3.53

vessels with gaseous carbon monoxide, has been reported.[99] Using a newly developed dedicated multimode microwave reactor (Anton Paar Synthos 3000) it is possible to perform reactions in heavy-walled quartz reaction vessels with operating limits of 80 bar. Reactions are performed using 1 mol.% of $Pd(OAc)_2$ or 0.01 mol.% of commercially available Pd solutions in a sealed tube pre-loaded with 14 bar CO, using Na_2CO_3 as base, in water at 165 °C for 20 min. A range of aryl iodides could be converted into the corresponding benzoic acid, including *ortho*-substituted examples (Scheme 3.53) but with this methodology the aryl bromides remained completely unreactive.

3.9 Representative Experimental Syntheses

3.9.1 Synthesis of 4,5-Dimethoxy-2-vinyl-2′-pivaloylamino-biphenyl (Scheme 3.14)[28]

A mixture of 1-bromo-4,5-dimethoxy-2-vinylbenzene (0.25 mmol), 2-pivaloyl-aminoophenyl-boronic acid (0.325 mmol), $NaHCO_3$ (0.75 mmol, 3.0 equiv.) and $Pd(Ph_3P)_4$ (5 mol.%) was placed in DMF (1.5 mL) and water (1.5 mL) and then irradiated for 15 min at 150 °C. After the reaction, the vial was cooled to 60 °C by air jet cooling. The crude mixture was partitioned between diethyl ether (25 mL) and water (25 mL) and the aqueous layer was extracted with diethyl ether. The combined organic layers were dried with $MgSO_4$ and solvents were removed under reduced pressure to yield the crude product as a yellow oil that was purified by column chromatography on gel silica (heptane–diethyl ether) to afford the desired product (93% yield).

3.9.2 Synthesis of 4-Phenyltoluene (Scheme 3.24)[43]

A solution of 4-bromotoluene (1 mmol), Ph_4BNa (0.25 mmol), K_2CO_3 (3.5 mmol), TBAB (0.3 mmol), PS-supported palladium catalyst 3 (1 mol.%) and 2 mL H_2O was placed in a 10-mL glass tube that was sealed and placed into the microwave. The reaction mixture was then heated to 120 °C and held at this temperature for 18 min. After cooling to room temperature, the organic material was dissolved in ethyl acetate–acetone (5 : 1 v/v, 10 mL). The reaction mixture was filtered and polymeric catalyst was extracted twice with ethyl acetate–acetone. The water layer was separated and the organic phase was washed with saturated NaCl and then dried with magnesium sulfate.

The solvent was removed under reduced pressure and the crude product was recrystallized from ethanol or purified by column chromatography on silica gel using light petroleum–ethyl acetate (30 : 1 v/v).

3.9.3 Synthesis of *m*-Carboxycinnamic Acid in Homogeneous Heck Reactions in D_2O (Scheme 3.36)[62]

A mixture of 3-bromobenzoic acid (0.5 mmol), acrylic acid (1.0 mmol) and K_2CO_3 (1.5 mmol) was added to 2.5 mL of D_2O in a 10 mL microwave vial. Then 60 μL of a 8.4 mM (5×10^{-7} mol, 0.1 mol.%) solution of **11** in D_2O was added. The mixture was then irradiated at 30 W with the maximum temperature set to 150 °C. After 10 min, the resulting mixture was quickly cooled using the cooling gas (N_2) feature on the microwave. The percent conversion was determined by ^1H NMR spectroscopy of the reaction mixture, monitoring the disappearance of 3-bromobenzoic acid and the appearance of product peaks. When the reaction was complete, the solution was neutralized with 6 M H_2SO_4 and the precipitate was collected and washed with water. The product was then dried and the structures of the isolated products were confirmed by ^1H and ^{13}C NMR spectroscopy.

3.9.4 Synthesis of Ethyl 3-(6-Methoxy-2-naphthyl)propanoate (Scheme 3.37)[63]

A mixture of 6-methoxy-2-bromonaphthalene (1 mmol), acrolein diethyl acetal (1.5 mmol), dicyclohexylmethylamine (1.5 mmol), tetrabutylammonium bromide (1 mmol) and 1 mol.% of **12** in a mixture of *N,N*-dimethylacetamide–water (4 : 1, 5 mL) was stirred at 120 °C for 10 min. After the reaction was complete, the resulting solution was cooled, poured into ethyl acetate (20 mL) and washed successively with 2 M HCl and H_2O. After drying over Na_2SO_4, the product was isolated by evaporating the solvent under vacuum and subjecting the residue to chromatography.

3.9.5 Synthesis of 5-Chloro-2-phenylethynyl-pyridine (Scheme 3.40)[61]

5-Bromo-2-chloropyridine (1 mmol), phenylacetylene (1.5 mmol), TBAB (0.6 mmol), sodium hydroxide (3 mmol) and precatalyst **10** (0.7 mol.%) were mixed in 3 mL of water in a properly capped process vial and thereafter the mixture was subjected to microwave irradiation at 160 °C for 15 min. The reaction mixture was cooled to room temperature and the solid catalyst was removed by filtration, washed with water and ethyl acetate, and the combined washings were added to the filtrate, which was then extracted with ethyl acetate. Purification of the product by flash chromatography (1 : 25, ethyl acetate–light petroleum) afforded the desired product in 82% yield.

3.9.6 Synthesis of 4-Methoxybenzonitrile (Scheme 3.47)[91]

A mixture of 4-iodoanisole (1 mmol), potassium hexacyanoferrate(II) (0.22 mmol), sodium fluoride (1 mmol), tetrabutylammonium bromide (1 mmol) and Pd(OAc)$_2$ (10 mg, 5 mol.%) was placed in a 10 mL glass tube. After adding water (2 mL), the vessel was sealed with a septum and an initial microwave irradiation of 150 W was used, the temperature being ramped from room temperature to the desired temperature of 150 °C. The reaction mixture was then held at 150 °C until a total time of 20 min had elapsed. After allowing the mixture to cool to room temperature, water (30 mL) and diethyl ether (30 mL) were added. The ether layer was separated and washed with magnesium sulfate and then isolated and characterized by comparison of NMR data.

3.9.7 Aminocarbonylation and Hydroxycarbonylation

3.9.7.1 General Procedure for Aminocarbonylation in Water (Scheme 3.49)[95]

Amine (1.0 mmol), Mo(CO)$_6$ (132 mg, 0.5 mmol), palladacycle **13** (47 mg, 0.05 mmol), Na$_2$CO$_3$ (318 mg, 3 mmol), aryl bromide (3 equiv.) and water (2 mL) were charged into a microwave vessel and the resulting mixture was heated for 10 min at 170 °C. After cooling, the mixture was extracted with ethyl acetate and purified by column chromatography on gel silica (CH$_2$Cl$_2$, then EtOAc–hexane) to give the corresponding amide.

3.9.7.2 Typical Procedure for the Aminocarbonylation of Trifluoromethanesulfonic Acid p-Nitrophenyl Ester[98]

A mixture of palladium diacetate (0.4 mmol), trifluoromethanesulfonic acid 4-nitrophenyl ester (4 mmol), 1,1′-bis(diphenylphosphino)ferrocene (0.4 mmol), pyridine (1.5 mL) and water (12 mL) was charged into a microwave vessel. Hexacarbonylmolybdenum (2 mmol) was then added and the resultant mixture was immediately heated by microwave at 150 °C for 20 min, under stirring. A solution of aqueous 6 M HCl (10 mL) was added very carefully to the reaction mixture and the resulting solution was filtered through Celite and then extracted with diethyl ether. The combined organic phases were then extracted with 2 M NaOH. After acidification of the combined basic aqueous phases with 6 M HCl, the pure carboxylic acid was recovered by filtration of the precipitate or extraction with Et$_2$O and evaporation of the solvent under reduced pressure.

3.9.7.3 Typical Procedure for the Hydroxycarbonylation of 4-Iodoanisole (Scheme 3.53)[99]

A mixture 4-iodoanisole (2 mmol), palladium acetate (0.02 mmol), sodium carbonate (7.4 mmol) and water (10 mL) was placed in an 80-mL quartz tube. The vessel was sealed and loaded onto the rotor then pressurized to 14 bar with

CO. The loaded rotor was subjected to a maximum of 1000 W microwave power in a ramp to 165 °C and then held at this temperature until a total reaction time of 20 min had elapsed. Upon completion, the reaction vessel was allowed to cool to 50 °C. The contents were then acidified with HCl to pH 1–3. The aqueous solution was extracted with diethyl ether. The organic washings were combined, dried over magnesium sulfate and the diethyl ether evaporated. The crude product was isolated and the yield determined by NMR spectroscopy using an internal standard.

3.10 Conclusion

Microwave irradiation has become very important in organic reactions in the last decade and it is reasonable to assert that there are now very few areas of organic chemistry that have not been shown to be enhanced using microwave heating. Many Pd-catalyzed reactions have been developed over the past three decades using this technique and the growing number of scientific publications in this area shows its high demand. Suzuki, Heck, Sonogashira, Stille, Hiyama, cyanation and carbonylation are among the predominantly used reactions that are catalyzed by palladium using MW. In addition, the use of MW and water as solvent in cross-coupling reactions has profoundly changed organic synthesis. Certainly, microwave-assisted coupling reactions in aqueous media will continue to be a fast-moving topic for the next several years and it will not be surprising to find microwave technology as one of the most preferred methods in cross-coupling reactions in the near future.

References

1. For a discussion, see (a) V. V. Grushin and H. Alper, in *Activation of Unreactive Bonds and Organic Synthesis,* ed. S. Murai, Springer, Berlin, 1999, pp. 193–226; (b) V. V. Grushin and H. Alper, *Chem. Rev.*, 1994, **94**, 1047–1062.
2. V. Polshettiwar, C. Len and A. Fihri, *Coord. Chem. Rev.*, 2009, **253**, 2599–2626.
3. L. Yin and J. Liebscher, *Chem. Rev.*, 2007, **107**, 133–173.
4. V. Polshettiwar and R. S. Varma, *Chem. Soc. Rev.*, 2008, **37**, 1546–1557.
5. D. Dallinger and C. O. Kappe, *Chem. Rev.*, 2007, **107**, 2563–2591.
6. N. Miyaura, T. Yanagi and A. Suzuki, *Synth. Commun.*, 1981, **11**, 513–519.
7. N. Miyaura and A. Suzuki, *Chem. Rev.*, 1995, **95**, 2457–2483.
8. N. Miyaura, *Cross-Coupling Reactions: A Practical Guide,* Springer, New York, 2002.
9. M. Larhed and A. Hallberg, *J. Org. Chem.*, 1996, **61**, 9582–9584.
10. M. Larhed, G. Lindeberg and A. Hallberg, *Tetrahedron Lett.*, 1996, **37**, 8219–8222.
11. N. E. Leadbeater and M. Marco, *Org. Lett.*, 2002, **4**, 2973–2976.

12. N. E. Leadbeater and M. Marco, *J. Org. Chem.*, 2003, **68**, 888–892.
13. A. F. Littke and G. C. Fu, *Angew. Chem. Int. Ed.*, 2002, **41**, 4176–4211.
14. G. Miao, P. Ye, L. Yu and C. M. Baldino, *J. Org. Chem.*, 2005, **70**, 2332–2334.
15. R. K. Arvela and N. E. Leadbeater, *Org. Lett.*, 2005, **7**, 2101–2104.
16. K. W. Anderson and S. L. Buchwald, *Angew. Chem. Int. Ed.*, 2005, **44**, 6173–6177.
17. P. Čapek, R. Pohl and M. Hocek, *Org. Biomol. Chem.*, 2006, **4**, 2278–2284.
18. P. Čapek, M. Vrábel, Z. Hasník, R. Pohl and M. Hocek, *Synthesis*, 2006, 3515–3526.
19. R. B. Bedford, C. P. Butts, T. E. Hurst and P. Lidström, *Adv. Synth. Catal.*, 2004, **346**, 1627–1630.
20. I. D. Kostas, G. A. Heropoulos, D. Kovala-Demertzi, P. N. Yadav, J. P. Jasinski, M. A. Demertzis, F. J. Andreadaki, G. Vo-Thanh, A. Petit and A. Loupy, *Tetrahedron Lett.*, 2006, **47**, 4403–4407.
21. X. Mei, Rhia M. Martin and C. Wolf, *J. Org. Chem.*, 2006, **71**, 2854–2861.
22. G. Cravotto, M. Beggiato, A. Penoni, G. Palmisano, S. Tollari, J.-M. Lévêque and W. Bonrath, *Tetrahedron Lett.*, 2005, **46**, 2267–2271.
23. M. Nishimura, M. Ueda and N. Miyaura, *Tetrahedron*, 2002, **58**, 5779–5787.
24. A. Konovets, A. Penciu, E. Framery, N. Percina, C. Goux-Henry and D. Sinou, *Tetrahedron Lett.*, 2005, **46**, 3205–3208.
25. E. J. García Suarez, A. Ruiz, S. Castillón, W. Oberhauser, C. Bianchini and C. Claver, *Dalton Trans.*, 2007, 285–2861.
26. R. Lépine and J. Zhu, *Org. Lett.*, 2005, **7**, 2981–2984.
27. J. Wannberg, Y. A. Sabnis, L. Vrang, B. Samuelsson, A. Karlén, A. Hallberg and M. Larhed, *Bioorg. Med. Chem.*, 2006, **14**, 5303–5315.
28. P. Appukkuttan, W. Dehaen and E. Van der Eycken, *Org. Lett.*, 2005, **7**, 2723–2726.
29. P. Appukkuttan, W. Dehaen and E. Van der Eycken, *Chem. Eur. J.*, 2007, **13**, 6452–6460.
30. P. Appukkuttan, A. B. Orts, R. P. Chandran, J. L. Goeman, J. Van der Eycken, W. Dehaen and E. Van der Eycken, *Eur. J. Org. Chem.*, 2004, 3277–3285.
31. M. D. Crozet, C. Castera-Ducros and P. Vanelle, *Tetrahedron Lett.*, 2006, **47**, 7061–7065.
32. J. V. Metzger, in *Thiazole and its Derivatives,* John Wiley & Sons, Inc., New York, 1979.
33. A. A. Kiryanov, P. Sampson and A. J. Seed, *J. Org. Chem.*, 2001, **66**, 7925–7929.
34. T. Bach and S. Heuser, *Tetrahedron Lett.*, 2000, **41**, 1707–1710.
35. A. Cohen, M. D. Crozet, P. Rathelot and P. Vanelle, *Green Chem.*, 2009, **11**, 1736–1742.
36. E. De Clercq, *Nat. Rev. Drug Discovery*, 2002, **1**, 13–25.
37. E. De Clercq, *Antiviral Res.*, 2005, **67**, 56–75.
38. R. Zhu, F. Qu, G. Quéléver and L. Peng, *Tetrahedron Lett.*, 2007, **48**, 2389–2393.

39. Y. Gong, J. K. Barbay, A. B. Dyatkin, T. A. Miskowski, E. S. Kimball, S. M. Prouty, M. C. Fisher, R. J. Santulli, C. R. Schneider, N. H. Wallace, S. A. Ballentine, W. E. Hageman, J. A. Masucci, B. E. Maryanoff, B. P. Damiano, P. Andrade-Gordon, D. J. Hlasta, P. J. Hornby and W. He, *J. Med. Chem.*, 2006, **49**, 3402–3411.
40. R. K. Arvela, N. E. Leadbeater, T. L. Mack and C. M. Kormos, *Tetrahedron Lett.*, 2006, **47**, 217–220.
41. R. L. Harker and R. D. Crouch, *Synthesis*, 2007, **1**, 25–27.
42. G. W. Kabalka, L. Li Zhou and A. Naravane, *Tetrahedron Lett.*, 2006, **47**, 6887–6889.
43. L. Baia and J.-X. Wanga, *Adv. Synth. Catal.*, 2008, **350**, 315–320.
44. F.-X. Felpin, T. Ayad and S. Mitra, *Eur. J. Org. Chem.*, 2006, 2679–2690.
45. N. T. S. Phan, M. Van Der Sluys and C. W. Jones, *Adv. Synth. Catal.*, 2006, **348**, 609–679.
46. W. Solodenko, U. Schön, J. Messinger, A. Glinschert and A. Kirschning, *Synlett*, 2004, 1699–1702.
47. K. M. Dawood, *Tetrahedron*, 2007, **63**, 9642–9651.
48. S.-S. Yi, D.-H. Lee, E. Sin and Y.-S. Lee, *Tetrahedron Lett.*, 2007, **48**, 6771–6775.
49. J. Ku Cho, R. Najman, T. W. Dean, O. Ichihara, C. Muller and M. Bradley, *J. Am. Chem. Soc.*, 2006, **128**, 6276–6277.
50. C. G. Blettner, W. A. König, W. Stenzel and T. Schotten, *J. Org. Chem.*, 1999, **64**, 3885–3890.
51. W. Zhang, C. Hiu-Tung Chen, Y. Lu and T. Nagashima, *Org. Lett.*, 2004, **6**, 1473–1476.
52. A. M. Trzeciak and J. J. Ziolkowski, *Coord. Chem. Rev.*, 2005, **249**, 2308–2322.
53. T. Mizoroki, K. Mori and A. Ozaki, *Bull. Chem. Soc. Jpn.*, 1971, **44**, 581.
54. R. F. Heck and J. P. Nolly, *J. Org. Chem.*, 1972, **37**, 2320–2322.
55. A. Eisenstadt and D. J. Ager, in *Fine Chemicals through Heterogeneous Catalysis*, ed. R. A. Sheldon and H. van Bekkum, Wiley-VCH, Verlag GmbH, Weinheim, 2001, p. 576.
56. A. Zapf and M. Beller, *Top. Catal.*, 2002, **19**, 101–109.
57. J.-X. Wang, Z. Liu, Y. Hu, B. Wei and L. Bai, *Synth. Commun.*, 2002, **32**, 1607–1614.
58. R. K. Arvela and N. E. Leadbeater, *J. Org. Chem.*, 2005, **70**, 1786–1790.
59. R. K. Arvela, S. Pasquini and M. Larhed, *J. Org. Chem.*, 2007, **72**, 6390–6396.
60. J. Gil-Moltó, S. Karlström and C. Nájera, *Tetrahedron*, 2005, **61**, 12168–12176.
61. K. M. Dawood, W. Solodenko and A. Kirschning, *ARKIVOC*, 2007 (v) 104–124.
62. D. E. Bergbreiter and S. Furyk, *Green Chem.*, 2004, **6**, 280–285.
63. C. Nájera and L. Botella, *Tetrahedron*, 2005, **61**, 9688–9695.
64. M. Zhu, Y. Song and Y. Cao, *Synthesis*, 2007, **6**, 853–856.

65. I. Paterson, R. D. Davies and R. Marquez, *Angew. Chem., Int. Ed.*, 2001, **40**, 603–607.
66. M. Toyota, C. Komori and M. Ihara, *J. Org. Chem.*, 2000, **65**, 7110–7113.
67. K. C. Nicolaou and W.-M. Dai, *Angew. Chem., Int Ed. Engl.*, 1991, **30**, 1387–1416.
68. R. Wu, J. S. Schumm, D. L. Pearson and J. M. Tour, *J. Org. Chem.*, 1996, **61**, 6906–6921.
69. U. H. F. Bunz, *Chem. Rev.*, 2000, **100**, 1605–1644.
70. K. Sonogashira, Y. Tohda and N. Hagihara, *Tetrahedron Lett.*, 1975, **16**, 4467–4470.
71. M. Erdélyi and A. Gogoll, *J. Org. Chem.*, 2001, **66**, 4165–4169; some earlier single results are known; see, for example, G. W. Kabalka, L. Wang, V. Namboodiri and R. M. Pagni, *Tetrahedron Lett.*, 2000, **41**, 5151–5154. See also J. Li, A. W.-H. Mau and C. R. Strauss, *Chem. Commun.*, 1997, 1275–1276.
72. N. E. Leadbeater, M. Marco and B. J. Tominack, *Org. Lett.*, 2003, **5**, 3919–3922.
73. P. Appukkuttan, W. Dehaen and E. van der Eycken, *Eur. J. Org. Chem.*, 2003, 4713–4716.
74. R. K. Arvela, N. E. Leadbeater, M. Sangi, V. Williams, P. Granados and R. D. Singer, *J. Org. Chem.*, 2005, **70**, 161–168.
75. A. J. McCarroll, T. D. Bradshaw, A. D. Westwell, C. S. Matthews and M. F. G. Stevens, *J. Med. Chem.*, 2007, **50**, 1707–1710.
76. D. Milstein and J. K. Stille, *J. Am. Chem. Soc.*, 1978, **100**, 3636–3638.
77. J. K. Stille, *Angew. Chem. Int. Ed. Engl.*, 1986, **25**, 508–524.
78. P. Espinet and A. M. Echavarren, *Angew. Chem. Int. Ed.*, 2004, **43**, 4704–4734.
79. N. Kaval, K. Bisztray, W. Dehaen, C. O. Kappe and E. Van der Eycken, *Mol. Diversity*, 2003, **7**, 125–133.
80. T. Hiyama and Y. Hatanaka, *Pure Appl. Chem.*, 1994, **66**, 1471–1478.
81. T. Hiyama, in *Metal-Catalyzed Cross-Coupling Reactions*, ed. F. Diederich, P.J. Stang, Wiley-VCH, New York, 1998, ch. 10.
82. C. J. Handy, A. S. Manoso, W. T. McElroy, W. M. Seganish and P. DeShong, *Tetrahedron*, 2005, **61**, 12201–12225.
83. T. Huang and C.-J. Li, *Tetrahedron Lett.*, 2002, **43**, 403–405.
84. C. Wolf and R. Lerebours, *Org. Lett.*, 2004, **6**, 1147.
85. E. Alacid and C. Nájera, *Adv. Synth. Catal.*, 2006, **348**, 2085–2091.
86. D. E. Uehling, S. S. Nanthakumar, D. Croom, D. L. Emerson, P. P. Leitner, M. J. Luzzio, G. McIntyre, B. Morton, S. Profeta, J. Sisco, D. D. Sternbach, W.-Q. Tong, A. Vuong and J. M. Besterman, *J. Med. Chem.*, 1995, **38**, 1106–1118.
87. K. Friedrich and K. Wallenfels, in *The Chemistry of the Cyano Group*, Z. Rappoport (ed.), Interscience, London, 1970, p. 67.
88. G. A. Ellis and T. M. Romney-Alexander, *Chem. Rev.*, 1987, **87**, 779–794.
89. M. Alterman and A. Hallberg, *J. Org. Chem.*, 2000, **65**, 7984–7989.

90. R. K. Arvela, N. E. Leadbeater, H. M. Torenius and H. Tye, *Org. Biomol. Chem.*, 2003, **1**, 1119–1121.
91. S. Velmathi and N. E. Leadbeater, *Tetrahedron Lett.*, 2008, **49**, 4693–4694.
92. *Applied Homogeneous Catalysis with Organometallic Compounds: A Comprehensive Handbook in Three Volumes, 2nd, Completely Revised and Enlarged Edition*, ed. B. Cornils and W. A. Herrmann, Wiley-VCH, 2002.
93. N.-F. K. Kaiser, A. Hallberg and M. Larhed, *J. Comb. Chem.*, 2002, **4**, 109–111.
94. J. Georgsson, A. Hallberg and M. Larhed, *J. Comb. Chem.*, 2003, **5**, 350–352.
95. X. Wu and M. Larhed, *Org. Lett.*, 2005, **7**, 3327–3329.
96. W. A. Herrmann, V. P. W. Böhm and C.-P. Reisinger, *J. Organomet. Chem.*, 1999, **576**, 23–41.
97. X. Wu, J. K. Ekegren and M. Larhed, *Organometallics*, 2006, **25**, 1434–1439.
98. G. Lesma, A. Sacchetti and A. Silvani, *Synthesis*, 2006, 594–596.
99. C. M. Kormos and N. E. Leadbeater, *Synlett*, 2006, 1663–1666.

CHAPTER 4
Microwave-assisted Synthesis of Bio-active Heterocycles in Aqueous Media

VIVEK POLSHETTIWAR[†] AND RAJENDER S. VARMA

Sustainable Technology Division, National Risk Management Research Laboratory, U. S. Environmental Protection Agency, 26 W. Martin Luther King Dr., MS 443, Cincinnati, Ohio, 45268, USA

4.1 Introduction

Human health, especially in an aging population, mostly depends on various medicines, and researchers are combating emerging diseases with new drug discovery. Heterocyclic compounds hold a special place among pharmaceutically active natural products as well as synthetic compounds.[1,2] The outstanding ability of heterocyclic nuclei to act as biomimetics and pharmacophores has contributed extensively to their unique value as traditional key elements of numerous drugs. Heterocycles have the enormous ability to selectively influence the activity of biological systems. Thus, in both lead identification and optimization processes, there is an acute need for new heterocycle-based active molecules. However, pathways currently used for drug discovery are not sustainable in terms of the time required for the synthesis of new bioactive molecules. Conventional methods of organic synthesis are orders of magnitude too slow to

---

[†] Current address: KAUST Catalysis Center (KCC), King Abdullah University of Science and Technology, Thuwal 23955, Kingdom of Saudi Arabia.

RSC Green Chemistry No. 7
Aqueous Microwave Assisted Chemistry: Synthesis and Catalysis
Edited by Vivek Polshettiwar and Rajender S. Varma
© Royal Society of Chemistry 2010
Published by the Royal Society of Chemistry, www.rsc.org

satisfy the demand for the preparation of new compounds. Combinatorial and automated medicinal chemistry sub-disciplines have emerged to meet the increasing requirement of new compounds for drug discovery, where a high turnover is the essence.[3] While most of these techniques are rapid and productive, they generate considerable quantities of chemical waste. Synthetic chemists have been under increasing pressure to produce, in an environmentally benign fashion, the series of bioactive heterocyclic-based novel drugs required to protect our heath in a short period of time. One alternative is to use non-conventional microwave (MW) heating technology. The efficiency of MW flash-heating has resulted in a dramatic reduction in reaction times – from days to minutes – which is potentially important in traditional medicinal chemistry and the assembly of heterocyclic systems.[4-6]

In recent years, MW-assisted chemistry has emerged into a mature and useful technique for various applications. Although MW-assisted reactions in conventional solvents have been developed rapidly, the main focus has now shifted to environmentally benign processes, which use greener solvents such as water.[7,8] As water is abundantly available in nature and devoid of any toxic properties, it is advantageous to carry out reactions in aqueous media. Reactions in aqueous media are environmentally benign, which will reduce the reliance of chemists on ecologically hazardous solvents. Interestingly, the combination of MW and aqueous media shows excellent benefits, such as shorter reaction times, homogeneous heating, and enhanced yields and selectivity.[9,10] There are many examples of the successful application of MW-assisted aqueous chemistry (the use of water as a non-toxic reaction medium, together with the deployment of microwave as a heating source) to organic synthesis and it seems to be a promising and enabling greener alternative.[11-14]

4.2 MW-assisted Nitrogen-containing Heterocycle Synthesis in Water

Nature is rich in nitrogen-containing heterocycles in the form of natural products such as vitamins, hormones, antibiotics, and alkaloids, as well as pharmaceuticals, herbicides, and dyes. These molecules play vital roles in drug discovery for almost every disease, making them one of the most important building blocks in medicinal chemistry. Rapid synthesis of these heterocycles using green chemistry principles is currently a real need, and aqueous microwave chemistry is one of the emerging techniques to achieve this goal.[1]

Nitrogen-heterocycles bearing triazine and tetrazole nuclei are an integral part of therapeutically attractive compounds showing varied biological activities. A series of these molecules have been designed and synthesised by first converting primary alcohols and aldehydes into respective nitriles, by treatment with iodine in aqueous ammonia under MW irradiation, which then underwent cycloaddition with dicyandiamide and sodium azides to produce triazines and tetrazoles, respectively, with high optical purity (Scheme 4.1).[15] For triazines, the reaction mixture was cooled and the precipitated product was then isolated.

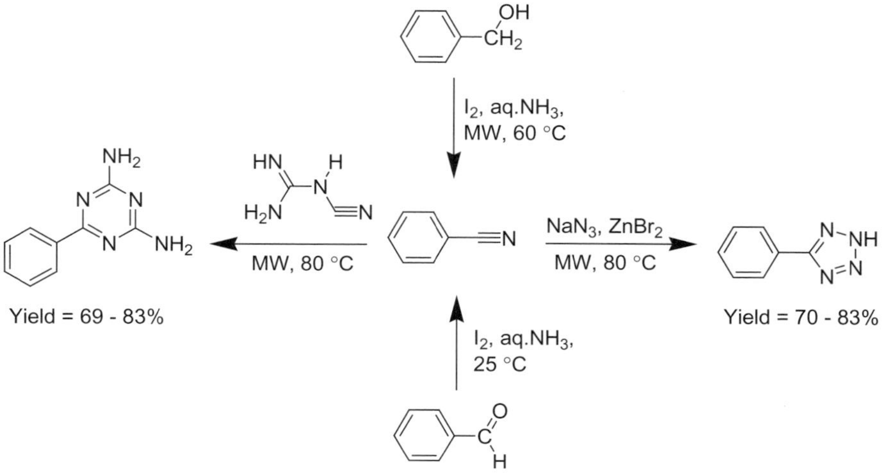

Scheme 4.1

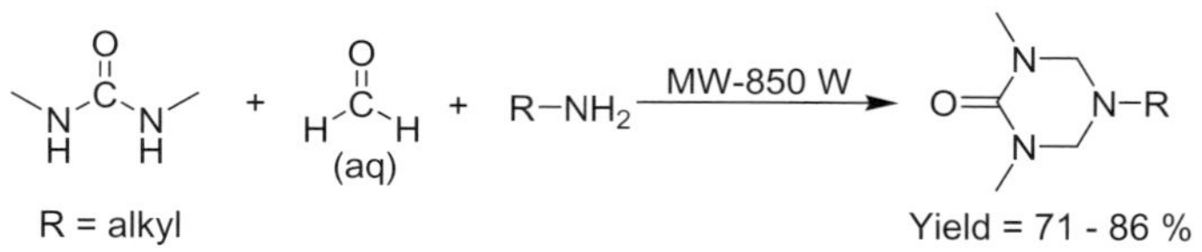

Scheme 4.2

However, in the case of tetrazoles, extraction with ethyl acetate was needed to isolate the desired product.

Balalaie and Shokrollahi have synthesized triazones by the three-component condensation reaction of N,N'-dimethylurea, formaldehyde, and primary amine in aqueous medium under microwave irradiation at 850 W for 2–4 min (Scheme 4.2). The final product was isolated by extraction with dichloromethane and purified by vacuum distillation.[16]

Triaza-benzo[b]fluoren-6-one derivatives exhibit excellent biological activity, including various central nervous system affecting properties. Tu and his co-workers[17] have synthesized these important molecules by the reaction of aldehyde, cyclohexane-1,3-dione, and 2-aminobenzimidazole in neat water using microwaves (Scheme 4.3). Microwave exposure for 2–4 min completed the reaction and the product was isolated by filtration in excellent yield.

The MW-assisted synthesis of nitrogen-containing heterocycles, such as azetidines, pyrrolidines, piperidines, azepanes, N-substituted 2,3-dihydro-1H-isoindoles, 4,5-dihydropyrazoles, pyrazolidines, and 1,2-dihydrophthalazines, has been conducted in water catalyzed by weak base. The reactions proceeded *via*

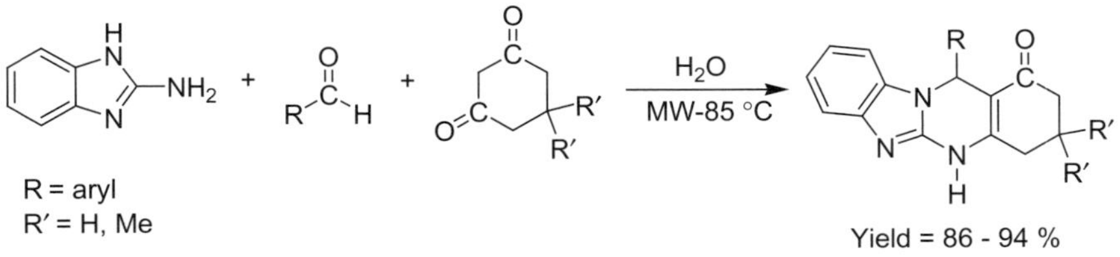

Scheme 4.3

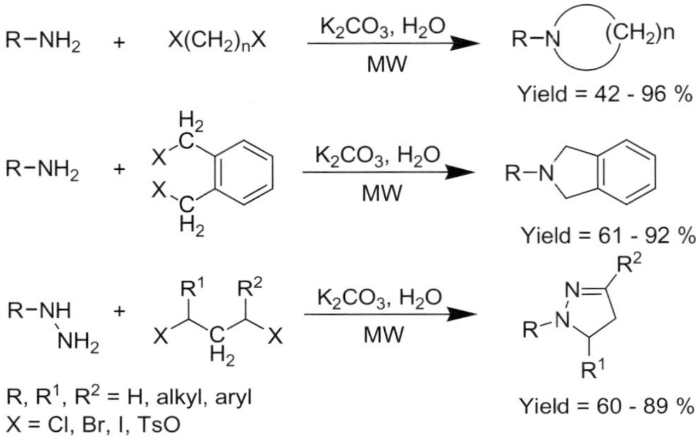

R = aryl
R' = H, Me

Yield = 86 - 94 %

Scheme 4.4

R, R¹, R² = H, alkyl, aryl
X = Cl, Br, I, TsO

Yield = 42 - 96 %
Yield = 61 - 92 %
Yield = 60 - 89 %

double *N*-alkylation of primary amines and hydrazine derivatives (Scheme 4.4) with alkyl dihalides, thus providing facile entry to important classes of building blocks in natural products and pharmaceuticals.[18–20]

The reaction time was significantly reduced and the protocol only utilizes easily available amines, hydrazines, alkyl dihalides, and ditosylates to make two C–N bonds in a S_N2-like sequential heterocyclization that is otherwise difficult to create under conventional conditions. This strategy also avoids multi-step reactions, functional group protection/deprotection sequences, and eliminates the use of expensive phase-transfer and transition metal catalysts. Phase separation of the desired product in either solid or liquid form from the aqueous media was done by simple filtration or decantation instead of tedious work-up processes, thereby circumventing the use of volatile organic solvents.[20]

The double alkylation of hydrazine was preferred by MW irradiation (Scheme 4.5) because of the increased polarity of **2** and **4**, which direct the reaction to form the desired product 1-phenylpyrazolidine, **5**.[18–20] The formation of a carbon–nitrogen double bond to generate 1-phenyl-4,5-dihydro-1*H*-pyrazole (**6**) was examined in two different experiments by introducing oxygen into the system or by using activated palladium on carbon as a

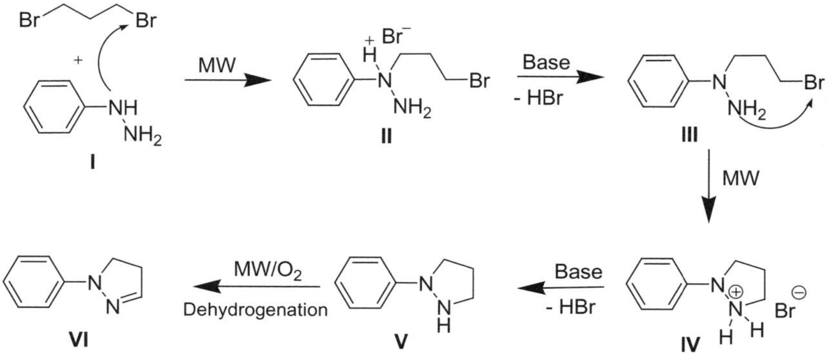

Scheme 4.5

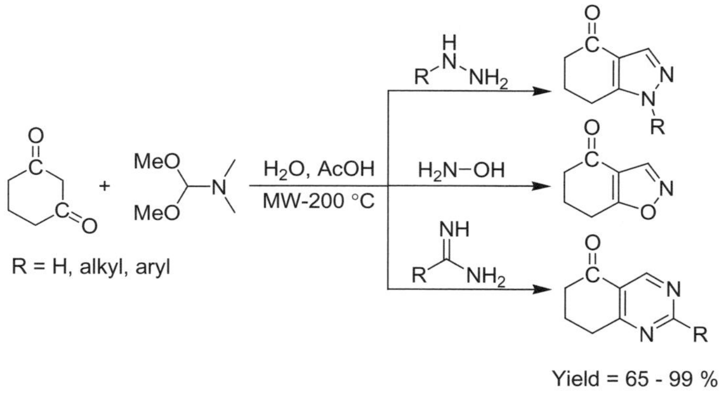

Scheme 4.6

dehydrogenation catalyst. The experimental observation that oxygen in the air plays a critical role in promoting the formation of a C=N double bond is consistent with the mechanistic hypothesis wherein the polar transition state of the reaction is favored by MW irradiation with respect to the dielectric polarization nature of MW energy transfer.

In situ synthesis of heterocycles is one of the excellent ways for the rapid generation of a library of bioactive molecules. Molteni and his colleagues[21] have developed a simple one-pot protocol for the conversion of enaminoketones into various heterocycles by employing different bidentate nucleophiles (Scheme 4.6) and using microwaves as a heating source and benign water as a reaction solvent. Advantages of this method are reduced reaction time and, in particular, the easy isolation of product by simple filtration, as it precipitated out from the reaction mixture.

Peptides encompassing spiro-amino acid units are helpful to restrict the flexibility of the peptide and also to provide information on the topographical requirements of peptide receptors. An aqueous protocol for the synthesis of spiro-2,5-diketopiperazines has been developed by cyclization of Boc-protected dipeptides containing spiro-amino acids using MW irradiation (Scheme 4.7).[22] After completion of the reaction, water was evaporated to isolate the product, which was further purified by column chromatography.

Schollkopf's bis-lactam ether chiral auxiliaries (3S)- and (3R)-3,6-dihydro-2,5-diethoxy-3-isopropylpyrazine have been designed and synthesized on a multi-gram scale, wherein the key step, the preparation of the 2,5-diketopiperazine derivative, was carried out by microwave-assisted heating in water (Scheme 4.8). This protocol avoided the use of any organic solvents and only inexpensive and easily available starting materials were used. The product was isolated by extraction with dichloromethane and purified by column chromatography.[23]

Polysubstituted benzo[b]pyrans and their derivatives have fascinated good medicinal chemists due to their wide range of pharmaceutical and biological activities. Haung, Lee, and co-workers[24] have designed a facile MW-assisted protocol for the synthesis of these potential cytotoxic agents by reaction of aromatic aldehyde, trimethoxyphenol, and malononitrile in aqueous medium, catalyzed by quaternary ammonium salts (Scheme 4.9). Reactions were carried out at 150 °C using the 300 W microwave power level and were completed in less than 10 min. After cooling the reaction mixture, products were isolated by filtration to yield pure benzopyran derivatives, which can be then evaluated for biological activity.

Benzopyrano[4,3-c]pyrazoles were also synthesized by heterocondensation between *in situ* generated 3-arylidene-2,4-chromanediones and an N-substituted hydrazine moiety, without use of organic solvents. This high yielding protocol for pharmaceutically useful molecules was archived out in water as a solvent under MW irradiation conditions (Scheme 4.10).[25]

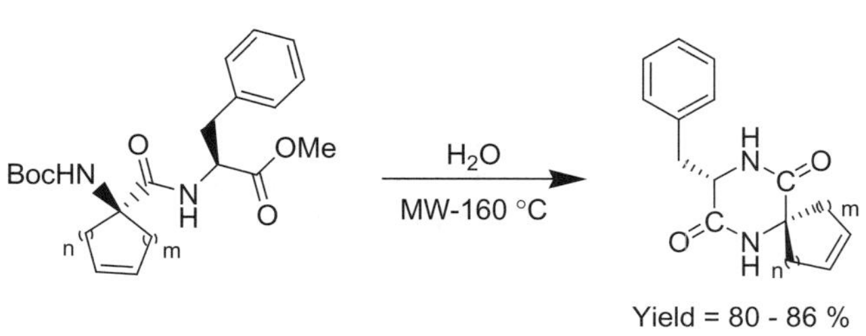

Scheme 4.7

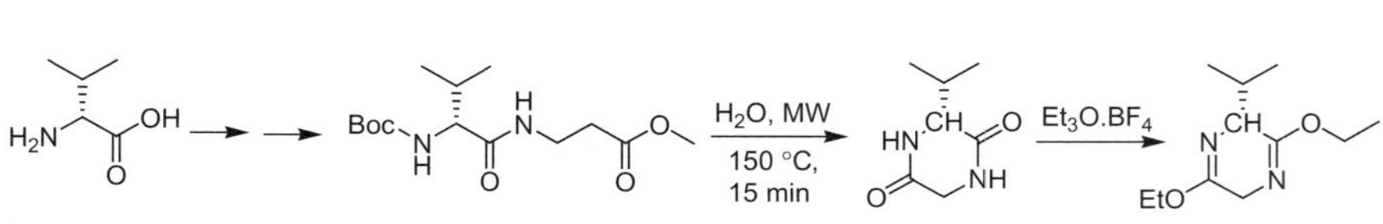

Scheme 4.8

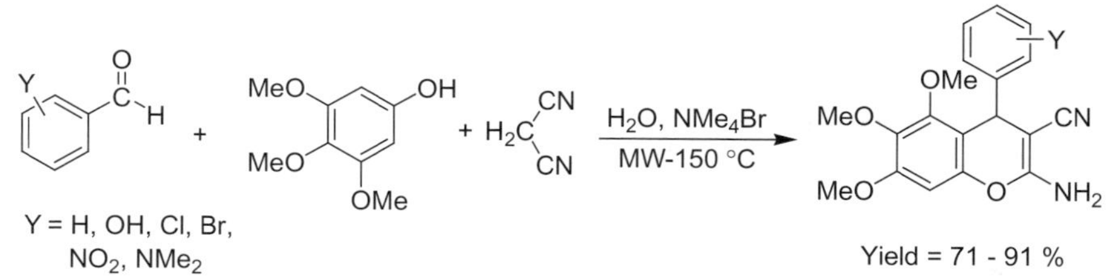

Scheme 4.9

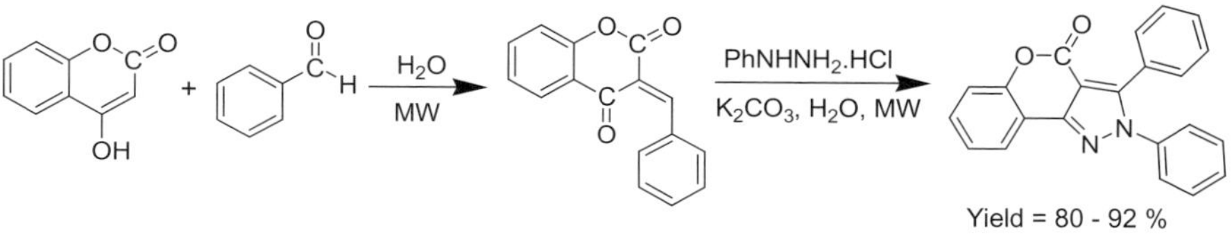

Scheme 4.10

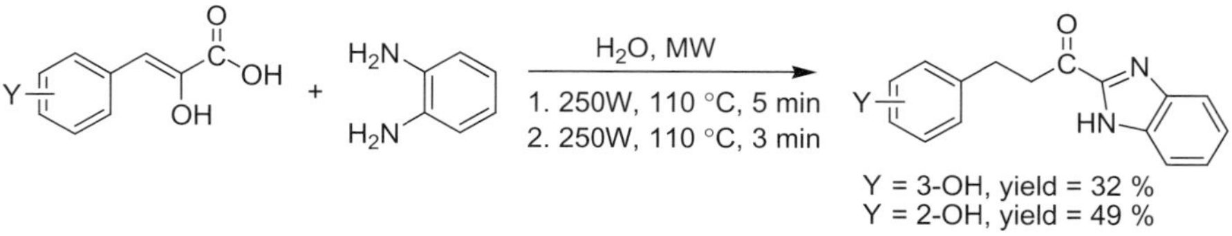

Scheme 4.11

Monforte et al. have synthesized 1-(1H-benzimidazol-2-yl)-2-(3- or 4-hydroxyphenyl)ethanone, a potential HIV-1 integrase inhibitor, in aqueous medium under MW irradiation (Scheme 4.11). Within 8 min of exposure, product was obtained in 32–49% yield. No organic solvent was used even during the work-up and the product was simply precipitated out by neutralization reaction and recrystallized to obtain pure product.[26]

Non-substituted tri-, tetra-, and pentapyrranes, essential precursors for the synthesis of porphyrins and their derivatives, have been designed and synthesized by MW-assisted one-step condensation of aqueous formaldehyde with pyrrole (Scheme 4.12). Shorter reaction times and the use of a green aqueous reaction medium are the main advantages of this protocol.[27] Product was isolated by extraction with dichloromethane to yield a brownish oil that was then separated and purified by column chromatography.

Dihydropyrimidinones, an important class of biologically important heterocyclic compounds, have been synthesized by the Biginelli protocol using polystyrene sulfonic acid (PSSA) as a catalyst under aqueous microwave

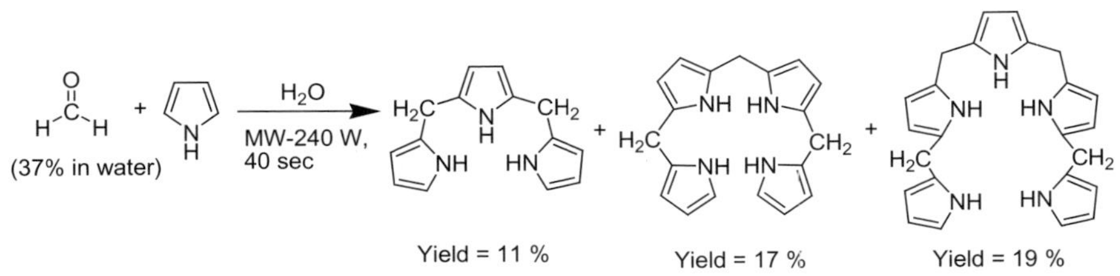

Scheme 4.12

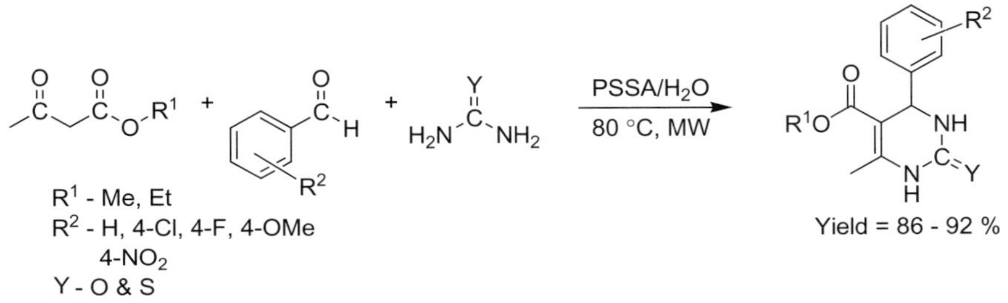

Scheme 4.13

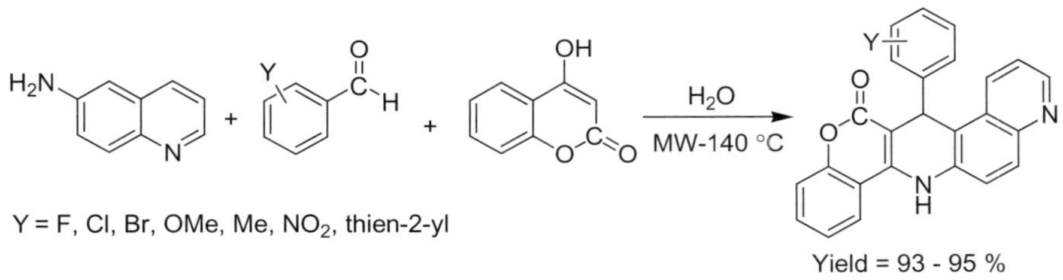

Scheme 4.14

condition (Scheme 4.13).[28] The use of polymer-supported, low toxic, and inexpensive PSSA as a catalyst rendered this method eco-friendly, with a very simple isolation procedure that involved filtration of the precipitated products.

Chromenes and phenanthroline derivatives have high biological activities and the combination of these pharmacophoric moieties can provide a new series of active heterocycles for drug development. This has been carried out by MW-assisted reaction of aromatic aldehyde, 6-aminoquinoline, and 4-hydroxycoumarin (Scheme 4.14) in aqueous medium. The reaction was completed in 5–6 min at 140 °C, affording >90% product yield, and the crude product was isolated by filtration from the reaction mixture.[29]

The Hantzsch reaction is a sustainable tool to synthesize 1,4-dihydropyridine, an important building block for drug discovery. Ohberg and Westman[30] have accelerated this protocol for the synthesis of 4-aryl and 4-alkyl-2,6-dimethyl-1,4-dihydropyridine-3,5-dicarboxylates using microwave irradiation in aqueous ammonium hydroxide (Scheme 4.15). The reaction rate was increased, with good product yield and acceptable purity. After completion of reaction, the reaction mixture was evaporated to dryness and product was isolated by column chromatography.

Another important class of derivatives of 2-amino-4-aryl-6-ferrocenylpyridine, has been synthesized *via* reaction of aromatic aldehyde, malononitrile or ethyl cyanoacetate, acetylferrocene, and ammonium acetate. The reaction was performed in one-pot under catalyst-free and MW heating conditions in aqueous medium (Scheme 4.16).[31] This low cost protocol afforded higher product yield with reduced environmental impact.

Amino-substituted 2-pyridone derivatives, because of their attractive features as building blocks for medicinal chemistry, have garnered the attention of synthetic chemists. These molecules have been designed and prepared in water medium under microwave irradiation by a one-pot coupling reaction of aldehydes and malononitrile, catalyzed by sodium hydroxide (Scheme 4.17).[32]

Although water is one of the best green solvents, execution of many organic reactions in water is not simple due to the inherent solubility limitation of non-polar reactants in polar aqueous medium. Peng, Song, and Dou[33] have demonstrated excellent strategies for aqueous organic reactions using the synergistic effect of microwave and ultrasound irradiation, termed CMUI. They observed a dramatic increase in the rate of reactions, which may be due to

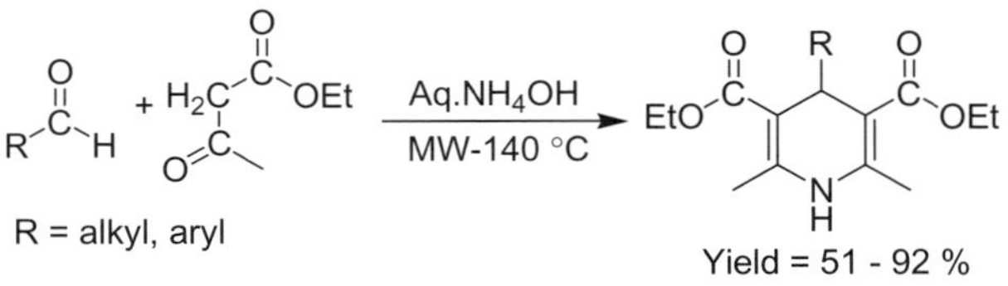

Scheme 4.15

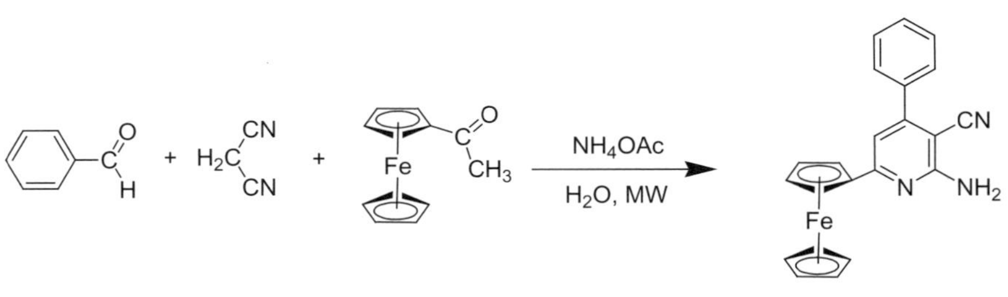

Scheme 4.16

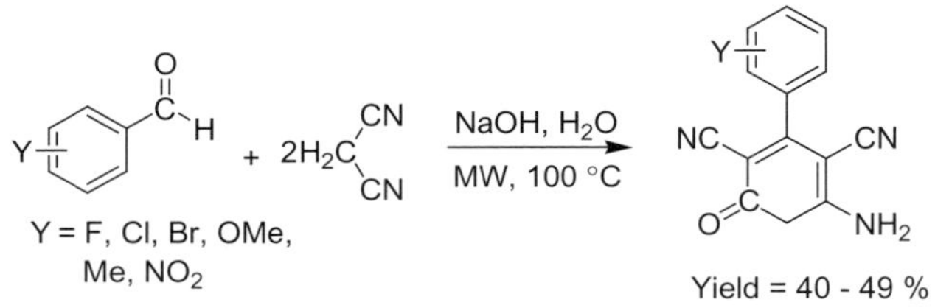

Scheme 4.17

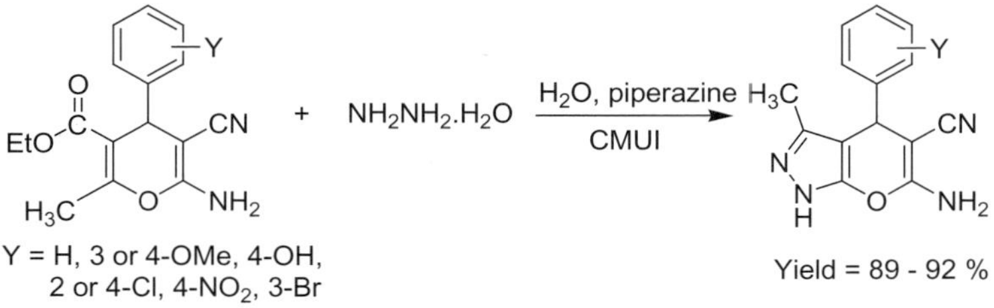

Scheme 4.18

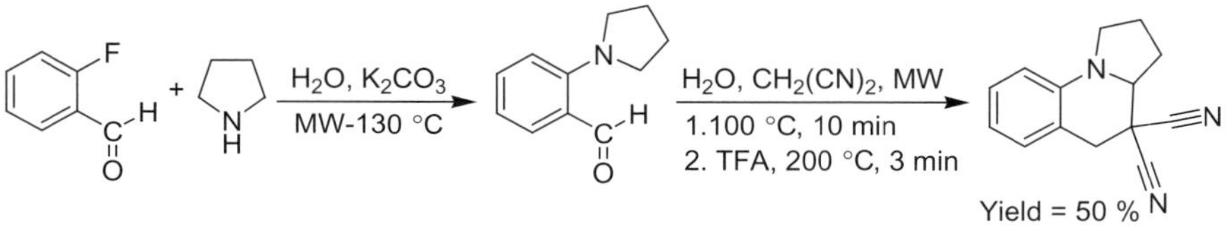

Scheme 4.19

the efficient removal of a passivation coating on the substrate particles, which enhances the mass and heat transfer. To prove the concept they synthesized, in good yield, a series of 4*H*-pyrano[2,3-*c*]pyrazole derivatives in aqueous media using the CMUI strategy (Scheme 4.18). Reactants were treated under combined MW and ultrasound irradiation and precipitated products were isolated by filtration, and were then purified by recrystallization from ethanol.

Van der Eycken and co-workers[34] have synthesized biologically important pyrido-fused heterocycles under microwave conditions. They synthesized a tri-cyclic heterocycle in a two-step procedure starting from *ortho*-fluorobenzaldehyde (Scheme 4.19). In the initial step, substitution of the fluorine with pyrrolidine was carried out by microwave irradiation of aqueous pyrrolidine and *ortho*-fluorobenzaldehyde in the presence of potassium carbonate in water. In the next step, after 10 min of microwave irradiation of the mixture of the pyrrolidine substituted aldehyde and malonitrile at 100 °C, trifluoroacetic acid

(TFA) (one drop) was added to the reaction mixture and irradiation was continued at 200 °C for 3 min. The desired compound was isolated by extraction with diethyl ether and the crude product was recrystallized from methanol.

Bicyclo[2.2.2]oct-7-enes, bio-active molecules found in the skeleton of *Kopsia* alkaloids, are well known multifunctional building blocks in drug discovery. Kočevar *et al.* have developed a microwave process for the synthesis of *N*-aminosuccinimide derivatives of bicyclo[2.2.2]oct-7-enes from fused anhydrides and different hydrazines (Scheme 4.20).[35] The reaction was carried out in neat water medium under neutral conditions, without the use of any catalyst.[53] This method has been extended to acetyl-containing derivatives of bicyclo[2.2.2]oct-7-enes with a range of amines (Scheme 4.21), generating succinimide derivatives chemoselectively.[35]

Indenoquinoline derivatives exhibit several useful biological properties, including anti-inflammatory activity, antitumor agents, inhibitors, and many more. Tu and co-workers[36] have designed and prepared these poly-substituted indeno[1,2-*b*]quinolines by the reaction of aldehydes, 1,3-indanedione, and enaminones under aqueous microwave conditions, using *p*-toluene sulfonic acid (*p*-TsOH) as a catalyst (Scheme 4.22). Several poly-substituted indeno[1,2-*b*]quinolines were synthesized successfully; however, the reactions were

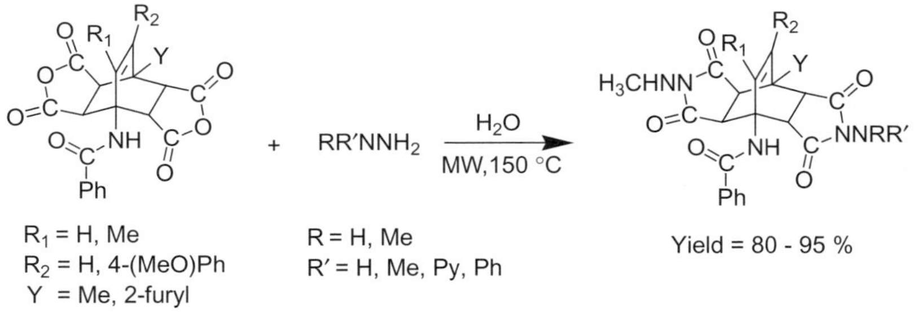

Scheme 4.20

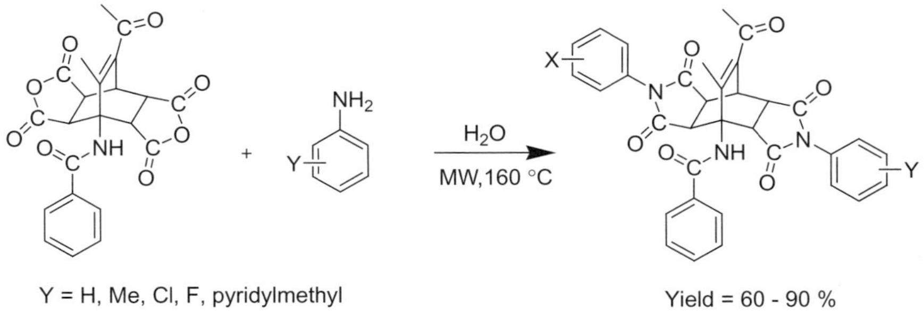

Scheme 4.21

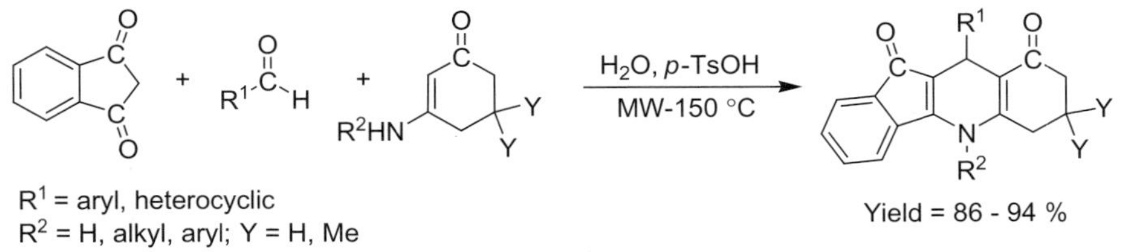

Scheme 4.22

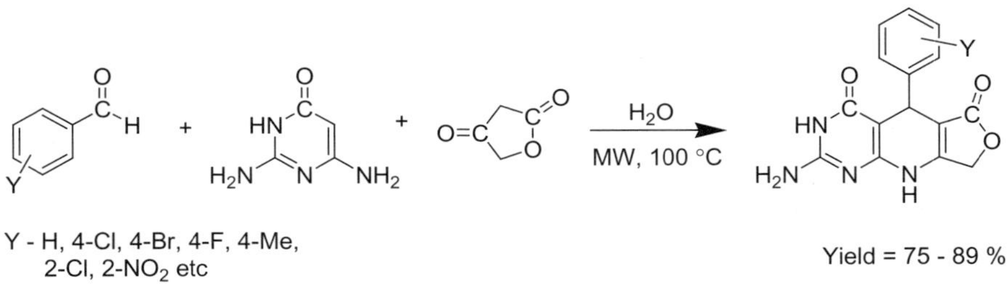

Scheme 4.23

conducted at high temperature (150 °C). Products were isolated by cooling the neutralized reaction mixture, followed by the filtration of solids.

Similarly, a series of related furo[3′,4′:5,6]pyrido[2,3-d]pyrimidine derivatives have also been achieved by reactions between aldehyde, 2,6-diaminopyrimidine-4(3H)-one, and tetronic acid/indane-1,3-dione, without using any catalyst (Scheme 4.23).[37]

Furopyridine is a well known bio-active scaffold used in drug discovery, and compounds based on this heterocyclic system displaying a diverse set of pharmacological activities. These heterocycles have been assembled by using the three-component reaction above between aldehyde, 6-aminoquinoline, and tetronic acid or 1,3-indanedione under aqueous microwave conditions (Scheme 4.24).[38,39] Tu and co-workers have extended this strategy for the synthesis of furo[3,4-e]pyrazole[3,4-b]pyridine derivatives (Scheme 4.25),[40] benzo[4,5]imidazo[1,2-a]pyrimidine derivatives (Scheme 4.26),[41] and pyrazolo[4′,3′:5,6]pyrido[2,3-d]pyrimidine derivatives (Scheme 4.27).[42] These protocols were operationally simple, needed less than 10 min for completion, and products were isolated by simple filtration of the reaction mixture, without using any organic solvent either as a reaction medium or in the work-up stage, thus making these protocols really green and sustainable.

Isoxazole derivatives are full of pharmacological as well as agrochemical properties such as herbicidal and soil fungicidal activity. Several polycyclic-fused isoxazolo[5,4-b]pyridines have been designed and prepared by microwave-assisted one-pot tandem reactions in water as a reaction solvent (Scheme 4.28). This protocol did not need any catalyst or additives.[43]

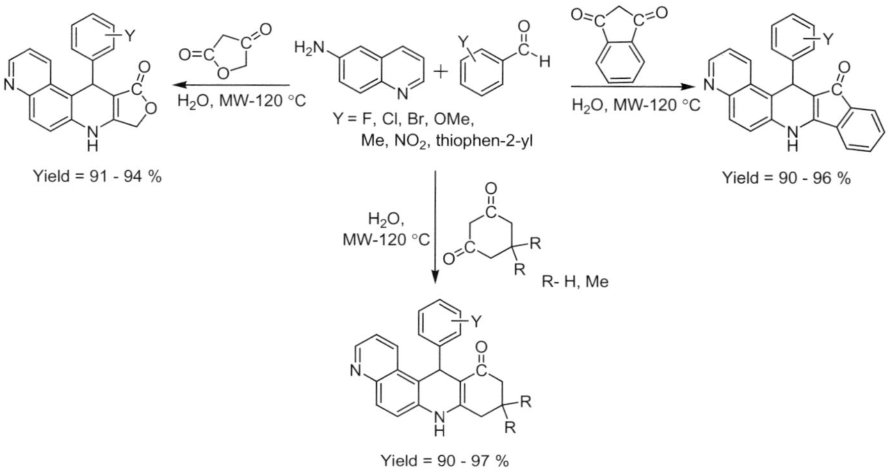

Scheme 4.24

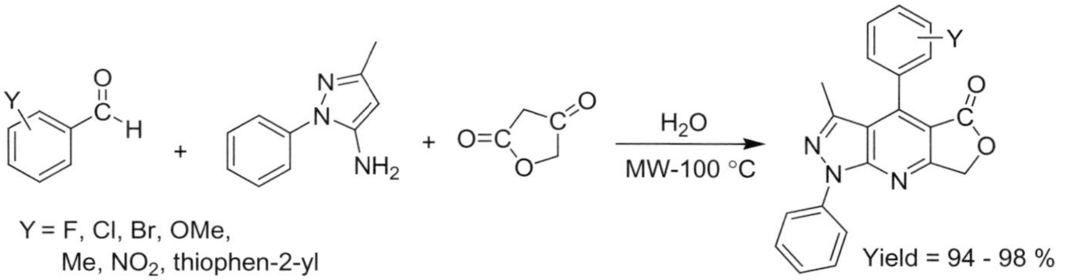

Scheme 4.25

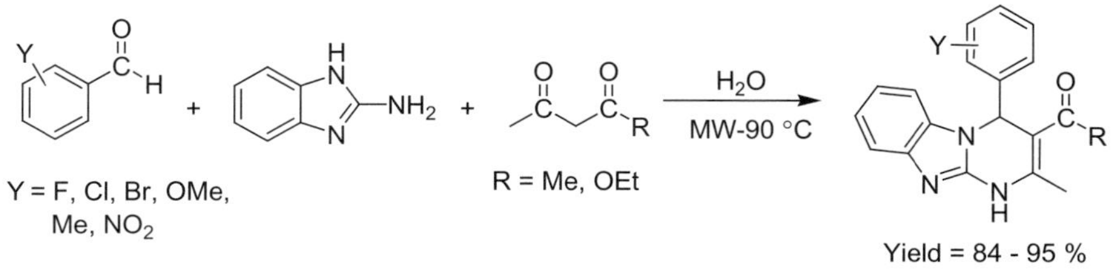

Scheme 4.26

Luthman *et al.* have reported the sustainable synthesis of various structurally diverse 2,5-diketopiperazines from dipeptide methyl esters (Scheme 4.29), using both thermal and microwave-assisted heating.[44] Microwave irradiation in aqueous medium proved to be the most proficient method of cyclization with

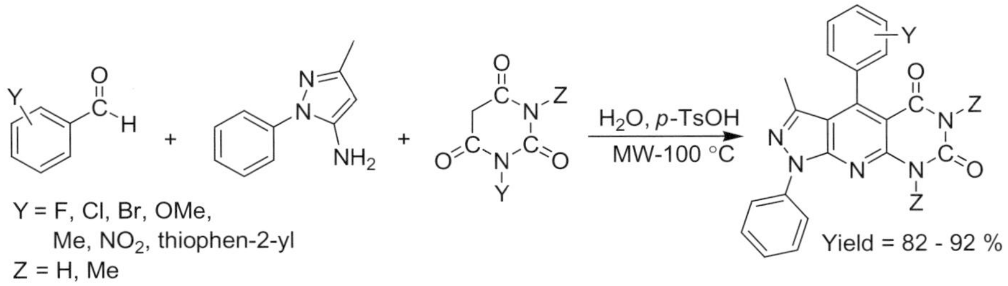
H2O, p-TsOH
MW-100 °C
Y = F, Cl, Br, OMe,
Me, NO2, thiophen-2-yl
Z = H, Me
Yield = 82 - 92 %

Scheme 4.27

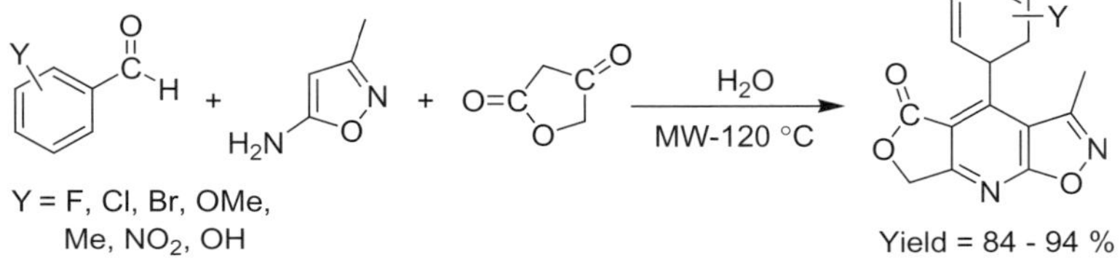
H2O
MW-120 °C
Y = F, Cl, Br, OMe,
Me, NO2, OH
Yield = 84 - 94 %

Scheme 4.28

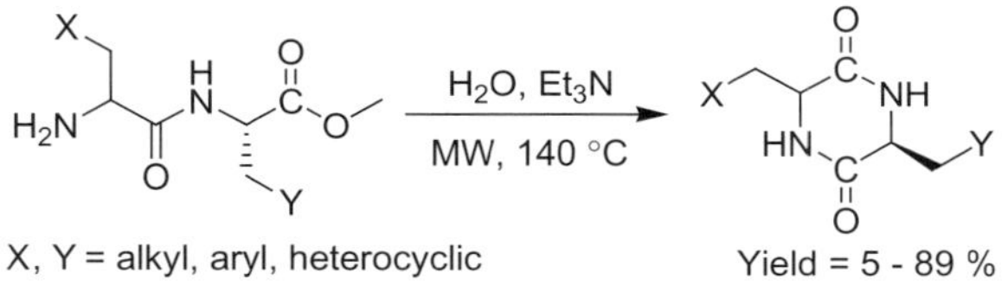
H2O, Et3N
MW, 140 °C
X, Y = alkyl, aryl, heterocyclic
Yield = 5 - 89 %

Scheme 4.29

good yields, and the outcome of this method is independent of the amino acid sequence. The crude product precipitated spontaneously, and was then simply isolated by filtration.[44]

1,4-Dihydropyridines, an important class of calcium channel blockers and one of the known classes of drugs for the treatment of cardiovascular diseases, have been synthesized under aqueous microwave conditions by the reaction of aldehydes, alkyl acetoacetates, and ammonium acetate using a phase-transfer catalyst (Scheme 4.30) in good to excellent yield in less than 10 min. Bifunctional dihydropyridines starting from dialdehyde precursors are also accessible *via* this general approach. The reactions were carried out in a domestic microwave oven in a flask connected to a reflux condenser. After completion of the reaction, product was isolated by extraction with ethyl acetate and purified by column chromatography.[45]

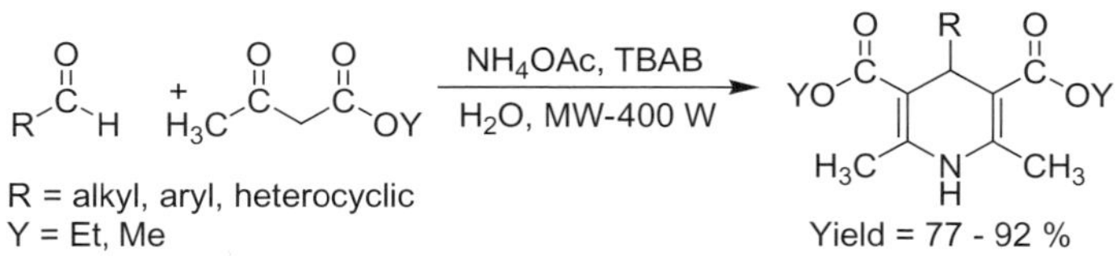

Scheme 4.30

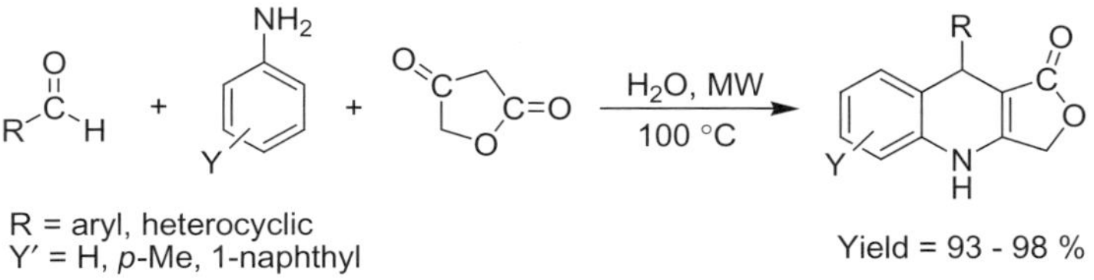

R = alkyl, aryl, heterocyclic
Y = Et, Me

Yield = 77 - 92 %

Scheme 4.31

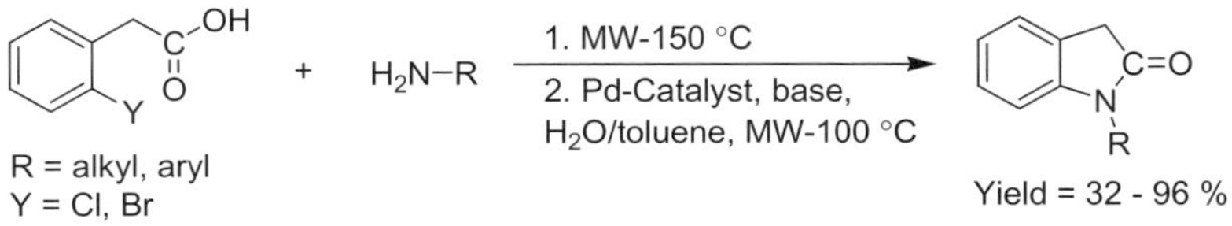

R = aryl, heterocyclic
Y' = H, p-Me, 1-naphthyl

Yield = 93 - 98 %

Scheme 4.32

R = alkyl, aryl
Y = Cl, Br

Yield = 32 - 96 %

Podophyllotoxin derivatives act as anti-tumor lignans that inhibit microtubule assembly. To obtain more potent and less toxic anticancer agents, many structural modifications have been performed on this skeleton. Importantly, 4-aza-podophyllotoxin was designed and prepared in water by microwave-assisted three-component reaction of an aldehyde, aromatic amine, and tetronic acid or 1,3-indanedione (Scheme 4.31). Reactions using water as a solvent resulted in higher yields and shorter reaction times than those using other organic solvents. Product was precipitated out by cooling the reaction mixture, and was further purified by recrystallization.[46]

Oxindoles, an important class of natural alkaloids, are very useful heterocycles in medicinal chemistry for novel drug development. Poondra and Turner have developed a two-step method for the synthesis of various N-substituted oxindole derivatives (Scheme 4.32). It needs amide bond formation between 2-halo-arylacetic acids and alkylamines and anilines, followed by a palladium-catalyzed intramolecular amidation under microwave conditions. Although the authors claimed it as an aqueous protocol, they actually conducted the reaction in a 1 : 1 mixture of water and toluene. In the case of aromatic amines, after initial amide formation step, it was necessary to isolate the crude product, to be used for second step. Notably, there is no need to isolate the intermediate amide in the case of alkylamines. Product was isolated by passing the reaction mixture (diluted by ethyl acetate) through a plug of Celite and purified by column chromatography.[47]

Vanelle and co-workers have synthesized a series of quinazoline-based anticancer agents. They used a MW-assisted reaction (Scheme 4.33) in water medium, and in most cases the product precipitated out from the reaction mixture and was isolated by simple filtration.[48]

4.3 MW-assisted Synthesis of Oxygen- and Sulfur-containing Heterocycles in Water

Oxygen- and sulfur-containing heterocycles are an important and well known class of building blocks in synthetic medicinal chemistry. Over the years, a variety of these heterocycles has attracted the attention of chemists for new drug discovery.[1,2]

Tetrahydrobenzo[b]pyran derivatives are an essential class of bio-active molecules that show various activities like anticancer, anti-schistosomal agent, etc. Hagiwara and co-workers[49] have reported a one-pot sequential synthesis of these heterocycles by three-component coupling of aromatic aldehydes, cyanoacetate, and dimedone (Scheme 4.34). The reaction proceeded well in neat water under MW condition. Notably, N,N-diethylamino-propylated silica was used as a solid catalyst, which could be re-used several times. Reactions, carried out in a domestic microwave oven, needed only 10 min for completion and the product was isolated by extraction with ethyl acetate.

Derivatives of isochromenone systems are known for various insecticidal natural and medicinal products. Bryson et al.[50] have synthesized a series of isochromenones using a copper-catalyzed microwave protocol in near critical water (Scheme 4.35). They observed that the cleavage of an acyl group in the water medium was the main driving force for this synthesis.

Kidwai and co-workers[51] have reported the synthesis of 2-aminochromenes, widely used in cosmetics and potential biodegradable agrochemicals, using aqueous microwave conditions. Reactions were carried out in a domestic microwave oven by exposing a mixture of aldehyde, malononitrile, and resorcinol or β-naphthol in a saturated aqueous solution of water at 560 W for 2–4 min (Scheme 4.36). The reaction mixture was then cooled, triturated by cold water, and solid product was isolated by simple filtration. In this method, only water was used in both the reaction step and workup, thus rendering the entire procedure a real green protocol.

The same reaction was also studied by Pasha and Jayashankara, wherein they replaced potassium carbonate with relatively inexpensive tetrabutylammonium bromide (TBAB).[52] Although the reaction rate and product yield were similar, this protocol required solvent extraction with ethyl acetate for isolation of the product. Thus, one can infer that Kidwai's protocol is much greener than Pasha's method.

Coumarin derivatives show several important biological activities and are an important class of heterocycles for the development of new medicinal compounds. A catalyst-free MW protocol for the synthesis of these derivatives

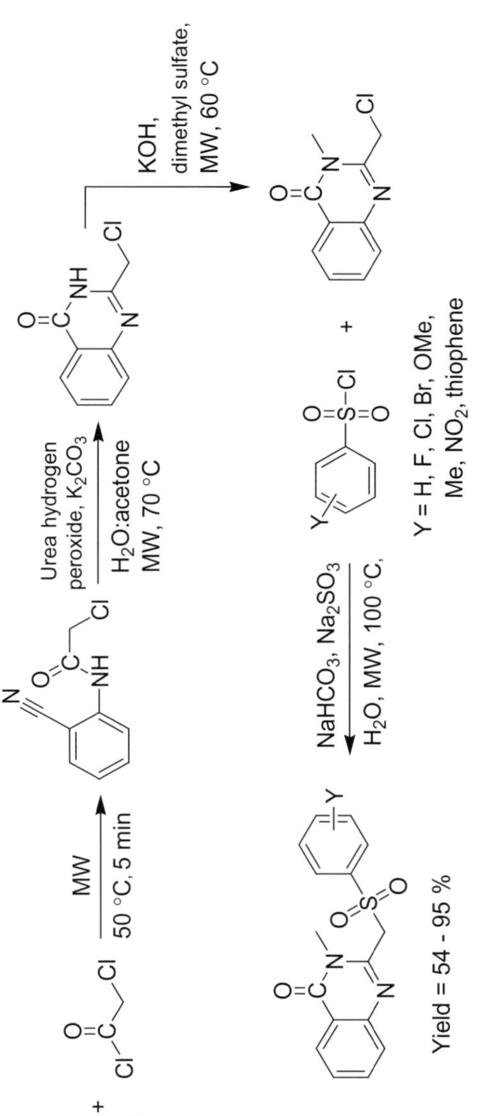
MW
50 °C, 5 min
Urea hydrogen peroxide, K_2CO_3
H_2O:acetone
MW, 70 °C
KOH, dimethyl sulfate,
MW, 60 °C
$NaHCO_3$, Na_2SO_3
H_2O, MW, 100 °C,
Y = H, F, Cl, Br, OMe, Me, NO_2, thiophene
Yield = 54 - 95 %

Scheme 4.33

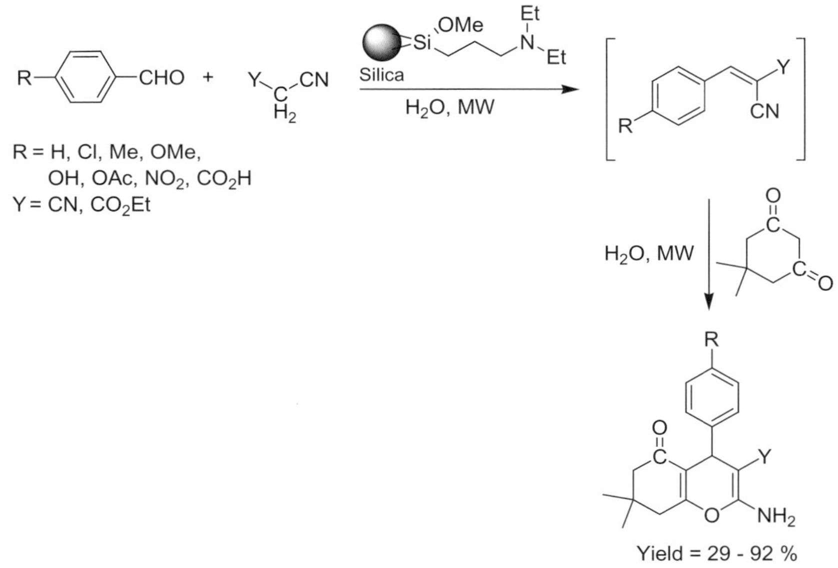

Scheme 4.34

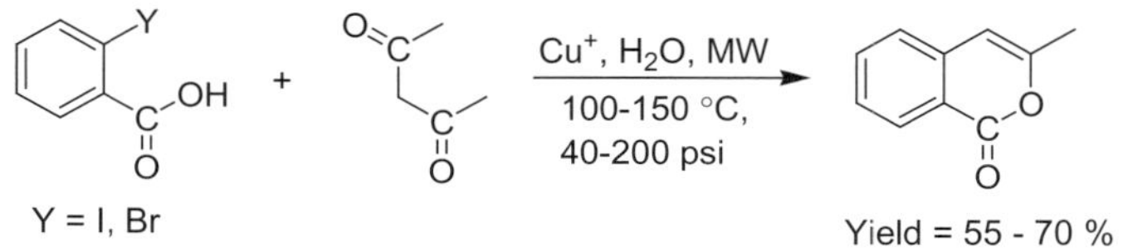

Scheme 4.35

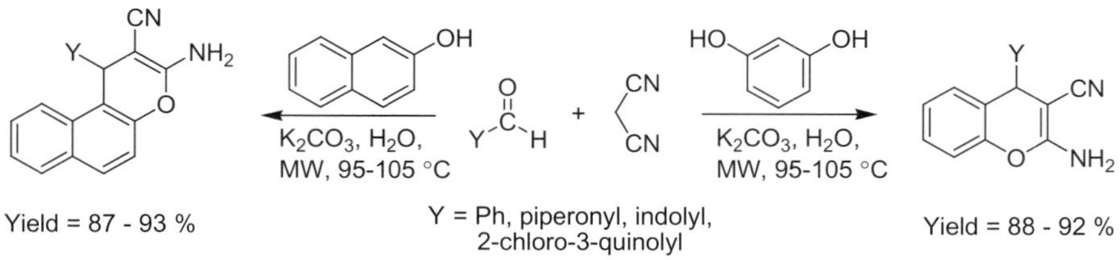

Scheme 4.36

was achieved by the reaction of 4-hydroxycoumarin with aromatic aldehydes in pure water medium (Scheme 4.37). The reaction was completed in 8–10 min and, after cooling the reaction mixture, precipitated solid was filtered off to isolate the crude product.[53]

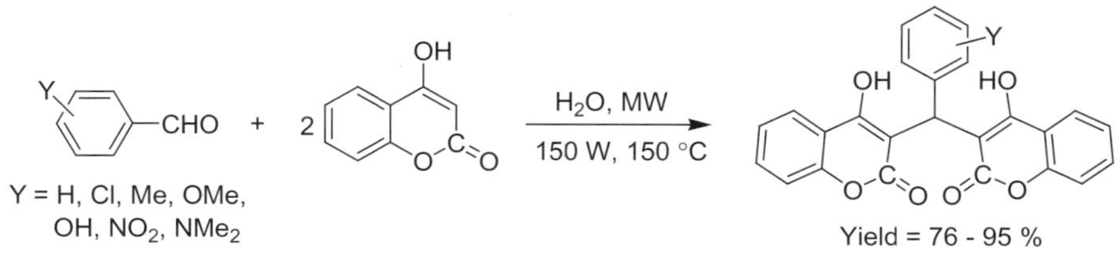

Scheme 4.37

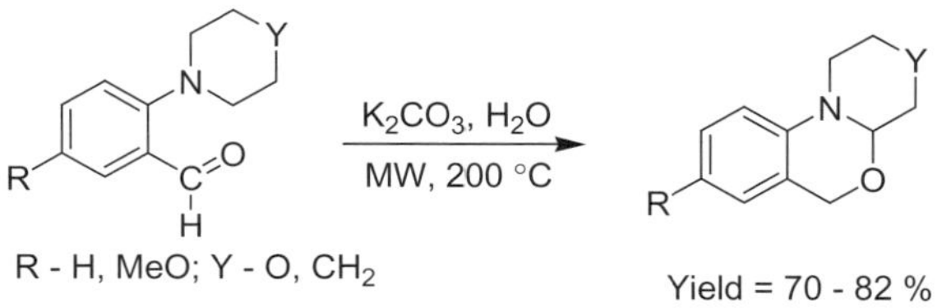

Y = H, Cl, Me, OMe, OH, NO_2, NMe_2

Yield = 76 - 95 %

Scheme 4.38

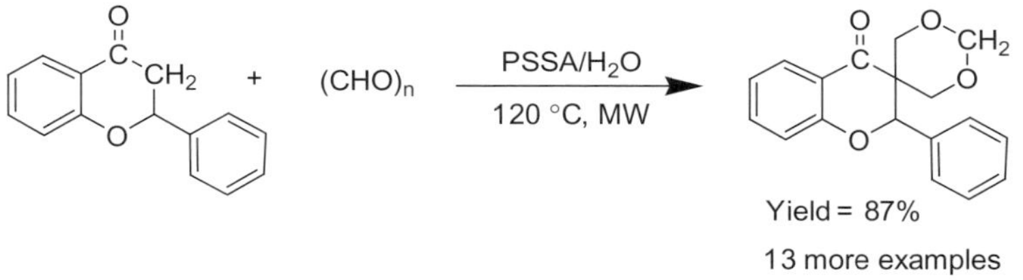

R - H, MeO; Y - O, CH_2

Yield = 70 - 82 %

Scheme 4.39

Yield = 87%

13 more examples

Benzoxazines, an important class of bio-active heterocycles, have been synthesized using MW irradiation (Scheme 4.38). This operationally sustainable protocol proceeded in pure water medium with good yield.[54]

Dioxane-functionalized molecules are well known for their potential as drug candidates; their one-pot synthesis has been achieved by microwave-assisted tandem bis-aldol reaction of ketones with paraformaldehyde. This approach provides a convenient and flexible method to attach dioxane arms to various ketones for further elaboration in drug design. The reaction was conducted in aqueous media and catalyzed by polymer (polystyrene) supported sulfonic acid (PSSA) (Scheme 4.39).[55] Remarkably, these reactions proceeded in an aqueous medium without using any phase-transfer catalyst, which clearly signals the selective absorption of microwaves by reactants, intermediates, and mainly the polar aqueous medium, thus proving the superiority of the concept of fusing microwave heating and water as a medium. In most cases, after reaction was completed, the phase of the desired product separated out from the aqueous media (under hot conditions), which facilitated isolation of the crude product by simple decantation, thus avoiding the use of organic solvents for product extraction.

This PSSA-catalyzed tandem bis-aldol reaction of ketone with paraformaldehyde in water may proceed *via* the following mechanism (Scheme 4.40). Initially, the addition of protonated formaldehyde molecule to the enol form of the ketone yields β-hydroxy ketone **7**, then the addition of another protonated formaldehyde molecule to **7** to give diol **8**, which in turn attacks the third formaldehyde molecule to produce adduct **9**, which after dehydration yields the final product 1,3-dioxane **10**.

Oxygen-heterocycles bearing hydrazone moieties have been found to be useful as anti-malaria drugs and as inhibitors of macrophage migration inhibitory factor (MIF) tautomerase activity. A MW protocol for the synthesis of these heterocyclic hydrazones using PSSA as a catalyst has also been developed by Polshettiwar and Varma (Scheme 4.41). The reaction proceeded

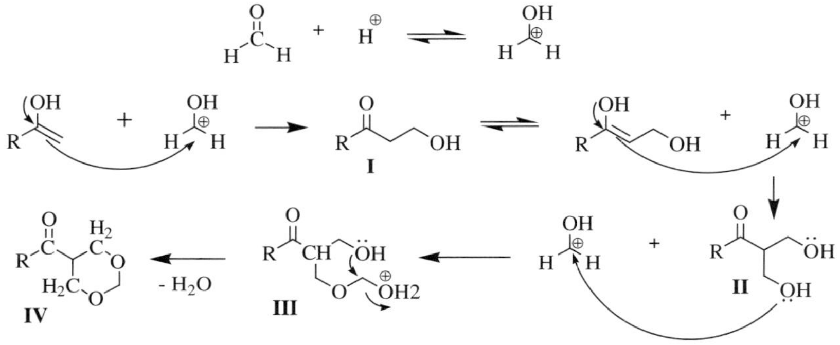

Scheme 4.40

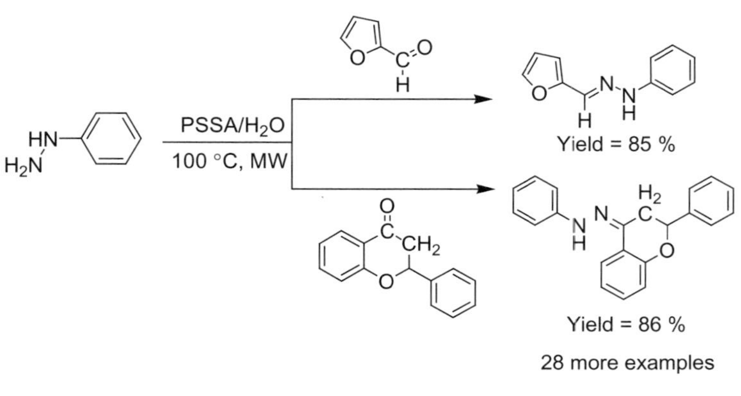

Scheme 4.41

efficiently in pure water in the absence of any organic solvent and involves basic filtration as the product isolation step.[56]

Sulfur-containing heterocycles are valued because of their rich and extraordinary bio-chemistry and also in view of the associated important biological properties. The scope of organosulfur chemistry has increased tremendously as sulfur-containing groups continue to serve an important auxiliary function in synthetic sequences.[57]

Benzothiazepinones and their fused derivatives, potential therapeutic medicines for type II diabetes, have attracted substantial synthetic attempts. Tu and co-workers[58] have designed and synthesised benzothiazepinones and thiazolidinones via a three-component reaction involving an aromatic aldehyde, aniline, and mercaptoacetic acid (Scheme 4.42). This one-pot protocol was carried out in pure water at 110 °C using 100–200 W microwave power. The solid product appeared after cooling the reaction mixture was crystallized to obtain pure product.

Another important class of sulfonyl-benzothiazole based bio-active compounds has been generated under microwave heating conditions. All the reactions were carried out in water by exposing three equivalents of sulfinic acid sodium salts with 2-chloromethyl-6-nitrobenzothiazole for 30 min at 800 W power (Scheme 4.43). After completion of the reaction, precipitates were filtered off to isolate the crude product.[59]

Rhodanine derivatives are compounds known for their exceptional biological activities, including hepatitis C virus protease inhibitors. A series of benzylidenerhodanine derivatives have been synthesized by MW-assisted crossed aldol condensation of aromatic aldehydes and rhodanine, catalyzed by tetrabutylammonium bromide (TBAB) in water (Scheme 4.44). The reactions

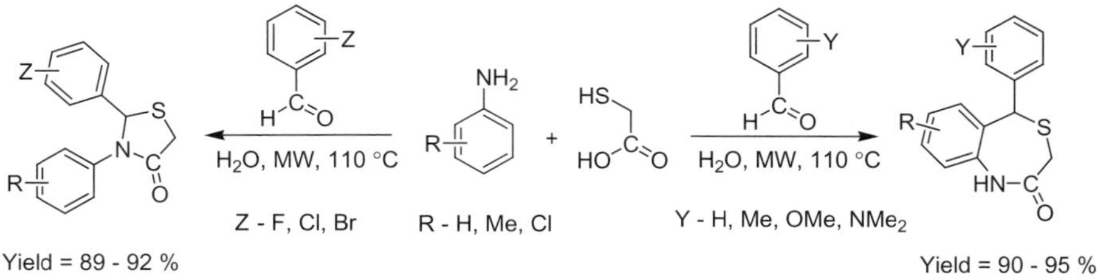

Scheme 4.42

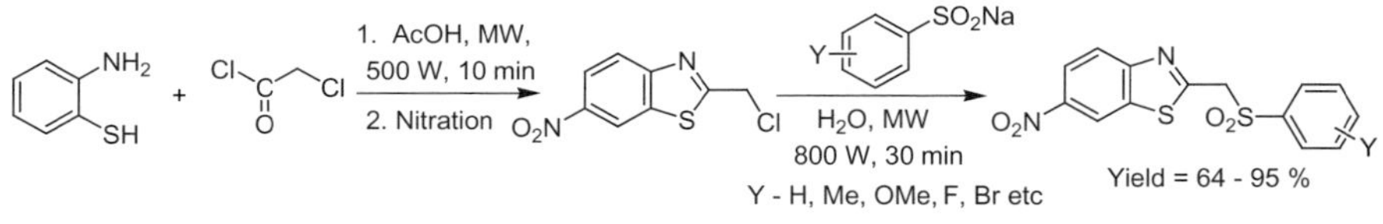

Scheme 4.43

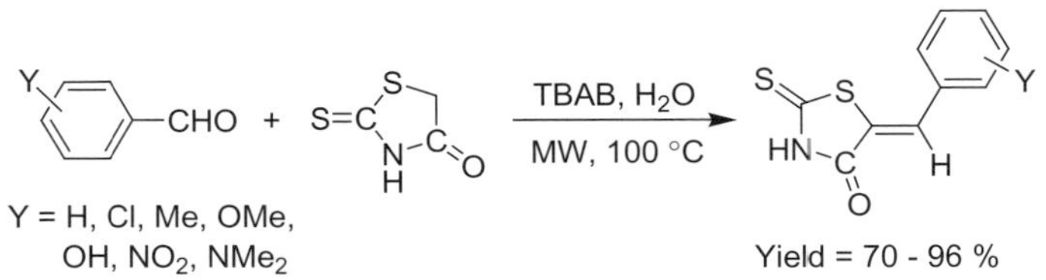

Scheme 4.44

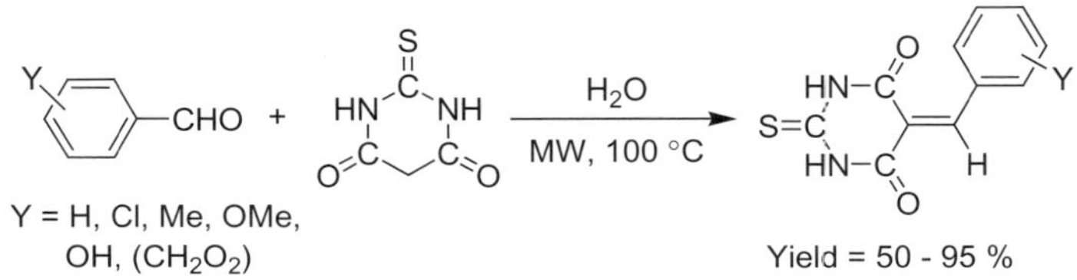

Y = H, Cl, Me, OMe, OH, NO$_2$, NMe$_2$

Yield = 70 - 96 %

Scheme 4.45

Y = H, Cl, Me, OMe, OH, (CH$_2$O$_2$)

Yield = 50 - 95 %

were completed in less than 10 min and solid product was isolated by filtration in good to excellent yield.[60]

A similar coupling reaction of aldehyde with thiobarbituric acid has been reported by Lu *et al.* (Scheme 4.45). They performed the reaction, without using any catalyst, in water under microwave irradiation conditions.[61]

4.4 MW-assisted Miscellaneous Reactions in Water

β-Hydroxy sulfides and β-hydroxy sulfoxides, an important class of intermediates in synthetic chemistry, have been generated by a MW-assisted one-pot reaction protocol in water (Scheme 4.46).[62] The reactions proceeded in a short period of time with good yield of sulfides, without using any metal catalyst. The *in situ* oxidation of these sulfides, mediated by *tert*-butyl hydroperoxide, yielded the desired β-hydroxy sulfoxides.

The Kröhnke reaction for the synthesis of 4′-aryl-2,2′:6′,2″-terpyridines has also been conducted in a clean aqueous medium *via* a one-pot reaction of 2-acetylpyridine with aromatic aldehyde and ammonium acetate under MW irradiation (Scheme 4.47).[63]

MW-assisted decarboxylation of substituted α-phenylcinnamic acids has been achieved, wherein a remarkable synergism between methylimidazole and aqueous NaHCO$_3$ produced the corresponding *para/ortho*-hydroxylated (*E*)-stilbenes in a benign manner (Scheme 4.48). The critical role of water in facilitating the decarboxylation is an important aspect to the synthetic utility of water-mediated organic reactions and needs more in-depth study.[64]

β-Amino carbonyl compounds are essential intermediates for the synthesis of various fine chemicals and pharmaceuticals products, and the Mannich

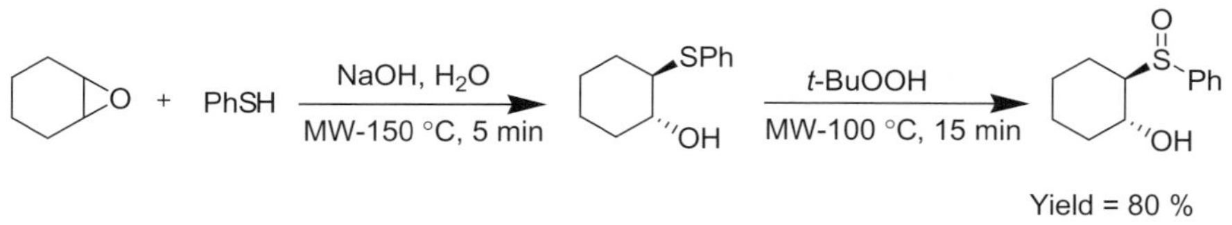

Scheme 4.46

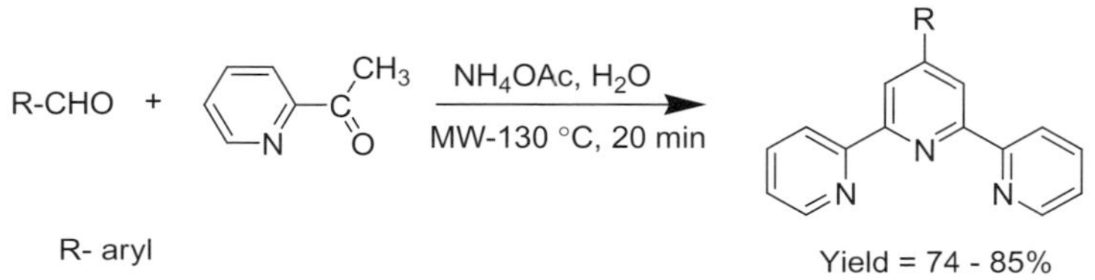

Scheme 4.47

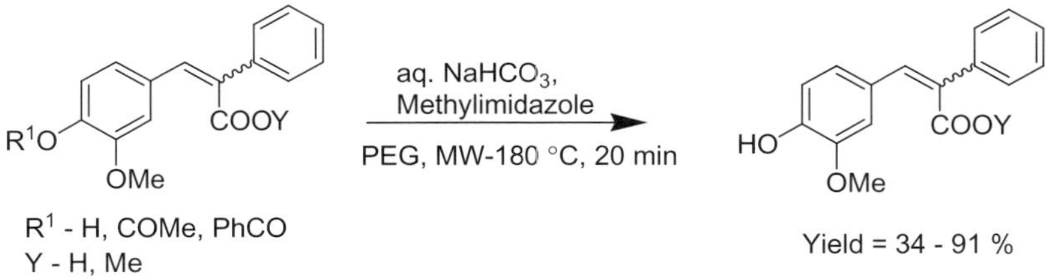

Scheme 4.48

reaction is one of the best protocols for their synthesis. Tu[65] synthesized several new β-aminoketones in water under microwave heating (Scheme 4.49). After completion of reaction, the mixture was simply poured into cold water and the solid that precipitated was isolated by filtration and purified by recrystallization. They achieved good stereoselectivity by controlling the steric hindrance of the reactants. This process has several advantages, such as the absence of toxic organic solvent, high speed, and operational ease with least environmental impact.

Tandem bis-aza-Michael addition of amines catalyzed by PSSA has been developed by Polshettiwar and Varma (Scheme 4.50).[66] This operationally simple, high yielding MW-assisted synthetic protocol proceeded in neat water in the absence of any organic solvent. A series of diamines were synthesized by changing the relative mole ratio of the reactants. One mole equivalent of diamine with two mole equivalents of Michael acceptor afforded di-substituted diamines, without any tri- or tetra-substituted products, whereas using four

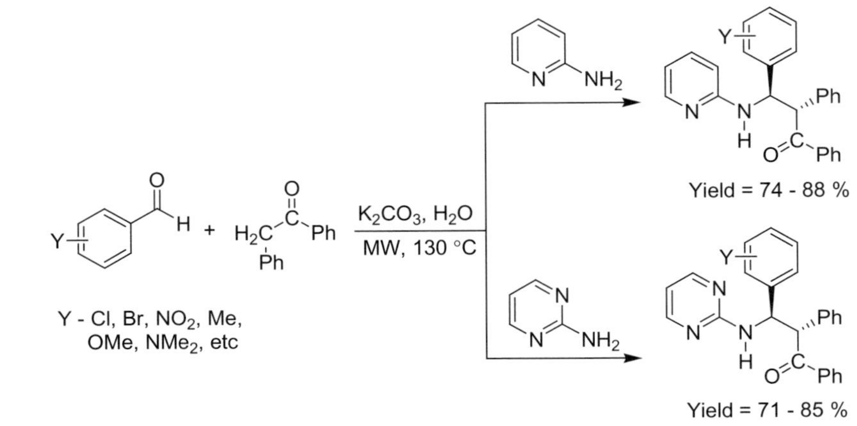

Scheme 4.49

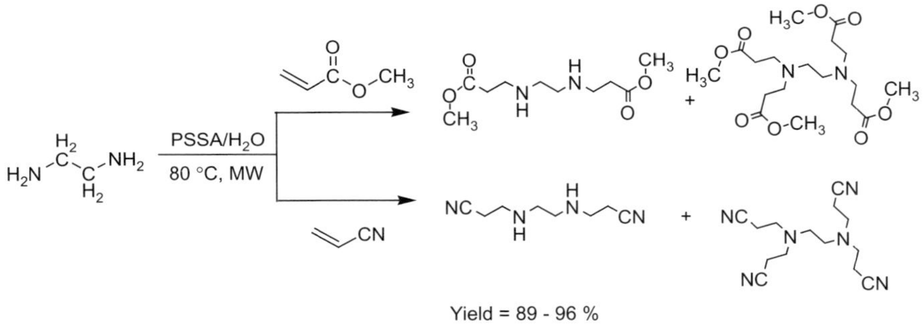

Scheme 4.50

equivalents of Michael acceptor gave exclusively the tetra-substituted diamine product.

Copper-catalyzed aqueous *N*-arylations of amines, amides, imides, and β-lactams with aryl halides under MW irradiation conditions have been reported (Scheme 4.51).[67] These reactions were performed at 85–90 °C in aqueous media, as well as under solvent-free conditions. However, under solvent-free conditions, lower yields were obtained. Notably this method was also successfully applicable to intramolecular *N*-arylation of β-lactam derivatives (Scheme 4.52). Most other methods were unsuccessful and led to decomposition of the starting material because of the presence of a base.

Nucleosides play significant roles in various biological processes and a range of modified nucleoside analogues have been known for their antiviral and anticancer activities. Kinase inhibitors, C6-cyclo amine-substituted purines and their analogues, have been synthesized *via* a mild aqueous MW protocol that

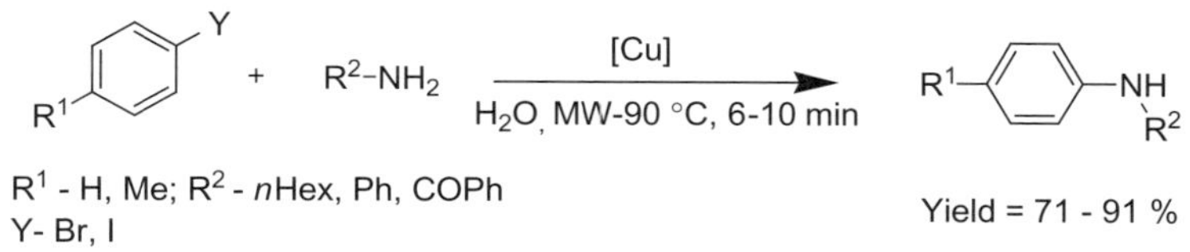

Scheme 4.51

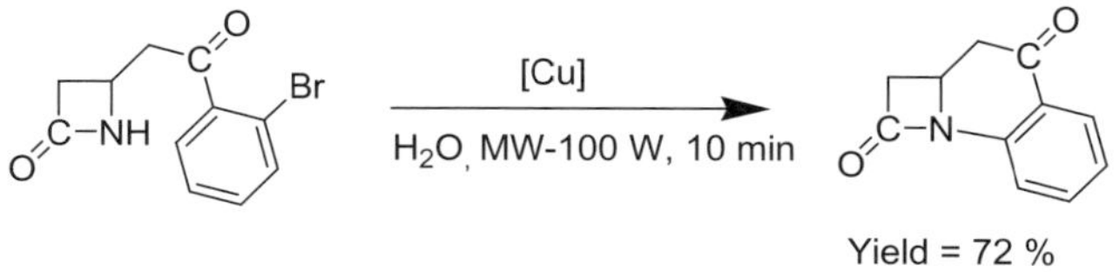

R^1 - H, Me; R^2 - nHex, Ph, COPh
Y - Br, I

Yield = 71 - 91 %

Scheme 4.52

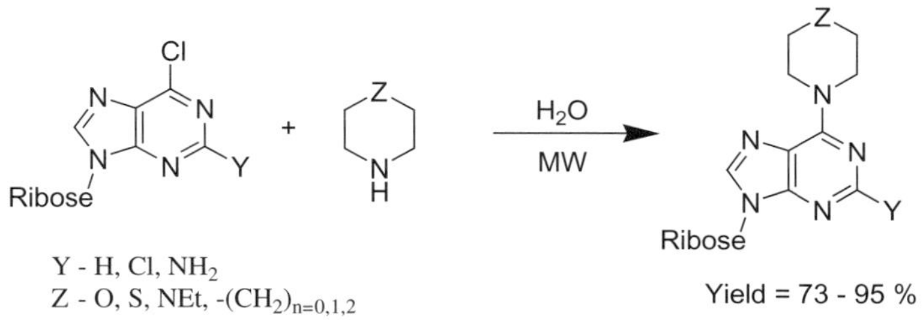

Yield = 72 %

Scheme 4.53

Y - H, Cl, NH$_2$
Z - O, S, NEt, -(CH$_2$)$_{n=0,1,2}$

Yield = 73 - 95 %

afforded the desired compounds in higher purity and yield, making this methodology suitable for rapid drug discovery (Scheme 4.53).[68]

Similarly, Guo and co-workers[69] have designed and synthesised 6-[N,N-bis(2-hydroxyethyl)amino]purine nucleosides by nucleophilic substitution reaction of 6-chloropurine nucleosides with diethanolamine using aqueous microwave chemistry (Scheme 4.54). Reactions occurred under microwave irradiation in less than 10 min and most of the products were crystallized from the solution; the pure products were isolated in excellent yields by simple filtration and washing.

Acetals and ketals are often used as carbonyl protecting groups in organic synthesis, and several methods have also been developed for their deprotection. Procopio et al. have conducted these deprotection protocols in water under MW-assisted catalyst-free conditions (Scheme 4.55).[70]

The selective transformation of amines into ketones is an important reaction in total organic synthesis. Olah et al. have designed a MW-assisted Pd-catalyzed retro-reductive amination reaction for direct conversion of amines

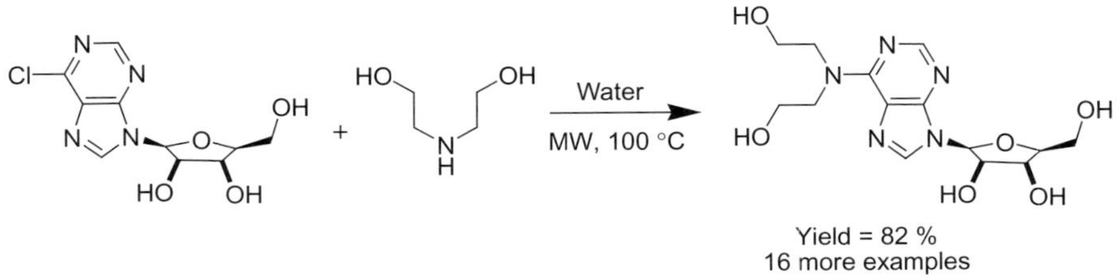

Scheme 4.54

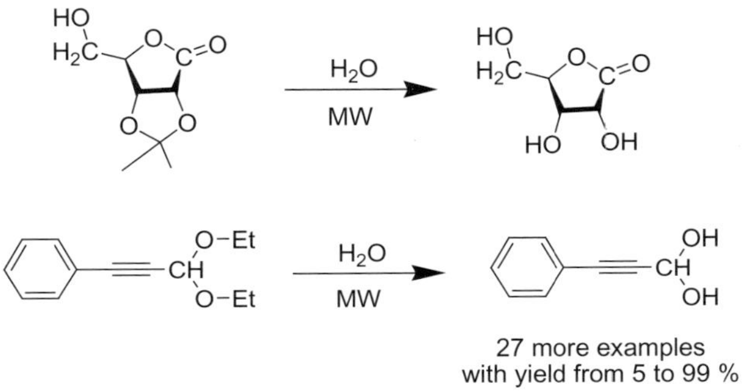

Scheme 4.55

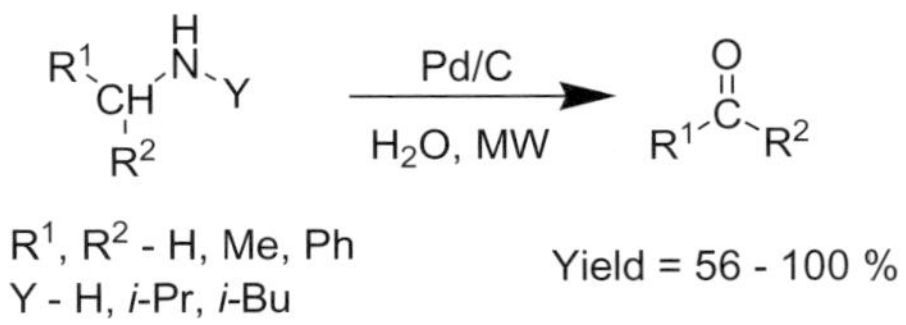

Scheme 4.56

into ketones in water medium. (Scheme 4.56).[71] This quick and selective reaction proceeded smoothly without any metal-based oxidants or volatile organic solvents.

Another excellent way to convert amines into carbonyl compounds (amides) is aminocarbonylation reactions. Wu and Larhed[72] rapidly converted a series of amines into the corresponding benzamides in neat water medium, using $Mo(CO)_6$ under MW heating (Scheme 4.57). Interestingly despite the use of water as solvent, aminocarbonylation strongly dominates over hydroxycarbonylation, providing good yields of both secondary and tertiary benzamides.

Deamination of aryl 3-amino-4(3*H*)-quinazolinone derivatives using oxidant (potassium permanganate) has been achieved in neat water medium using MW

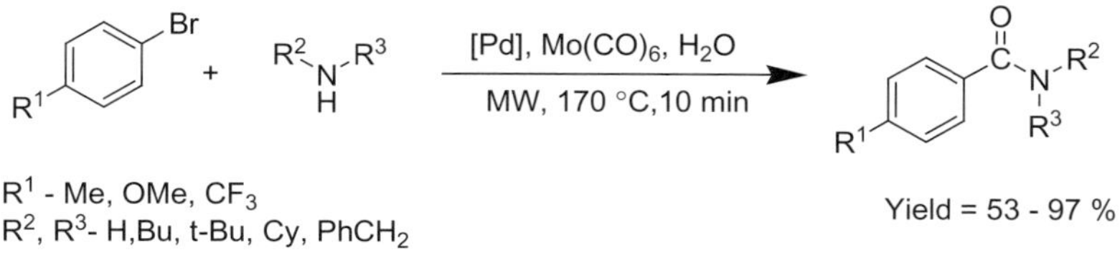
Br
R^2 N R^3 H
[Pd], Mo(CO)6, H2O
MW, 170 °C,10 min
O C N R^2 R^3
R^1
R^1 - Me, OMe, CF3
R^2, R^3- H,Bu, t-Bu, Cy, PhCH2
Yield = 53 - 97 %

Scheme 4.57

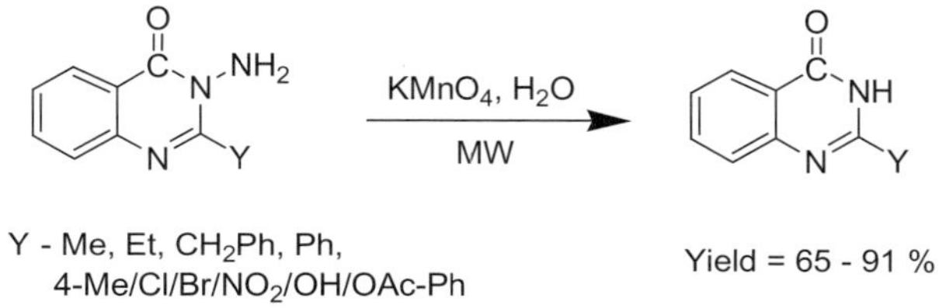
O C N NH2
N Y
KMnO4, H2O
MW
O C NH
N Y
Y - Me, Et, CH2Ph, Ph,
4-Me/Cl/Br/NO2/OH/OAc-Ph
Yield = 65 - 91 %

Scheme 4.58

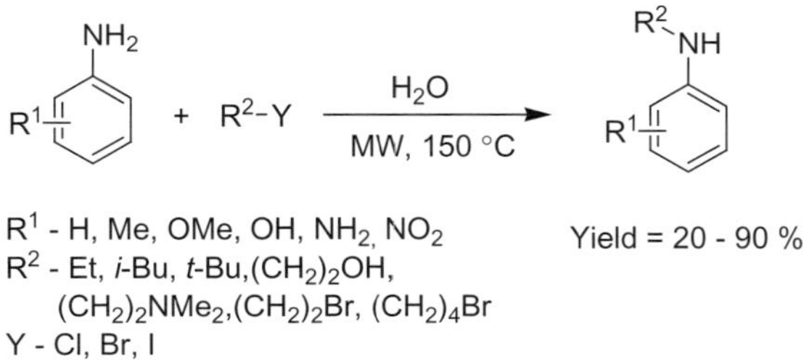
NH2
R^1
+ R^2-Y
H2O
MW, 150 °C
R^2 NH
R^1
R^1 - H, Me, OMe, OH, NH2, NO2
R^2 - Et, i-Bu, t-Bu,(CH2)2OH,
(CH2)2NMe2,(CH2)2Br, (CH2)4Br
Y - Cl, Br, I
Yield = 20 - 90 %

Scheme 4.59

conditions, which provided considerably higher yields than obtained using conventional oil bath heating (Scheme 4.58).[73] However, after completion of the reaction, the water was evaporated completely and the mixture was extracted with dichloromethane to isolate the product, thus losing the greener characteristics.

Selective mono-alkylation of anilines plays a major role in organic chemistry due to the extensive use of such compounds as intermediates for new drug discovery as well as in fine chemical industries. Chilin and co-workers[74] have reported the direct mono-*N*-alkylation of aromatic amines by alkyl halides using microwave irradiation without using any catalyst or organic solvents (Scheme 4.59). Reactions were completed in 20 min at 150 °C with good yields in water. Interestingly, when organic solvents were used instead of water the reaction did not occur in the absence of bases and/or catalysts, indicating the unusual catalytic effect of water during the reaction.

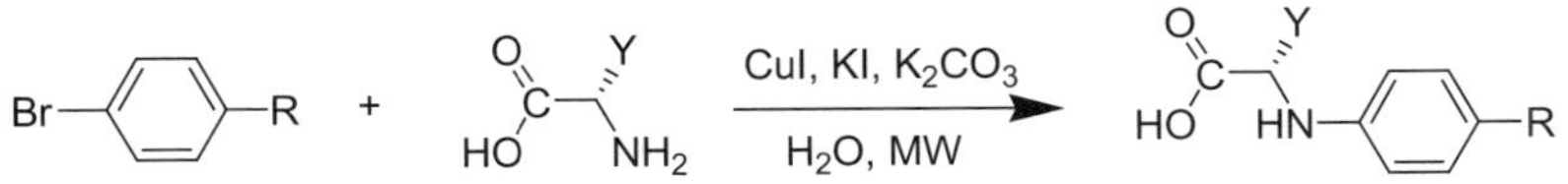

Scheme 4.60

Similarly, Larhed et al.[75] have developed copper-catalyzed aqueous MW protocols for the N-arylation of amino acids using various substituted aryl bromides. Reactions were completed in 40 min with good yields of non-protected N-arylated amino acids with slight racemization (Scheme 4.60).

4.5 Representative Experimental Procedures

4.5.1 Synthesis of Dihydropyrimidinones (Scheme 4.13)[28]

Alkyl acetoacetate (1 mmol), aldehyde (1 mmol), and urea/thiourea (1.2 mmol) were dissolved in 20% aqueous PSSA solution (three times the weight of aldehyde) in a 10 mL glass tube. The reaction tube was then placed inside a microwave oven, operated at 80 °C (power 40–100 W) for 20 min. After completion of reaction, all solid products were isolated by simple filtration and recrystallized to afford pure derivatives of 3,4-dihydropyrimidin-2($1H$)-ones.

4.5.2 Synthesis of 2-Aminochromene Derivatives (Scheme 4.36)[51]

Aldehyde (1 equiv.), malononitrile (1 equiv.), and either resorcinol or β-naphthol (1 equiv.) were mixed in a saturated aqueous solution of potassium carbonate (10 mL). The reaction mixture was exposed to microwaves (560 W) for 2–4 min and then cooled and triturated with ice cold water (2–3 mL). The solid product that appeared was filtered off, washed with cold water, dried, and recrystallized from ethanol to furnish the pure 2-aminochromene derivative.

4.5.3 Synthesis of Dioxane-functionalized Molecules (Scheme 4.39)[55]

Ketones (5 mmol) and paraformaldehyde (20 mmol) were dissolved in a 20% aqueous PSSA solution (five times the weight of ketone) in a 10 mL reaction tube. The capped reaction tube was then exposed to microwaves (40–140 W) at 120 °C for 30 min. Phase separation of the desired product from the aqueous medium occurred (under hot conditions) and the product – isolated by simple decantation – was subjected to column chromatography to afford pure dioxane-functionalized molecules.

4.5.4 Synthesis of Sulfonyl-benzothiazole-based Bio-active Compounds (Scheme 4.43)[59]

2-Chloroacetyl chloride (1.35 g, 11.89×10^{-3} mol) was added drop-wise to 2-aminobenzenethiol (1 g, 7.93×10^{-3} mol) in 15 mL of acetic acid. This reaction mixture was then exposed to microwaves (500 W) for 10 min, cooled, and poured onto 100 mL of basified (using 5 M NaOH) crushed ice. The product was extracted with chloroform (3×50 mL), dried over magnesium sulfate, and purified by column chromatography (silica gel, chloroform), yielding 1.97 g (90%) of 2-chloromethylbenzothiazole as a yellow solid, which was subsequently nitrated conventionally using concentrated sulfuric and fuming nitric acid to yield 2-chloromethyl-6-nitrobenzothiazole. This product (0.2 g, 8.75×10^{-3} mol) was then added to an aqueous solution (30 mL) of the sodium salt of sulfinic acid (26.25×10^{-3} mol). This reaction mixture was irradiated in a microwave oven (800 W) for 30 min and the precipitated product (sulfonyl-benzothiazole derivative) was filtered off, washed with water (3×20 mL), vacuum dried, and recrystallized from ethanol.

4.5.5 Synthesis of a Series of new β-Aminoketones (Scheme 4.49)[65]

Water (1 mL) was added to a mixture of aldehyde (2.0 mmol), pyridin-2-amine or pyrimidin-2-amine (3 mmol), and 1,2-diphenylethanone (2.2 mmol) in a 10 mL reaction tube. The capped reaction tube was irradiated with microwaves (100–250 W) for 10–28 min at 130 °C. The reaction mixture was then cooled and poured into cold water. The solid that appeared was filtered off, washed with water, and purified by recrystallization from ethanol to yield β-aminoketones.

4.6 Conclusions

The synthesis of bio-active heterocycles and fine chemicals in aqueous media is one of the best solutions for the development of green and sustainable protocols. However, execution of many organic reactions in water is not simple due to the inherent limitation of solubility of non-polar reactants in a polar aqueous medium, which can be overcome by using MW irradiation conditions. Thus, the *fusion* of benign water medium and non-conventional MW heating seems to be the preeminent way to develop the next generation of highly efficient processes. The true potential of this concept in various processes has not been fully explored yet, and further progress is expected in the future.

References

1. L. Garuti, M. Roberti and D. Pizzirani, *Mini Rev. Med. Chem.*, 2007, **7**, 481–489.

2. J. B. Sperry and D. L. Wright, *Curr. Opin. Drug Discovery Dev.*, 2005, **8**, 723–740.
3. C. O. Kappe, *Curr. Opin. Chem. Biol.*, 2002, **6**, 314–320.
4. V. Polshettiwar and R. S. Varma, *Pure Appl. Chem.*, 2008, **80**, 777–790.
5. V. Polshettiwar and R. S. Varma, *Curr. Opin. Drug Discovery Devel.*, 2007, **10**, 723–737.
6. V. Polshettiwar and R. S. Varma, *Acc. Chem. Res.*, 2008, **41**, 629–639.
7. C. J. Li and L. Chen, *Chem. Soc. Rev.*, 2006, **35**, 68–82.
8. J. H. Clark and S. J. Tavener, *Org. Process Res. Dev.*, 2007, **11**, 149–156.
9. V. Polshettiwar and R. S. Varma, *Chem. Soc. Rev.*, 2008, **37**, 1546–1557.
10. D. Dallinger and C. O. Kappe, *Chem. Rev.*, 2007, **107**, 2563–2591.
11. P. T. Anastas and J. C. Warner, *Green Chemistry: Theory and Practice*, Oxford University Press, Oxford, 2000.
12. J. H. Clark, *Green Chem.*, 2006, **8**, 17–21.
13. M. Poliakoff and P. Licence, *Nature*, 2007, **450**, 810–812.
14. J. H. Clark, L. Summerton and L. Nattrass, *Chem. Ind.*, 2008 (26th May), 16–17.
15. J.-J. Shie and J.-M. Fang, *J. Org. Chem.*, 2007, **72**, 3141–3144.
16. S. Balalaie and A. Shokrollahi, *Indian J. Chem. B.*, 2001, **40**, 612–613.
17. Q. Shao, S. Tu, C. Li, L. Cao, D. Zhou, Q. Wang, B. Jiang, Y. Zhang and W. Hao, *J. Heterocycl. Chem.*, 2008, **45**, 411–416.
18. Y. Ju and R. S. Varma, *Tetrahedron Lett.*, 2005, **46**, 6011–6014.
19. Y. Ju and R. S. Varma, *Org. Lett.*, 2005, **7**, 2409–2411.
20. Y. Ju and R. S. Varma, *J. Org. Chem.*, 2006, **71**, 135–141.
21. V. Molteni, M. M. Hamilton, L. Mao, C. M. Crane, A. P. Termin and D. M. Wilson, *Synthesis*, 2002, 1669–1674.
22. F. Jam, M. Tullberg, K. Luthman and M. Grøtli, *Tetrahedron*, 2007, **63**, 9881–9889.
23. A.-C. Carlsson, F. Jam, M. Tullberg, A. Pilotti, P. Ioannidis, K. Luthman and M. Grotli, *Tetrahedron Lett.*, 2006, **47**, 5199–5201.
24. C.-M. Huang, L.-Y. Su, W.-H. Huang and A.-R. Lee, *Chin. Pharm. J.*, 2005, **57**, 1–6.
25. M. Kidwai, P. Priya, K. Singhal and S. Rastogi, *Heterocycles*, 2007, **71**, 569–576.
26. S. Ferro, A. Rao, M. Zappala, A. Chimirri, M. L. Barreca, M. Witvrouw, Z. Debyser and P. Monforte, *Heterocycles*, 2004, **63**, 2727–2734.
27. I. Saltsman and Z. Gross, *Tetrahedron Lett.*, 2008, **49**, 247–249.
28. V. Polshettiwar and R. S. Varma, *Tetrahedron Lett.*, 2007, **48**, 7343–7346.
29. Q. Zhuang, D. Zhou, S. Tu, C. Li, L. Cao and Q. Shao, *J. Heterocycl. Chem.*, 2008, **45**, 831–835.
30. L. Ohberg and J. Westman, *Synlett*, 2001, 1296–1298.
31. Q. Zhuang, R. Jia, S. Tu, J. Zhang, B. Jiang, Y. Zhang and C. Yao, *J. Heterocycl. Chem.*, 2007, **44**, 895–900.
32. R. Jia, S. Tu, Y. Zhang, B. Jiang, J. Zhang, C. Yao and F. Shi, *J. Heterocycl. Chem.*, 2007, **44**, 1177–1180.
33. Y. Peng, G. Song and R. Dou, *Green Chem.*, 2006, **8**, 573–575.

34. N. Kayal, W. Dehaen, P. Matyus and E. van der Eycken, *Green Chem.*, 2004, **6**, 125–127.
35. J. Hren, K. Kranjc, S. Polanc and M. Kočevar, *Synthesis*, 2008, **1**, 452–458.
36. S.-J. Tu, B. Jiang, J.-Y. Zhang, R.-H. Jia, Y. Zhang and C.-S. Yao, *Org. Biomol. Chem.*, 2006, **4**, 3980–3985.
37. S.-J. Tu, Y. Zhang, H. Jiang, B. Jiang, J.-Y. Zhang, R.-H. Jia and F. Shi, *Eur. J. Org. Chem.*, 2007, 1522–1528.
38. F. Shi, D. Zhou, C. Li, Q. Shao, L. Cao and S. Tu, *J. Heterocycl. Chem.*, 2008, **45**, 405–410.
39. F. Shi, D. Zhou, S. Tu, Q. Shao, C. Li and L. Cao, *J. Heterocycl. Chem.*, 2008, **45**, 1065–1070.
40. F. Shi, Q. Wang, S. Tu, J. Zhou, B. Jiang, C. Li, D. Zhou, Q. Shao and L. Cao, *J. Heterocycl. Chem.*, 2008, **45**, 1103–1108.
41. S. Tu, Q. Shao, D. Zhou, L. Cao, F. Shi and C. Li, *J. Heterocycl. Chem.*, 2007, **44**, 1401–1406.
42. F. Shi, D. Zhou, S. Tu, C. Li, L. Cao and Q. Shao, *J. Heterocycl. Chem.*, 2008, **45**, 1305–1310.
43. S.-J. Tu, X.-H. Zhang, Z.-G. Han, X.-D. Cao, S.-S. Wu, S. Yan, W.-J. Hao, G. Zhang and N. Ma, *J. Comb. Chem.*, 2009, **11**, 428–432.
44. M. Tullberg, M. Grotli and K. Luthman, *Tetrahedron*, 2006, **62**, 7484–7491.
45. H. Salehi and Q.-X. Guo, *Synth. Commun.*, 2004, **34**, 4349–4357.
46. S. Tu, Y. Zhang, J. Zhang, B. Jiang, R. Jia, J. Zhang and S. Ji, *Synlett*, 2006, 2785–2790.
47. R. R. Poondra and N. J. Turner, *Org. Lett.*, 2005, **7**, 863–866.
48. Y. Kabri, A. Gellis and P. Vanelle, *Green Chem.*, 2009, **11**, 201–208.
49. H. Hagiwara, A. Numamae, K. Isobe, T. Hoshi and T. Suzuki, *Heterocycles*, 2006, **68**, 889–895.
50. T. A. Bryson, J. J. Stewart, J. M. Gibson, P. S. Thomas and J. K. Berch, *Green Chemistry*, 2003, **5**, 174–176.
51. M. Kidwai, S. Saxena, M. K. R. Khan and S. S. Thukral, *Bioorg. Med. Chem. Lett.*, 2005, **15**, 4295–4298.
52. M. A. Pasha and V. P. Jayashankara, *Indian J. Chem. B*, 2007, **46**, 1328–1331.
53. G.-X. Gong, J.-F. Zhou, L.-T. An, X.-L. Duan and S.-J. Ji, *Synth. Commun.*, 2009, **39**, 497–505.
54. N. Kaval, B. Halasz-Dajka, G. Vo-Thanh, W. Dehaen, J. V. Eycken, P. Matyus, A. Loupy and E. van der Eycken, *Tetrahedron*, 2005, **61**, 9052–9057.
55. V. Polshettiwar and R. S. Varma, *J. Org. Chem.*, 2007, **72**, 7420–7422.
56. V. Polshettiwar and R. S. Varma, *Tetrahedron Lett.*, 2007, **48**, 5649–5652.
57. V. Polshettiwar and M. P. Kaushik, *J. Sulfur Chem.*, 2006, **27**, 353.
58. S.-J. Tu, X.-D. Cao, W.-J. Hao, X.-H. Zhang, S. Yan, S.-S. Wu, Z.-G. Han and F. Shi, *Org. Biomol. Chem.*, 2009, **7**, 557–563.
59. A. Gellis, N. Boufatah and P. Vanelle, *Green Chem.*, 2006, **8**, 483–487.
60. J.-F. Zhou, F.-X. Zhu, Y.-Z. Song and Y.-L. Zhu, *ARKIVOC*, 2006, **14**, 175–480.

61. J. Lu, Y. Li, Y. Bai and M. Tian, *Heterocycles*, 2004, **63**, 583–589.
62. V. Pironti and S. Colonna, *Green Chem.*, 2005, **7**, 43–45.
63. S. Tu, R. Jia, B. Jiang, J. Zhang, Y. Zhang, C. Yaoa and S. Jib, *Tetrahedron*, 2007, **63**, 381–388.
64. V. Kumar, A. Sharma, A. Sharma and A. K. Sinha, *Tetrahedron*, 2007, **63**, 7640–7646.
65. W.-J. Hao, B. Jiang, S.-J. Tu, X.-D. Cao, S.-S. Wu, S. Yan, X.-H. Zhang, Z.-G. Han and F. Shi, *Org. Biomol. Chem.*, 2009, **7**, 1410–1414.
66. V. Polshettiwar and R. S. Varma, *Tetrahedron Lett.*, 2007, **48**, 8735–8738.
67. L. D. S. Yadav, B. S. Yadav and V. K. Rai, *Synthesis*, 2006, **1**, 1868–1872.
68. G.-R. Qu, L. Zhao, D.-C. Wang, J. Wu and H.-M. Guo, *Green Chem.*, 2008, **10**, 295–297.
69. G.-R. Qu, J. Wu, Y.-Y. Wu, F. Zhang and H.-M. Guo, *Green Chem.*, 2009, **11**, 760–762.
70. A. Procopio, M. Gaspari, M. Nardi, M. Oliverio, A. Tagarellib and G. Sindona, *Tetrahedron Lett.*, 2007, **48**, 8623–8627.
71. A. Miyazawa, K. Tanaka, T. Sakakura, M. Tashiro, H. Tashiro, G. K. Surya Prakash and G. A. Olah, *Chem. Commun.*, 2005, 2104–2106.
72. X. Wu and M. Larhed, *Org. Lett.*, 2005, **7**, 3327–3329.
73. M. Arfan, R. Khan, S. Anjum, S. Ahmad and M. I. Choudhary, *Chin. Chem. Lett.*, 2008, **19**, 161–165.
74. G. Marzaro, A. Guiotta and A. Chilin, *Green Chem.*, 2009, **11**, 774–776.
75. S. Röttger, P. J. R. Sjöberg and M. Larhed, *J. Comb. Chem.*, 2007, **9**, 204–209.

CHAPTER 5
Microwave-assisted Enzymatic Reactions in Aqueous Media

HUA ZHAO

Chemistry Program, Savannah State University, Savannah, GA 31404, USA

5.1 Introduction

In contrast to conventional heating methods, microwave irradiation (0.3–300 GHz) can serve as an alternative energy source for initiating chemical reactions.[1] The electromagnetic field of microwaves induces rotation of dipolar molecules. However, since the molecular rotation is much slower than the change of electric field ($2.45 \times 10^9 \, \text{s}^{-1}$, the frequency of commercial microwave ovens), the resulting intermolecular friction converts some electromagnetic energy into thermal energy. The dielectric constant (ε_s), the ability of a species to be polarized by an electric field, determines the capacity of a medium to absorb microwave energy. Thus, microwave irradiation is often referred as dielectric heating.[2–4] Polar solvents, such as water ($\varepsilon_s = 80.1$ at 20 °C, all ε_s data from ref. 5 except those noted otherwise), methanol (33.0 at 20 °C), and dimethylformamide (38.25 at 20 °C), can be heated rapidly by microwaves, whereas hexane (1.89 at 20 °C), toluene (2.38 at 23 °C), and diethyl ether (4.27 at 20 °C) are basically "inert" to microwave irradiation. On the other hand, a solvent with a low heat capacity (C_p) can also be heated relatively easily by microwaves, even if it has a low dielectric constant. For example, 1-propanol ($\varepsilon_s = 20.8$ at 20 °C) can be heated 1.7 times faster than water ($\varepsilon_s = 80.1$ at 20 °C) mainly due to the low C_p of 1-propanol ($2.45 \, \text{J} \, \text{g}^{-1} \, \text{K}^{-1}$) compared with water ($4.18 \, \text{J} \, \text{g}^{-1} \, \text{K}^{-1}$).[6]

RSC Green Chemistry No. 7
Aqueous Microwave Assisted Chemistry: Synthesis and Catalysis
Edited by Vivek Polshettiwar and Rajender S. Varma
© Royal Society of Chemistry 2010
Published by the Royal Society of Chemistry, www.rsc.org

In terms of energy efficiency of microwave irradiation, Razzaq and Kappe[7] discovered recently that (i) in open-vessel reflux reaction setup, microwave heating required much more energy than conventional heating methods (such as oil baths and heating mantles); (ii) in closed-vessel systems considerable saving in energy was detected. However, the authors argued that such energy saving was mainly due to the reduction in reaction time in a closed vessel, and was not an intrinsic property of microwave heating. Another recent study[8] has confirmed the rather poor heating-efficiency of microwaves, as illustrated by the example of heating deionized water by monomode and multimode microwave synthesizers: maximum average heating efficiencies of 10% for small-scale vessels (5 mL), 20% for medium-size (50 mL), and 30% for large-scale (400 mL).

Microwaves can increase reaction rates through the thermal effect. Although many studies have addressed non-thermal effects (or so-called 'specific effects') of microwaves,[9–11] the controversy continues regarding the extent to which such non-thermal effects actually exist – in part because many earlier applications used household microwave ovens that did not allow a fine control of temperature. Currently, microwave reactors equipped with *in situ* temperature and pressure controls are readily available commercially. Many researchers have argued that reaction rates increase in response to microwave irradiation because microwaves can superheat solvents beyond their normal boiling-points (due to the retardation of nucleation under microwave heating).[1–3,12,13] Therefore, they suggest, there actually is no non-thermal effect.[1] However, the superheating effect is expected to disappear or be minimized when the reaction mixtures are well stirred under low microwave power.[13–15] Other investigators propose the existence of non-thermal effects since polar functional groups absorb more microwave energy than other chemical groups, subsequently releasing this energy into the surrounding solution. Therefore, the functional groups have higher reactivity with adjacent reactants under microwave irradiation conditions than under conventional heating conditions at the same temperature.[16]

Nevertheless, microwave irradiation has become a routine heating device employed in various chemical reactions,[3,9,17–20] which include limited studies on enzymatic syntheses.[21–23] Among those biological applications, microwave irradiation was initially adopted by the food industry, not to activate enzymes but rather to inactivate them for reducing food rancidity.[24–28] For this reason, microwave irradiation has been an effective tool in food and medical industries to destroy bacteria, preserve food, and sterilize medical products,[29,30,25–27] as well as in biomedical research [such as the rapid inactivation of brain enzymes[31–34] and microwave coagulation therapy (MCT)[35]]. Enzyme inactivation typically results from the thermal effect of microwaves in the presence of water, but the non-thermal effect also plays important roles in some cases.[26] It is not the main task of this chapter to focus on the inactivation of enzymes in food and medical applications; we are more interested in understanding how microwave exposure affects the enzyme activity in aqueous solutions, and how the enzyme activity can be improved through microwave irradiation in aqueous solutions.

Accelerating enzymatic reactions by microwaves often requires minimizing the water content in the system because enzymes have much higher thermal stabilities in non-aqueous media (such as organic solvents, especially hydrophobic ones). In some cases, no significant loss of enzyme activity in hydrophobic organic solvents has been observed, even at 100 °C for an extended period of time.[36,37] In addition, many important chemical and enzymatic transformations are conducted in organic solvents.[38–41] For these reasons, many microwave-assisted enzymatic syntheses have been carried out under nearly anhydrous conditions.[22,38] With temperature control, many studies have described higher enzyme activities and selectivities in organic solvents (usually containing little water) under microwave irradiation, than those under conventional heating. Examples of these reactions include protease- (subtilisin Carlsberg and α-chymotrypsin)[42] and lipase-[43–47] catalyzed esterification and/or transesterification, lipase- (porcine pancreas) catalyzed chemoselective reduction of organic azides[48] and acylation of alcohols,[49] cabbage chitinase-catalyzed the hydrolysis of chitin (chitin was pretreated by microwaves, not the reaction),[50] and glycosidase-catalyzed reversed hydrolysis and transglycosidations.[51] Therefore, microwave irradiation is an effective means to activate enzymes in hydrophobic solvents. Microwave-assisted biocatalysis in non-aqueous solvents has been reviewed recently.[21–23] On the other hand, this chapter will focus on enzymatic reactions in aqueous solutions under microwave irradiation.

5.2 Microwave-assisted Enzymatic Reactions in Water (or Aqueous Buffer)

Much early research (Section 5.2.1) focused on the microwave effect on enzyme activities in aqueous buffer solutions, where many studies have found no specific microwave effect on enzymes but some researchers have indicated the existence of a non-thermal effect of microwaves. Recent advances in the field have concentrated on the enzymatic hydrolysis of proteins irradiated by microwaves, where a rapid proteolysis was often observed (Section 5.2.2).

5.2.1 Effect of Microwave Irradiation on Enzyme Activity in Aqueous Solutions

Early studies on the influence of microwaves on enzymes in aqueous solutions can be generally divided into two categories: (i) enzyme activity was measured after microwave exposure (*i.e.*, the enzyme in buffer solution was exposed to microwave irradiation at a fixed temperature and, periodically, enzyme samples were taken from the microwave field and then assayed[52]), and (ii) enzyme activity was measured during microwave irradiation (*i.e.*, the enzymatic reaction in aqueous solution was subject to microwave heating, and the reaction mixture was periodically withdrawn and analyzed[53]). Some of these studies were conducted with the control of microwave power output, but without the control of temperature during microwave irradiation. Representative examples are compiled in Table 5.1, and discussed in detail below.

Table 5.1 Effect of microwave irradiation on enzyme activity in aqueous solutions.

Enzyme	Frequency (GHz)	MW Power ($W\,g^{-1}$)	T (°C)	MW effect	Ref.
Lactate dehydrogenase, acid phosphatase, and alkaline phosphatase	2.8	400–1000 $mW\,cm^{-2}$	37–50 (fixed)	No non-thermal effect, but thermally induced inactivation	62
Glucose 6-phosphate dehydrogenase, adenylate kinase, cytochrome c reductase	2.45	0.042	25–60	No effect	54
Horseradish peroxidase	2.45	62.5 to 375	25 (fixed)	No effect at low power density, but inactivation at high power density	56
Lysozyme and trypsin	2.45	0.1–0.6	30–95	No effect	55
Acetylcholinesterase	2.45		37	No direct effect unless thermal inactivation	63
Lactate dehydrogenase	3	30–1800	25–60	No effect but thermal activation	64
Alcohol dehydrogenase	40 to 115	10 $mW\,cm^{-2}$		No effect	57
Lactate dehydrogenase, glutamic oxaloacetic transaminase, and creatine phosphokinase	2.45	0.004–0.0162	37.5	No effect	58
Acetylcholinesterase, creatine kinase	2.45	0.001–0.1	37 (fixed)	No effect	59
Acetylcholinesterase	2.45		25	No effect	60
Cellulase from *Penicillinum funiculosum*	2.45		35	No effect	61
β-Galactosidase from *Bacillus acidocaldarius*	10.4	1.1–1.7	70 (fixed)	Inactivation at low enzyme concentration, but no effect at high enzyme concentration	67

Enzyme				Observation	Ref.
Trichoderma reesei cellulose	2.45	3	45–55	Higher initial rate, but lower yield under microwaves	70
Thermophilic (S)-adenosylhomocysteine hydrolase and 5′-methylthioadenosine phosphorylase	10.4	1.5–3.1	70–90	Non-thermal, irreversible and time-dependent inactivation of both enzymes	52
γ-Amylase from Rhizopus mold	2.45	0–7.5	60 (fixed)	Higher enzyme activity under microwaves	65
Thermophilic alcohol dehydrogenase from Sulfolobus solfataricus	2.45			Non-thermal effect	66
Thermophilic β-galactosidase from Bacillus acidocaldarius	10.4	1.1–1.7	70 (fixed)	Irreversible inactivation at low enzyme concentration, but no effect at high concentration	67
Lactate dehydrogenase	2.45			Non-thermal effect	68
Several lipases	2.45	~135		Faster reaction rates under microwave irradiation (30 s at 1.35 kW) than the pH stat method	71
Aspergillus carneus lipase	2.45		38–40, 90	Faster enzymatic hydrolysis under microwaves	69
Hyperthermophilic enzymes: Pyrococcus furiosus β-glucosidase (Pfu CelB), α-galactosidase from Thermotoga maritima (Tm GalA), and carboxylesterase from Sulfolobus solfataricus P1 (SsoP1 CE)	2.45	83–500	−20 to 40	Enzyme activation and non-thermal effect under microwaves	53

Many studies have suggested that microwave irradiation has no effect on the enzyme activity in aqueous solutions. For example, during the exposure to microwaves at an absorbed dose rate of $42\,W\,kg^{-1}$, the activities of three enzymes (*i.e.*, glucose 6-phosphate dehydrogenase from human red blood cells and yeast, adenylate kinase from rat liver mitochondria and rabbit muscle, and rat liver microsomal NADPH cytochrome *c* reductase) were simultaneously monitored spectrophotometrically, and exhibited no difference from non-irradiated enzyme samples at the same temperature.[54] Through controlling the reaction temperature during microwave irradiation, Yeargers *et al.*[55] found that the activities of lysozyme and trypsin were comparable with those under conventional heating, implying the enzyme activity in aqueous solutions was not a function of heating methods. Henderson *et al.*[56] exposed aqueous solutions of horseradish peroxidase to 2.45 GHz microwave irradiation with a fine control of temperature at 25 °C, and detected no considerable inactivation of the enzyme unless absorbed power densities were greater than $125\,W\,cm^{-3}$ for 20 min, or were above $60\,W\,cm^{-3}$ for more than 20 min (which implies that a high power density of microwaves might cause non-thermal effect). Tuengler *et al.*[57] investigated the microwave effect on ethanol reduction in the presence of alcohol dehydrogenase using an irradiation density of $10\,mW\,cm^{-2}$ while continuously varying the frequency from 40 to 115 GHz; they observed no microwave influence on the reaction rate. After microwave irradiation (5 or $20\,mW\,cm^{-2}$) on embryonic quail hearts, Galvin *et al.*[58] detected no significant differences in enzymatic activities of three enzymes (lactate dehydrogenase, glutamic oxaloacetic transaminase, and creatine phosphokinase) between the irradiated embryonated eggs and non-exposed groups. The same research group[59] also exposed the acetylcholinesterase and creatine kinase to microwaves at specific absorption rates of 1, 10, 50, and $100\,mW\,g^{-1}$, measured the enzyme activities during microwave irradiation, and concluded that the microwaves had no effect on the activities of both enzymes when the temperatures of the control and exposed samples were identical (37 °C). Millar *et al.*[60] found no effect of microwave irradiation on the activity of acetylcholinesterase despite the use of different power densities, pulse widths, repetition rates, and duty cycles. Kabza *et al.*[61] constructed a continuous-flow setup using a household microwave oven to enable the temperature control of reactions, and studied the hydrolysis of cellobiose in aqueous buffer catalyzed by cellulase from *Penicillinum funiculosum* at 35 °C. They found the reaction rates between microwave irradiation and conventional heating were comparable, suggesting no microwave enhancement when the reaction temperature was well controlled.

Other studies suggest that the thermal effect of microwave heating could denature enzymes, but no non-thermal effect could be detected. Three human serum enzymes (lactate dehydrogenase, acid phosphatase, and alkaline phosphatase) were treated with 2.8 GHz microwaves at an incident power density of between 400 and $1000\,mW\,cm^{-2}$ at fixed temperatures through a heat exchanger; thermally induced inactivation of all three enzymes was observed while no non-thermal effect was detected.[62] Olcerst and Rabinowitz[63] measured the activity of acetylcholinesterase in aqueous solution and rabbit blood during microwave exposure at a constant temperature of 37 °C, and concluded that microwave irradiation had no direct impact on the enzyme activity only when the temperature increase was so great as to denature the enzyme. Bini et al.[64] studied the interactions between microwave irradiation (3 GHz) and lactate dehydrogenase (LDH) through monitoring the enzymatic activity during irradiation in steady-state or dynamic conditions, and suggested that, at a fixed temperature, the enzymatic activity under microwaves was not different from that under regular thermal condition while the high enzyme activity was solely due to the thermal activation.

However, some studies have implied the existence of a non-thermal effect of microwave irradiation in aqueous solutions. Porcelli et al.[52] irradiated two thermophilic and thermostable enzymes [(S)-adenosylhomocysteine hydrolase and 5'-methylthioadenosine phosphorylase] at a concentration of 0.3 mg mL^{-1} in 10 mM Tris-HCl buffer (pH 7.4) with 10.4 GHz microwaves. The exposures were maintained at a constant temperature between 70–90 °C. A non-thermal, irreversible, and time-dependent inactivation of both enzymes was observed, and the inactivation rate was associated with the energy absorbed but was independent of enzyme concentration. Fluorescence and CD spectroscopic techniques suggested the conformational changes of (S)-adenosylhomocysteine hydrolase induced by microwaves were not associated with temperature. Gelo-Pujic et al.[65] carried out the hydrolysis of starch in water by Celite 545-supported γ-amylase at 60 °C (Scheme 5.1); they obtained a 98% yield of D-glucose in 15 min under microwave irradiation and only a 55% yield in 15 min in an oil bath. Another study[66] on thermophilic alcohol dehydrogenase exposed to microwave irradiation suggested the presence of non-thermal effect of microwaves on the structural and functional properties of the enzyme. Interestingly, microwave (10.4 GHz) irradiation of low concentrations of thermophilic β-galactosidase (from *Bacillus acidocaldarius*) in aqueous solutions suggested an irreversible inactivation of the enzyme at 70 °C; however, the microwave effects disappeared at high enzyme concentrations above 50 μg mL^{-1}.[67] Hategan et al.[68] irradiated lactate dehydrogenase in ammonium sulfate solution at 0 °C (ice-water bath) at 310, 400, and 550 W power levels for 0–40 s; they observed a uniform loss of enzyme activity with the increase of temperature at the 310 W level, but a nonlinear behavior with the increase of

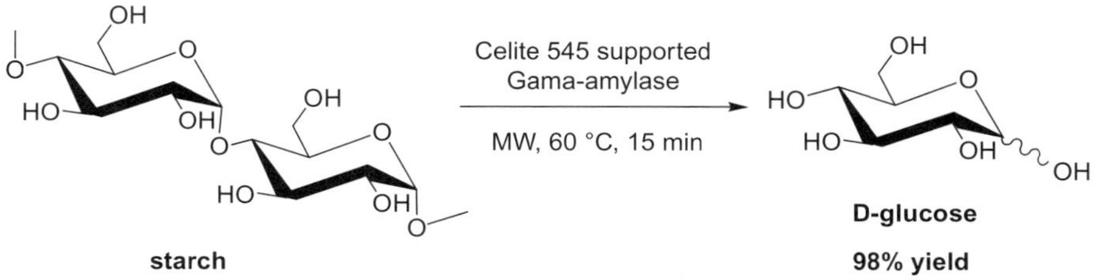

Scheme 5.1 Enzymatic hydrolysis of starch in water under microwave (MW) irradiation.

temperature at the 400 and 550 W levels. It was suggested that the nonlinear behavior was caused by the non-thermal effect of microwaves. The enzymatic hydrolysis of triolein through *Aspergillus carneus* lipase was performed under microwave irradiation at both low power mode (175 W, 38–40 °C) and high power mode (800 W, 90 °C); a complete hydrolysis of triolein was obtained in as short as 160 s at the low power level and 75 s at the high power level.[69] Recently, the Deiters' group[53] have demonstrated that microwave irradiation could activate hyperthermophilic enzymes [such as *Pyrococcus furiosus* β-glucosidase (Pfu CelB), α-galactosidase from *Thermotoga maritima* (Tm GalA), and carboxylesterase from *Sulfolobus solfataricus* P1 (SsoP1 CE)] in aqueous buffer solutions at temperatures far below their optimum conditions. Such a specific microwave effect was rationalized as the microwave-induced conformational flexibility of enzymes.

Sometimes, due to the lack of temperature control in domestic microwave ovens, the microwave effect on enzyme activity could not be well understood. Using a household microwave oven at a fixed power level (300 W), Zhu *et al.*[70] observed a higher initial rate of *Trichoderma reesei* cellulase-catalyzed hydrolysis of rice straw under microwave irradiation than that under conventional heating; however, the final yield under microwave irradiation was slightly lower. Therefore, the microwave effect on the enzymatic hydrolysis is not evident. Bradoo *et al.*[71] hydrolyzed triolein in aqueous buffers (10 mL) by several lipases, and reported 7–12-fold faster reaction rates under microwave irradiation (30 s at 1.35 kW) than the pH stat method (5 min at 37 °C for all lipases except at 50 °C for *Bacillus stearothermophilus* lipase). However, since the reaction temperature under microwave exposure remained unknown, it is not clear if the rate enhancement was due to a thermal or non-thermal effect.

5.2.2 Microwave-assisted Enzymatic Protein Digestion

Recombinant DNA technology has enabled the production of various therapeutic proteins. To analyze the structures of proteins, the enzymatic or chemical hydrolysis of proteins into smaller peptide fragments is routinely carried out in proteomics for biomarker discovery. Early studies on acid-catalyzed protein hydrolysis suggested the microwaves were able to reduce the reaction time to minutes from the 24 h of conventional heating.[72–75] The conventional method in enzymatic cleavage of proteins is also very time-consuming (typically hours, but could be up to 24 h for complex samples).[76,77] To overcome this bottleneck, several new approaches have been actively explored, which include heating, Microspin columns, ultrasonic energy, high pressure, infrared energy, microwave irradiation, alternating electric fields, and micro-reactors.[76] In particular, the use of microwaves in enzymatic protein digestion has gained some attention (see recent reviews[23,78,79]). Figure 5.1 shows the general procedures of microwave-assisted enzymatic protein digestion: after the protein is denatured, it is hydrolyzed in solutions or in gels under microwave irradiation, followed by the analysis *via* various instrumental methods [such as MALDI-MS

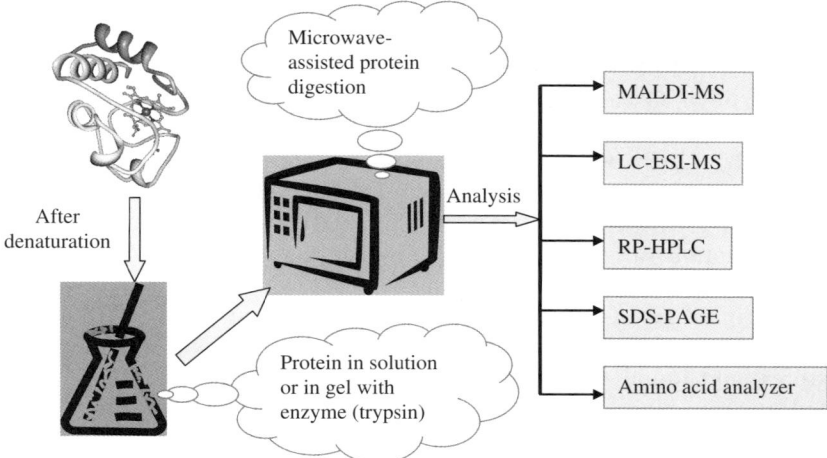

Figure 5.1 Microwave-assisted protein digestion catalyzed by the enzyme (typically trypsin). The protein shown is bovine heart cytochrome *c* from the RCSB Protein Data Bank: PDB ID: 2B4Z;[143] the protein structure was drawn using PDB Protein Workshop 3.5.[144]

(matrix-assisted laser desorption ionization-mass spectrometry),[77,80–83] LC-ESI-MS (liquid chromatography-electrospray ionization-mass spectrometry),[77,82] RP-HPLC (reversed-phase high-pressure liquid chromatography),[74,81,84–86] SDS-PAGE (sodium dodecyl sulfate polyacrylamide-gel electrophoresis),[84,85] and amino acid analyzer[75]]. The following is a review of recent advances in the field.

In 2002, Pramanik's group[77] reported the improved proteolytic cleavage of several proteins (cytochrome *c*, ubiquitin, myoglobin, lysozyme, and IFN α-2b) under controlled microwave irradiation (constant temperatures), and observed that the trypsin-catalyzed digestion of proteins finished in minutes compared to hours needed in conventional heating methods; on the other hand, their kinetic studies suggested that the acceleration in reaction rates was due to the rapid increase in temperature by microwaves, rather than the non-thermal effect. Further, Gómez and co-workers[84] investigated the microwave-assisted hydrolysis of β-lactoglobulin catalyzed by pronase or α-chymotrypsin, and obtained some interesting results: microwave irradiation improved the pronase's catalytic ability in digesting β-lactoglobulin, but had little effect on the performance of α-chymotrypsin in hydrolyzing the protein (all reactions were held at a constant temperature of 40 °C).

To accelerate the in-gel protein digestion during the identification of proteins using two-dimensional electrophoresis (2-DE) combined with mass spectrometry, Juan *et al.*[87] irradiated the trypsin-catalyzed in-gel digestion of lysozyme, albumin, conalbumin, and ribonuclease A with microwaves (195 or 325 W, no temperature control) and observed that the hydrolysis time was reduced to as little as 5 min from the traditional 16 h. Ho's group[88] have suggested that the digestion efficiencies of several proteins (myoglobin, cytochrome c, lysozyme, and ubiquitin) in aqueous solutions under microwave irradiation for 10 min at 60 °C were comparable with those without microwave heating for 6 h at 37 °C. As the microwave-assisted digestion proceeded at a higher temperature (60 °C) than that under regular heating (37 °C), the advantage of microwaves is not obvious. Vesper et al.[86] have demonstrated that the digestion efficiency of glycated hemoglobin HbA1c catalyzed by trypsin under microwave irradiation (at 50 °C) was 20% higher than that under conventional heating; however, such an enhancement effect was not achieved when the endoproteinase Glu-C was used as the enzyme. Sun et al.[80] extended the microwave-assisted enzymatic digestion of known proteins to complex protein mixtures, and found that the in-solution digestion of protein mixtures completed in 6 min while the in-gel digestion and the preparation time could be reduced to 25 min from overnight in the standard digestion protocol. Izquierdo et al.[85] conducted the microwave-assisted enzymatic digestion of a commercial bovine whey protein concentrate for 5 min at 40 or 50 °C; they observed faster proteolysis of protein under microwave irradiation than conventional heating for all seven enzymes investigated (especially pronase, papain, and Alcalase®). After immobilizing trypsin on magnetic silica nanoparticles (an excellent microwave absorber), Lin et al.[81,82] observed fast enzymatic digestion of bovine serum albumin (BSA), myoglobin, and cytochrome c under microwave irradiation; they reported that proteolysis could be completed in 15 s, compared with the conventional in-solution digestion time of 12 h. Similarly, Miao et al.[83] completed the enzymatic digestion of human lens tissue in 1 min (400 W and trypsin-to-protein ratio 1:5) under microwave irradiation when catalyzed by trypsin immobilized on amine-functionalized magnetic nanoparticles. Using a household microwave, Liu et al.[89] digested glycoproteins by pronase E, and completed the proteolytic cleavage of glycoproteins within 5 min under microwave irradiation. To further enhance the efficiency of protein digestion, Stencel et al.[90] demonstrated the use of a 48-position carbide plate for proteolytic hydrolysis of insulin chain B by trypsin under microwave irradiation, and observed a uniform and fast digestion of the polypeptide.

5.3 Microwave-assisted Enzymatic Reactions in Aqueous Solutions of Organic Solvents

Aqueous solutions of organic solvents[91,92] and ionic liquids (ILs)[93,94] are commonly used in enzymatic reactions to improve the substrate solubility, enzyme activity, and enantioselectivity, as well as to depress the water activity. However, little attention has been paid to performing biocatalysis in aqueous solutions of organic solvents and ILs under microwave irradiation. Lin et al.[88] conducted the trypsin-catalyzed protein digestion of myoglobin, cytochrome c, lysozyme, and ubiquitin in 50% (v/v) methanol (+50% water), or 30% (v/v) acetonitrile (+70% water), respectively, at 60 °C under microwave irradiation.

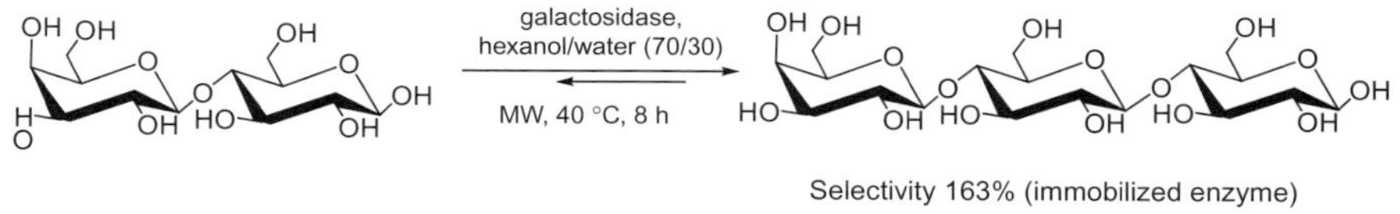

Scheme 5.2 Enzymatic transglycosylation of lactose in hexanol–water (70:30, v/v) under microwave irradiation.

The digestion efficiencies of proteins under microwave exposure for 10 min were higher than those under conventional heating at 37 °C for 6 h. However, the digestion efficiencies in aqueous solutions of organic solvents were generally lower than those in water, except in the case of ubiquitin. It was explained that organic solvents have two opposite effects on the protein hydrolysis: (i) they may unfold the target protein and increase the protein hydrolysis and (ii) they may denature the enzyme and slow down the protein digestion.

Most hydrolase-catalyzed reactions in aqueous solutions are hydrolysis processes. However, with the use of high concentrations of organic solvents, synthetic activities of enzymes may be achieved. Maugard et al.[95] performed the enzymatic transglycosylation of lactose in organic solvent–water (70:30, v/v) or in phosphate buffer (50 mM, pH 6.5) at 40 °C under microwave irradiation (Scheme 5.2). The reaction was catalyzed by free or immobilized β-galactosidase. A higher selectivity of galacto-oligosaccharides (GOS) was observed in hexanol–water (70 : 30) than in phosphate buffer, although the yields were comparable. The initial rates (within 1 h, by the immobilized enzyme) under microwave exposure and conventional heating were similar; however, microwave irradiation induced higher enzymatic selectivities, especially in the presence of hexanol (217-fold increase), and higher GOS yields at the end of the reaction (8 h).

5.4 Microwave-assisted Enzymatic Reactions in Ionic Liquids

Ionic liquids (ILs) consist of ions and are liquid at temperatures below 100 °C. Those salts with very low melting points, so-called room-temperature ionic liquids (RTILs), are most desirable as solvents. Compared with conventional organic solvents, ILs have many favorable properties, including extremely low vapor pressure, a wide liquid range, low-flammability, high ionic conductivity, high thermal conductivity, good dissolution power towards many substrates, high thermal/chemical stability, and a wide electrochemical potential window. Because of these unique properties, ILs have been widely recognized as solvents or co-catalysts in various applications, including organic catalysis,[96–104] inorganic synthesis,[105] biocatalysis,[102,106–111] polymerization,[112,113] and engineering fluids.[114–116] Typical IL cations are nitrogen containing (such as alkylammonium, N,N'-dialkylimidazolium, N-alkylpyridinium, and pyrrolidinium)

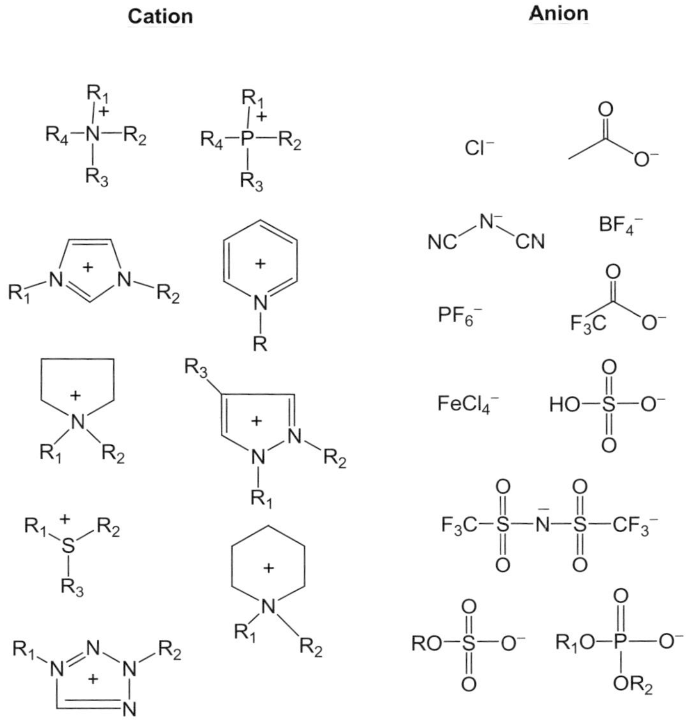

Figure 5.2 Structures of some cations and anions in ILs.

or phosphorous containing (such as alkylphosphonium). The common choices of anions include halides, BF_4^-, PF_6^-, $CH_3CO_2^-$, $CF_3CO_2^-$, NO_3^-, Tf_2N^- [$(CF_3SO_2)_2N^-$], [RSO_4]$^-$, and [R_2PO_4]$^-$. Figure 5.2 shows some representative cations and anions.

ILs are relatively polar, but not strongly polar. Their polarities are close to those of lower alcohols[117,118] or formamide.[119] The data presented in Table 5.2 indicate that ILs have moderate and adjustable dielectric constants (usually between 8 and 15). ILs have been shown to be efficient microwave absorbers and thus ideal solvents or co-solvents for many reactions.[120–122] In addition, ILs are safer reaction media for microwave-assisted reactions than volatile organic solvents. Since organic solvents usually possess high vapor pressures, microwave-assisted reactions in closed vessels result in high pressures, which is unsafe for preparative scale productions. In addition, most organic solvents are highly flammable, which imposes another safety hazard. In contrast, most ILs possess near-zero vapor pressures, and are usually much less flammable than organic solvents. Therefore, it is advantageous to replace organic solvents with ILs in microwave-irradiated enzymatic reactions.

As a result, ILs have been demonstrated more recently to be excellent solvents for microwave-assisted reactions, such as the Pd/C-catalyzed Heck

Table 5.2 Dielectric constants of some organic solvents and ionic liquids.

Solvent[a]	Dielectric constant, ε_s (temperature, °C)	Solvent[a]	Dielectric constant, ε_s (temperature, °C)
Water	80.1 (20)[b]	[EMIM][OTf]	15.2 (25)[c]
Methanol	33.0 (20)[b]	[EMIM][BF$_4$]	12.8 (25)[c]
Ethanol	25.3 (20)[b]	[EMIM][Tf$_2$N]	12.3 (25)[d]
Acetonitrile	36.64 (20)[b]	[EMIM][EtSO$_4$]	37.84 or 36.75[e]
DMSO	47.24 (20)[b]	[BMIM][BF$_4$]	11.7 (25)[c]
Dimethylformamide	38.25 (20)[b]	[BMIM][PF$_6$]	11.4 (25)[c] or (20)[f]
1-Butanol	17.84 (20)[b]	[BMIM][Tf$_2$N]	11.7 (25)[d]
Dichloromethane	8.93 (25)[b]	[HMIM][PF$_6$]	8.9 (25)[c]
THF	7.52 (22)[b]	Hexane	1.89 (20)[b]
Toluene	2.38 (23)[b]	Diethyl ether	4.27 (20)[b]

[a]Initialism of ionic liquids: EMIM = 1-ethyl-3-methylimidazolium, BMIM = 1-butyl-3-methylimidazolium, HMIM = 1-n-hexyl-3-methylimidazolium, OTf = trifluoromethylsulfonate, Tf$_2$N = bis(trifluoromethylsulfonyl)imide.
[b]Data from ref. 5.
[c]Ref. 139.
[d]Ref. 140.
[e]Calculated data from ref. 141 (this reference also calculated [EMIM][Tf$_2$N] ε_s = 15.76 or 14.01).
[f]Ref. 142.

reaction,[123] catalytic Beckmann rearrangement of ketoximes,[122] and nucleophilic substitution reactions of anilines or thiophenol with ethoxymethylene isopropylidene malonate,[124] and the synthesis of alkyl halides from alcohols and nitriles from aryl halides.[125] Further examples of microwave-assisted organic synthesis in ILs have been reviewed.[126]

However, although the catalytic properties of enzymes in organic solvents in response to microwave irradiation have been reported, much less is known about the enzyme activity in ILs under microwave irradiation. Lundell et al.[127] have investigated the use of PS-C II (lipase from *Burkholderia cepacia* immobilized on ceramic particles) in acylations of N-acylated 2-amino-1-phenylethanol and N-acylated norphenylephrine in imidazolium- and pyridinium-based ILs, *tert*-butyl methyl ether (TBME), and their mixtures. They observed similar substrate conversions and E values between microwave irradiation and classical heating. Our group[128] have recently examined the microwave effect on the enzymatic transesterification of ethyl butyrate and 1-butanol catalyzed by Novozym® 435 (immobilized *Candida antarctica* lipase B) in ILs and organic solvents at 40 °C. We observed that when the enzyme particles were surrounded by a layer of water molecules the enzyme surface might be over-heated to induce higher enzymatic activities. Figure 5.3(a) shows the enzyme particle surrounded by (at least) a layer of water molecules in a dried hydrophobic solvent and substrate environment. The solvent is hydrophobic so it has a low tendency to strip off the water layer. In such a scenario, the water layer has a much higher dielectric constant (80.1 at 20 °C) than the surrounding organic solvent or IL (in the range of 9–40, mostly 9–20). Meanwhile, the static

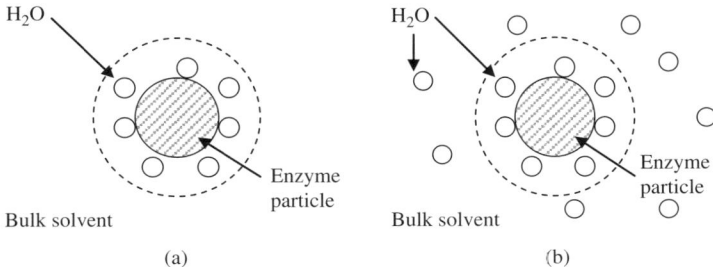

Figure 5.3 Illustrations of water-induced superheating: (a) the free or immobilized enzyme particle is surrounded by a layer of water molecules while the bulk hydrophobic solvent is dried; (b) the enzyme particle is surrounded by a layer of water molecules while the bulk hydrophobic solvent contains a small amount of dispersed water.

dielectric constant of dry protein powder is about 4,[129–132] and the immobilization support (acrylic resin) has a dielectric constant in the range 2.7–4.5[133] (most plastics are in the range 2–4). Consequently, under microwave exposure, the enzyme's surface is likely to have a higher temperature than the bulk solvent due to the superheating of the water layer. This might explain the higher reaction rates observed under microwaves when the enzyme was not dried but the solvent and substrates were dried. When the enzyme is also intensively dried, this layer of water is expected to disappear. In this scenario, there is no superheating of the enzyme; as a result, the reaction rates under both heating modes are very close. If additional water is added and dispersed (through vigorously stirring) into the bulk solvent (Figure 5.3b), the overall dielectric constant of the medium is increased and the superheating effect on the enzyme surface is reduced.

Although microwave-promoted enzymatic reactions in aqueous solutions of ILs have not been actively pursued, we can foresee some productive outcomes in the near future due to the unique combination of ILs and microwaves for enzyme stabilization and activation.

5.5 Non-thermal Effect of Microwave Irradiation on Enzymes

As discussed earlier, many studies have suggested that there is no specific microwave effect on an enzyme's activity in aqueous solutions, while the microwave-induced thermal effect may cause the inactivation of enzymes. A rationale of this observation can be understood from Figure 5.3(b): when the enzyme and substrate molecules are surrounded by aqueous solutions, the soluble enzyme and substrates are not superheated by microwave exposure since water has the highest dielectric constant and absorbs most of the microwave energy; if the reaction is well mixed and well controlled at a certain temperature, a higher enzymatic activity is not expected under microwaves than under conventional heating.

However, some researchers (see previous sections of this chapter) still observed faster enzymatic reactions under microwave exposure than under traditional heating. Some early studies carried out the enzymatic reactions using household microwave ovens with the control of microwave power input, but without a fine control of *in situ* temperature and the degree of agitation. Therefore, these studies leading to the conclusion of a non-thermal microwave effect may not be convincing. Despite these undesired comparisons, some studies (including a recent one[53]) reported the non-thermal effect on enzyme activities in aqueous solutions even with the control of reaction temperature. A reasonable explanation would be that proteins have high dipole moments, and thus are very susceptible to microwave irradiation.[22,23]

5.6 Prospects of Microwave Irradiation in Aqueous Phase Biocatalysis

Very recently, an even more attractive way to utilize microwave energy has been to apply the microwave irradiation with simultaneous cooling. This unique combination can increase the kinetic energy of reactant molecules through microwaves while the cooling keeps the temperature low by removing the unnecessary thermal energy from intermolecular friction. As a result, this method allows more microwave energy to be introduced into the reaction system for a faster reaction. Although only a few studies have dealt with this subject, most of them observed enhanced reaction rates through two methods: cooling with a stream of compressed air[134,135] or with a microwave transparent cooling liquid (such as liquid nitrogen, solid CO_2 in hexane, or cryogenic fluid).[53,136,137] The latter cooling method enables the reaction to be conducted at very low temperatures, which is especially favorable for improving enzyme stability and achieving high enantioselectivity. Currently, very limited studies on the low-temperature enzymatic reactions irradiated by microwaves have been performed in either organic solvents or ionic liquids.[138]

5.7 Experimental

5.7.1 Enzymatic Hydrolysis of Starch in Water under Microwave Irradiation (Scheme 5.1)[65]

Celite 545 (1.0 g) was added to a suspended solution of starch (0.5–1.0 g) and γ-amylase (0.25–0.5 g) in water (2.0 mL). The mixture was heated under microwave irradiation (between 0–15 w) at a constant temperature of 60 °C for 15 min. The D-glucose yield was 90–98%.

5.7.2 Enzymatic Transglycosylation of Lactose in Hexanol–Water (70:30, v/v) under Microwave Irradiation (Scheme 5.2)[95]

A solution of lactose (159 g L^{-1}), MgCl$_2$ (0.01 M) and free or immobilized β-galactosidase (80 U mL^{-1}) in phosphate buffer (50 mM, pH 6.5) was mixed with 70% (v/v) of hexanol (dried with 3Å molecular sieves). The final reaction volume was 5 mL. The reaction was maintained at 40 °C. After lactose was depleted, the reaction was stopped by heating it for 10 min at 100 °C. Galacto-oligosaccharides were analyzed by HPLC. Oligosaccharide analysis was performed with an Ultrasep ES 100 NH$_2$ column (250×4 mm, 6 μm) with elution of acetonitrile–water (80:20, v/v) at 2 mL min^{-1} at 70 °C. Hexyl-galactopyranoside, hexanediol-galactopyranoside and glycerol-galactopyranoside were analyzed with a Prontosyl C18 (250 × 4 mm, 5 μm) column with elution of water–methanol (50:50, v/v) at 0.6 mL min^{-1} at 70 °C.

5.7.3 Enzymatic Digestion of Proteins (Cytochrome c, Myoglobin, Lysozyme, and Ubiquitin)[77]

Digestion experiments were conducted in 350-μL Eppendorf tubes. The protein concentration was kept at 20 μM in 75 mM ammonium bicarbonate solution. A different ratio of trypsin-to-protein, ranging from 1:200 to 1:5 (w/w), was used. Samples were irradiated for 5, 10, 15, 20, 30, or 60 min with microwaves at 30% (144 W) of the maximum power setting. After irradiation, the sample was quenched by the addition of 0.1% trifluoroacetic acid solution and was stored in dry ice for mass spectrometric analysis.

Acknowledgements

The author thanks Drs Hubert W. Vesper (CDC at Atlanta) and Gary A. Baker and Sheila N. Baker (both at Oak Ridge National Laboratory) for providing reprints of papers.

References

1. D. Adam, *Nature*, 2003, **421**, 571–572.
2. D. R. Baghurst and D. M. P. Mingos, *J. Chem. Soc. Chem. Commun.*, 1992, 674–677.
3. L. Perreux and A. Loupy, *Tetrahedron*, 2001, **57**, 9199–9223.
4. C. O. Kappe and D. Dallinger, *Mol. Diversity*, 2009, **13**, 71–193.
5. D. R. Lide, *CRC Handbook of Chemistry and Physics*, CRC Press Inc., New York, 84th edn., 2003.
6. A. C. Metaxes and R. J. Meredith, *Industrial Microwave Heating*, Peter Perigrinus, London, 1983.

7. T. Razzaq and C. O. Kappe, *ChemSusChem*, 2008, **1**, 123–132.
8. R. Hoogenboom, T. F. A. Wilms, T. Erdmenger and U. S. Schubert, *Aust. J. Chem.*, 2009, **62**, 236–243.
9. S. Caddick, *Tetrahedron*, 1995, **51**, 10403–10432.
10. C. O. Kappe, *Chem. Soc. Rev.*, 2008, **37**, 1127–1139.
11. A. de La Hoz, A. Díaz-Ortiz and A. Moreno, *Chem. Soc. Rev.*, 2005, **34**, 164–178.
12. R. N. Gedye and J. B. Wei, *Can. J. Chem.*, 1998, **76**, 525–532.
13. F. Chemat and E. Esveld, *Chem. Eng. Technol.*, 2001, **24**, 735–744.
14. S. Raulta, A.-C. Gillarda, M.-P. Foloppea and M. Robba, *Tetrahedron Lett.*, 1995, **36**, 6673–6674.
15. A. Stadler and C. O. Kappe, *Eur. J. Org. Chem.*, 2001, 919–925.
16. H.-M. Yu, S.-T. Chen, P. Suree, R. Nuansri and K.-T. Wang, *J. Org. Chem.*, 1996, **61**, 9608–9609.
17. V. Sridar, *Curr. Sci.*, 1998, **74**, 446–450.
18. P. Lidström, J. Tierney, B. Watheyb and J. Westman, *Tetrahedron*, 2001, **57**, 9225–9283.
19. M. Nüchter, B. Ondruschka, W. Bonrath and A. Gum, *Green Chem.*, 2004, **6**, 128–141.
20. M. Nüchter, U. Müller, B. Ondruschka, A. Tied and W. Lautenschläger, *Chem. Eng. Technol.*, 2003, **26**, 1207–1216.
21. I. Roy and M. N. Gupta, *Curr. Sci.*, 2003, **85**, 1685–1693.
22. B. Rejasse, S. Lamare, M.-D. Legoy and T. Besson, *J. Enzyme Inhib. Med. Chem.*, 2007, **22**, 519–527.
23. J. M. Collins and N. E. Leadbeater, *Org. Biomol. Chem.*, 2007, **5**, 1141–1150.
24. R. Vetrimani, N. Jyothirmayi, P. Haridas Rao and C. S. Ramadoss, *Lebensm. Wiss. Technol.*, 1992, **25**, 532–535.
25. C. T. Ponne, A. C. Moller, L. M. M. Tijskens, P. V. Bartels and M. M. T. Meijer, *J. Agric. Food Chem.*, 1996, **44**, 2818–2824.
26. R. C. Anantheswaran and H. S. Ramaswamy, in *Handbook of Microwave Technology for Food Applications*, Food Science and Technology Series, Vol. **109**, ed. A. K. Datta and R. C. Anantheswaran, CRC Press, New York, 2001, pp. 191–213.
27. C. Devece, J. N. Rodriguez-Lopez, L. G. Fenoll, J. Tudela, J. M. Catala, E. de los Reyes and F. Garcia-Canovas, *J. Agric. Food Chem.*, 1999, **47**, 4506–4511.
28. S. Tajchakavit and H. S. Ramaswamy, *J. Microwave Power Electromag. Energy*, 1995, **30**, 141–148.
29. R. Lopez-Fandiño, M. Villamiel, N. Corzo and A. Olano, *J. Food Prot.*, 1996, **59**, 889–892.
30. L. Campanella, M. Cusano, R. Dragone, M. P. Sammartino and G. Visco, *Curr. Med. Chem.*, 2003, **10**, 663–669.
31. A. T. Modak, S. T. Weintraub, T. H. McCoy and W. B. Stavinoha, *J. Pharmacol. Exp. Ther.*, 1976, **197**, 245–252.
32. C. Galli and G. Racagni, *Methods Enzymol.*, 1982, **86**, 635–642.

33. Y. Ikarashi, Y. Maruyama and W. B. Stavinoha, *Jpn. J. Pharmacol.*, 1984, **35**, 371–387.
34. E. Marani, P. Bolhuis and M. E. Boon, *Histochem. J.*, 1988, **20**, 397–404.
35. T. Ozaki, K. Tabuse, T. Tsuji, Y. Nakamura, K. Kakudo and I. Mori, *Pathol. Int.*, 2003, **53**, 837–845.
36. A. Zaks and A. M. Klibanov, *Science*, 1984, **224**, 1249–1251.
37. B. Réjasse, T. Besson, M.-D. Legoy and S. Lamare, *Org. Biomol. Chem.*, 2006, **4**, 3703–3707.
38. M. N. Gupta and I. Roy, *Eur. J. Biochem.*, 2004, **271**, 2575–2583.
39. S. H. Krishna, *Biotechnol. Adv.*, 2002, **20**, 239–267.
40. Y. L. Khmelnitsky and J. O. Rich, *Curr. Opin. Chem. Biol.*, 1999, **3**, 47–53.
41. M. N. Gupta, *Eur. J. Biochem.*, 1992, **203**, 25–32.
42. I. Roy and M. N. Gupta, *Tetrahedron*, 2003, **59**, 5431–5436.
43. M. C. Parker, T. Besson, S. Lamare and M.-D. Legoy, *Tetrahedron Lett.*, 1996, **37**, 8383–8386.
44. J. R. Carrillo-Munoz, D. Bouvet, E. Guibé-Jampel, A. Loupy and A. Petit, *J. Org. Chem.*, 1996, **61**, 7746–7749.
45. M. Vacek, M. Zarevúcka, Z. Wimmer, K. Stránský, K. Demnerová and M.-D. Legoy, *Biotechnol. Lett.*, 2000, **22**, 1565–1570.
46. M. Gelo-Pujic, E. Guibé-Jampel, A. Loupy, S. Galema and D. Mathé, *J. Chem. Soc. Perkin Trans.*, 1996, **1**, 2777–2780.
47. G. D. Yadav and P. S. Lathi, *Synth. Commun.*, 2005, **35**, 1699–1705.
48. S. Mazumder, D. D. Laskar, D. Prajapati and M. K. Roy, *Chem. Biodiversity*, 2004, **1**, 925–929.
49. G. Lin and W.-Y. Lin, *Tetrahedron Lett.*, 1998, **39**, 4333–4336.
50. I. Roy, K. Mondal and M. N. Gupta, *Biotechnol. Prog.*, 2003, **19**, 1648–1653.
51. M. Gelo-Pujic, E. Guibé-Jampel, A. Loupy and A. Trincone, *J. Chem. Soc. Perkin Trans. 1*, 1997, **1**, 1001–1002.
52. M. Porcelli, G. Cacciapuoti, S. Fusco, R. Massa, G. d'Ambrosio, C. Bertoldo, M. De Rosa and V. Zappia, *FEBS Lett.*, 1997, **402**, 102–106.
53. D. D. Young, J. Nichols, R. M. Kelly and A. Deiters, *J. Am. Chem. Soc.*, 2008, **130**, 10048–10049.
54. T. R. Ward, J. W. Allis and J. A. Elder, *J. Microwave Power*, 1975, **10**, 315–320.
55. E. K. Yeargers, J. B. Langley, A. P. Sheppard and G. K. Huddleston, *Ann. N. Y. Acad. Sci.*, 1975, **247**, 301–304.
56. H. M. Henderson, K. Hergenroeder and S. S. Stuchly, *J. Microwave Power*, 1975, **10**, 27–35.
57. P. Tuengler, F. Keilmann and L. Genzel, *Z. Naturforsch., Teil C*, 1979, **34**, 60–63.
58. M. J. Galvin, D. I. McRee and M. Lieberman, *Bioelectromagnetics*, 1980, **1**, 389–396.
59. M. J. Galvin, D. L. Parks and D. I. McRee, *Radiat. Environ. Biophys.*, 1981, **19**, 149–156.

60. D. B. Millar, J. P. Christopher, J. Hunter and S. S. Yeandle, *Bioelectromagnetics*, 1984, **5**, 165–172.
61. K. G. Kabza, J. E. Gestwicki, J. L. McGrath and H. M. Petrassi, *J. Org. Chem.*, 1996, **61**, 9599–9602.
62. M. L. Belkhode, A. M. Muc and D. L. Johnson, *J. Microwave Power*, 1974, **9**, 23–29.
63. R. B. Olcerst and R. Rabinowitz, *Radiat. Environ. Biophys.*, 1978, **15**, 289–295.
64. M. Bini, A. Checcucci, A. Ignesti, L. Millanta, N. Rubino, G. Camici, G. Manao and G. Ramponi, *J. Microwave Power*, 1978, **13**, 95–100.
65. M. Gelo-Pujic, E. Guibé-Jampel and A. Loupy, *Tetrahedron*, 1997, **53**, 17247–17252.
66. F. La Cara, S. D'Auria, M. R. Scarfi, O. Zeni, R. Massa, G. d'Ambrosio, G. Franceschetti, M. De Rosa and M. Rossi, *Protein Peptide Lett.*, 1999, **6**, 155–162.
67. F. La Cara, M. R. Scarffi, S. D'Auria, R. Massa, G. d'Ambrosio, G. Franceschetti, M. Rossi and M. De Rosa, *Bioelectromagnetics*, 1999, **20**, 172–176.
68. A. Hategan, D. Martin, A. Popescu, C. Oproiu and I. Margaritescu, *Bulg. J. Phys.*, 2000, **27**, 203–206.
69. R. K. Saxena, J. Isar, S. Saran, R. Kaushik and W. S. Davidson, *Curr. Sci.*, 2005, **89**, 1000–1003.
70. S. Zhu, Y. Wu, Z. Yu, X. Zhang, H. Li and M. Gao, *Bioresour. Technol.*, 2006, **97**, 1964–1968.
71. S. Bradoo, P. Rathi, R. K. Saxena and R. Gupta, *J. Biochem. Biophys. Methods*, 2002, **51**, 115–120.
72. H.-M. Yu, S.-T. Chen, S.-H. Chiou and K.-T. Wang, *J. Chromatogr.*, 1986, **456**, 357–362.
73. S.-T. Chen, S.-H. Chiou, Y.-H. Chu and K.-T. Wang, *Int. J. Peptide Protein Res.*, 1987, **30**, 572–576.
74. S.-H. Chiou and K.-T. Wang, *J. Chromatogr.*, 1989, **491**, 424–431.
75. E. Marconi, G. Panfili, L. Bruschi, V. Vivanti and L. Pizzoferrato, *Amino Acids*, 1995, **8**, 201–208.
76. J. L. Capelo, R. Carreira, M. Diniz, L. Fernandes, M. Galesio, C. Lodeiro, H. M. Santos and G. Vale, *Anal. Chim. Acta*, 2009, **650**, 151–159.
77. B. N. Pramanik, U. A. Mirza, Y. H. Ing, Y.-H. Liu, P. L. Bartner, P. C. Weber and A. K. Bose, *Protein Sci.*, 2002, **11**, 2676–2687.
78. W. N. Sandoval, V. Pham, E. S. Ingle, P. S. Liu and J. R. Lill, *Comb. Chem. High Throughput Screening*, 2007, **10**, 751–765.
79. J. R. Lill, E. S. Ingle, R. S. Liu, V. Pham and W. N. Sandoval, *Mass Spectrom. Rev.*, 2007, **26**, 657–671.
80. W. Sun, S. Gao, L. Wang, Y. Chen, S. Wu, X. Wang, D. Zheng and Y. Gao, *Mol. Cell. Proteomics*, 2006, **5**, 769–776.
81. S. Lin, G. Yao, D. Qi, Y. Li, C. Deng, P. Yang and X. Zhang, *Anal. Chem.*, 2008, **80**, 3655–3665.

82. S. Lin, D. Yun, D. Qi, C. Deng, Y. Li and X. Zhang, *J. Proteome Res.*, 2008, **7**, 1297–1307.
83. A. Miao, Y. Dai, Y. Ji, Y. Jiang and Y. Lu, *Biochem. Biophys. Res. Commun.*, 2009, **380**, 603–608.
84. F. J. Izquierdo, I. Alli, R. Gómez, H. S. Ramaswamy and V. Yaylayan, *Food Chem.*, 2005, **92**, 713–719.
85. F. J. Izquierdo, E. Penas, M. L. Baeza and R. Gomez, *Int. Dairy J.*, 2008, **18**, 918–922.
86. H. W. Vesper, L. Mi, A. Enada and G. L. Myers, *Rapid Commun. Mass Spectrom.*, 2005, **19**, 2865–2870.
87. H.-F. Juan, S.-C. Chang, H.-C. Huang and S.-T. Chen, *Proteomics*, 2005, **5**, 840–842.
88. S.-S. Lin, C.-H. Wu, M.-C. Sun, C.-M. Sun and Y.-P. Ho, *J. Am. Soc. Mass Spectrom.*, 2005, **16**, 581–588.
89. X. Liu, K. Chan, I. K. Chu and J. Li, *Carbohydr. Res.*, 2008, **343**, 2870–2877.
90. L. M. Stencel, C. M. Kormos, K. B. Avery and N. E. Leadbeater, *Org. Biomol. Chem.*, 2009, **7**, 2452–2457.
91. S.-T. Chen, S.-Y. Chen, S.-C. Hsiao and K.-T. Wang, *Biotechnol. Lett.*, 1991, **13**, 773–778.
92. T. Kijima, K. Ohshima and H. Kise, *J. Chem. Technol. Biotechnol.*, 1994, **59**, 61–65.
93. H. Zhao, *J. Mol. Catal. B: Enzym.*, 2005, **37**, 16–25.
94. H. Zhao, O. Olubajo, Z. Song, A. L. Sims, T. E. Person, R. A. Lawal and L. A. Holley, *Bioorg. Chem.*, 2006, **34**, 15–25.
95. T. Maugard, D. Gaunt, M. D. Legoy and T. Besson, *Biotechnol. Lett.*, 2003, **25**, 623–629.
96. C. M. Gordon, *Appl. Cat. A*, 2001, **222**, 101–117.
97. S. Houlton, *Chem. Week*, 2004 (Feb 25), s10–s11.
98. K. R. Seddon, *J. Chem. Technol. Biotechnol.*, 1997, **68**, 351–356.
99. T. Welton, *Chem. Rev.*, 1999, **99**, 2071–2083.
100. H. Zhao and S. V. Malhotra, *Aldrichim. Acta.*, 2002, **35**, 75–83.
101. M. Earle, A. Forestier, H. Olivier-Bourbigou and P. Wasserscheid, in *Ionic Liquids in Synthesis,* ed. P. Wasserscheid and T. Welton, Wiley-VCH Verlag GmbH, Weinheim, 2003, pp. 174–288.
102. N. Jain, A. Kumar, S. Chauhan and S. M. S. Chauhan, *Tetrahedron*, 2005, **61**, 1015–1060.
103. W. Wasserscheid and W. Keim, *Angew. Chem. Int. Ed.*, 2000, **39**, 3772–3789.
104. P. Wasserscheid and T. Welton (eds), *Ionic Liquids in Synthesis,* Wiley-VCH Verlag GmbH, Weinheim, 2nd ed., 2008.
105. F. Endres and T. Welton, in *Ionic Liquids in Synthesis,* ed. P. Wasserscheid and T. Welton, Wiley-VCH Verlag GmbH, Weinheim, 2003, pp. 289–318.
106. T. L. Husum, C. T. Jorgensen, M. W. Christensen and O. Kirk, *Biocatalysis Biotransform.*, 2001, **19**, 331–338.

107. U. Kragl, M. Eckstein and N. Kaftzik, *Curr. Opin. Biotechnol.*, 2002, **13**, 565–571.
108. S. Park and R. J. Kazlauskas, *Curr. Opin. Biotechnol.*, 2003, **14**, 432–437.
109. R. A. Sheldon, R. M. Lau, M. J. Sorgedrager, F. van Rantwijk and K. R. Seddon, *Green Chem.*, 2002, **4**, 147–151.
110. F. van Rantwijk, R. Madeira Lau and R. A. Sheldon, *Trends Biotechnol.*, 2003, **21**, 131–138.
111. F. van Rantwijk and R. A. Sheldon, *Chem. Rev.*, 2007, **107**, 2757–2785.
112. P. Kubisa, *Prog. Polym. Sci.*, 2004, **29**, 3–12.
113. A. J. Carmichael and D. M. Haddleton, in *Ionic Liquids in Synthesis*, ed. P. Wasserscheid and T. Welton, Wiley-VCH Verlag GmbH, Weinheim, 2003, pp. 319–335.
114. J. F. Brennecke and E. J. Maginn, US Patent, 6,579,343, 2003.
115. H. Zhao, S. Xia and P. Ma, *J. Chem. Technol. Biotechnol.*, 2005, **80**, 1089–1096.
116. H. Zhao, *Chem. Eng. Commun.*, 2006, **193**, 1660–1677.
117. A. J. Carmichael and K. R. Seddon, *J. Phys. Org. Chem.*, 2000, **13**, 591–595.
118. S. N. V. K. Aki, J. F. Brennecke and A. Samanta, *Chem. Commun.*, 2001, 413–414.
119. C. Reichardt, *Chem. Rev.*, 1994, **94**, 2319–2358.
120. N. E. Leadbeater and H. M. Torenius, *J. Org. Chem.*, 2002, **67**, 3145–3148.
121. J. Hoffmann, M. Nuechter, B. Ondruschka and P. Wasserscheid, *Green Chem.*, 2003, **5**, 296–299.
122. J. K. Lee, D.-C. Kim, C. E. Song and S.-G. Lee, *Synth. Commun.*, 2003, **33**, 2301–2307.
123. X. Xie, J. Lu, B. Chen, J. Han, X. She and X. Pan, *Tetrahedron Lett.*, 2004, **45**, 809–811.
124. S.-R. Guo and Y.-Q. Yuan, *Synth. Commun.*, 2006, **36**, 1479–1484.
125. N. E. Leadbeater, H. M. Torenius and H. Tye, *Tetrahedron*, 2003, **59**, 2253–2258.
126. N. E. Leadbeater, H. M. Torenius and H. Tye, *Comb. Chem. High Throughput Screening*, 2004, **7**, 511–528.
127. K. Lundell, T. Kurki, M. Lindroos and L. T. Kanerva, *Adv. Synth. Catal.*, 2005, **347**, 1110–1118.
128. H. Zhao, G. A. Baker, Z. Song, O. Olubajo, L. Zanders and S. M. Campbell, *J. Mol. Catal. B: Enzym.*, 2009, **57**, 149–157.
129. S. T. Bayley, *Trans. Faraday Soc.*, 1951, **47**, 509–517.
130. S. Maricic, G. Pifat and V. Parvdic, *Biochim. Biophys. Acta - Specialized Sect. Biophys. Subjects*, 1964, **79**, 293–300.
131. S. Takashima and H. P. Schwan, *J. Phys. Chem.*, 1965, **69**, 4176–4182.
132. A. Tanaka and Y. Ishida, *J. Polym. Sci., Polym. Phys. Ed.*, 1973, **11**, 1117–1138.
133. S. Oshima, K. Ehata and T. Tomioka, *IEICE Trans. Electron.*, 2000, **E83-C**, 2–6.

134. R. K. Arvela and N. E. Leadbeater, *Org. Lett.*, 2005, **7**, 2101–2104.
135. J. J. Chen and S. V. Deshpande, *Tetrahedron Lett.*, 2003, **44**, 8873–8876.
136. J. Kurfürstová and M. Hájek, *Res. Chem. Intermed.*, 2004, **30**, 673–681.
137. B. K. Singh, P. Appukkuttan, S. Claerhout, V. S. Parmar and E. Van der Eycken, *Org. Lett.*, 2006, **8**, 1863–1866.
138. N. E. Leadbeater, L. M. Stencel and E. C. Wood, *Org. Biomol. Chem.*, 2007, **5**, 1052–1055.
139. C. Wakai, A. Oleinikova, M. Ott and H. Weingärtner, *J. Phys. Chem. B*, 2005, **109**, 17028–17030.
140. C. Wakai, A. Oleinikova and H. Weingärtner, *J. Phys. Chem. B*, 2006, **110**, 5824.
141. T. Köddermann, C. Wertz, A. Heintz and R. Ludwig, *Angew. Chem. Int. Ed.*, 2006, **45**, 3697–3702.
142. S. N. Baker, G. A. Baker, M. A. Kane and F. V. Bright, *J. Phys. Chem. B*, 2001, **105**, 9663–9668.
143. N. Mirkin, J. Jaconcic, V. Stojanoff and A. Moreno, *Proteins Struct., Funct., Bioinf.*, 2008, **70**, 83–92.
144. J. L. Moreland, A. Gramada, O. V. Buzko, Q. Zhang and P. E. Bourne, *BMC Bioinf.*, 2005, **6**, 21.

CHAPTER 6
Microwave-assisted Synthesis of Polymers in Aqueous Media

CATHERINE MARESTIN AND RÉGIS MERCIER

Laboratoire des Matériaux Organiques à Propriétés Spécifiques (LMOPS), CNRS/Université de Savoie-UMR 5041, Chemin du Canal – 69 360 Solaize, France

6.1 Introduction

Owing to the significant advantages of microwave-assisted processes for organic reactions, polymer synthesis under microwave irradiation has recently appeared as an emerging field of research (Figure 6.1) and microwave irradiation has become a well-established technique to promote the synthesis and modification of many polymers.

Most approaches involve organic solvents because solid-state polymer synthesis might be limited by melted state viscosity problems. Taking into account environmental as well as safety considerations, attempts to replace toxic and volatile organic solvents by safer and more environmentally-friendly media represent a significant and attractive challenge. In this context, water combines many interesting and valuable properties (abundant, non-toxic, *etc.*). In addition, thanks to its high dielectric constant, water is particularly adapted for microwave heating. The synthesis of polymers in aqueous medium and under microwave irradiation therefore appears as an innovative and promising area of research. Although this field is only in its infancy, the benefits already reported for the synthesis of different type of polymers highlight its great

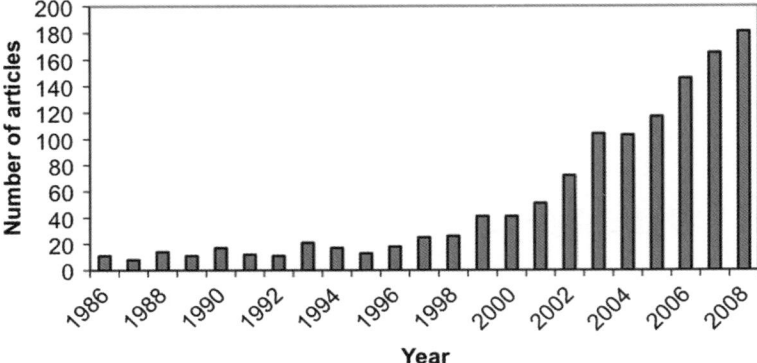

Figure 6.1 Microwave-assisted polymer synthesis; number of articles per year.

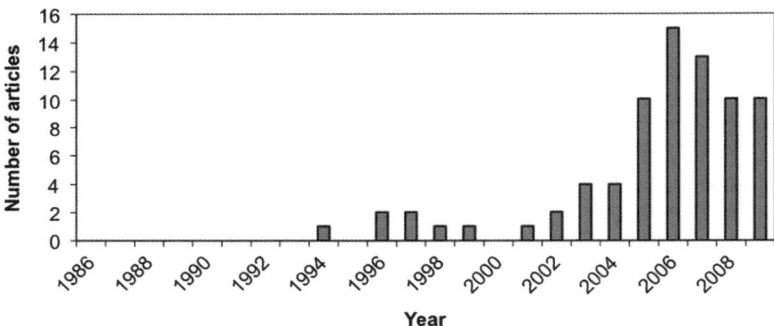

Figure 6.2 Microwave-assisted polymer synthesis in aqueous medium; number of articles per year.

potential and explains the increasing interest of both the academic and the industrial communities (Figure 6.2).

This chapter surveys recent developments in this field.

6.2 Radical Polymerization in Aqueous Medium

6.2.1 Free-Radical Polymerization

6.2.1.1 Radical Homogeneous Polymerization in Water

The polymerization under microwave irradiation of acrylamide and acrylic acid has been investigated with the purpose of developing polyelectrolytes for wastewater treatment.[1] A typical formulation of the aqueous reaction medium consists of a mixture of acrylamide, acrylic acid monomers, ethylenediamine tetraacetate (EDTA) as complexing agent, sodium formate and sodium persulfate as, respectively, chain-transfer agent and initiator. The polymerization performed at 60 °C under microwave irradiation (115–140 s at 250 W) afforded high molecular weight polymers ranging from 3 to 5 10^6 g mol^{-1}. The reaction time was 50–100 times lower under microwaves than under conventional operating conditions.

Microwave irradiation has also used by Xu and co-workers[2] to synthesize superabsorbent systems based on starch, 2-acrylamido-2-methyl-propanosulfonic acid (AMPS), and sodium acrylate. They observed for these superabsorbent materials higher swelling rates than those obtained by conventional heating. Interestingly, SEM analyses showed well-dispersed even pores for materials prepared under microwaves, whereas more irregular pores were observed for conventionally prepared materials. To account for this morphology difference, it was suggested that a higher temperature in the microwave-assisted process would be responsible for an *in situ* drying of the material, which should favor the formation of bubbles by rapid evaporation.

Other superabsorbent materials have been synthesized by grafting sodium acrylate onto starch, using poly(ethylene glycol diacrylate) as crosslinker.[3] In contrast to the conventional heating process, microwave-assisted polymerization does not require the removal of the monomer inhibitor and, furthermore, the presence of residual oxygen does not have a detrimental effect. Materials with high swelling ratio and low solubility can be obtained after 10 min irradiation at 85–90 W.

6.2.1.2 Macromolecular Chain Grafting – Copolymer Synthesis

Most studies related to macromolecular chains grafting onto polymer backbone using microwave irradiation in aqueous media concern the polymerization of vinyl monomers on polysaccharide substrates. Depending on the specific structure of the substrate, vinyl monomers can be chemically linked to a polysaccharide backbone through hydroxyl or amino groups. For Singh *et al.*[4] the grafting process should involve a free-radical polymerization reaction (Scheme 6.1).

In addition to the propagation and termination steps of the grafting reaction, undesired homopolymerization also occurs. To characterize the copolymers both the grafting percentage (%G) and the grafting efficiency (%E) are then determined.

The development of new efficient processes involving the graft copolymerization of vinyl monomers onto polysaccharides under microwave irradiation and in aqueous medium has recently attracted great attention. Various initiation systems have been tested.

Kaith *et al.*[5] have reported the microwave-assisted grafting of poly(methyl methacrylate) (PMMA) onto acetylated *Saccharum spontaneum* L fiber, using ferrous ammonium sulfate–potassium persulfate redox initiator (FAS-KPS). Different experimental parameters were tested (reaction time, pH of the medium, dilution effect, initiator molar ratio, *etc.*) to optimize the grafting yield.

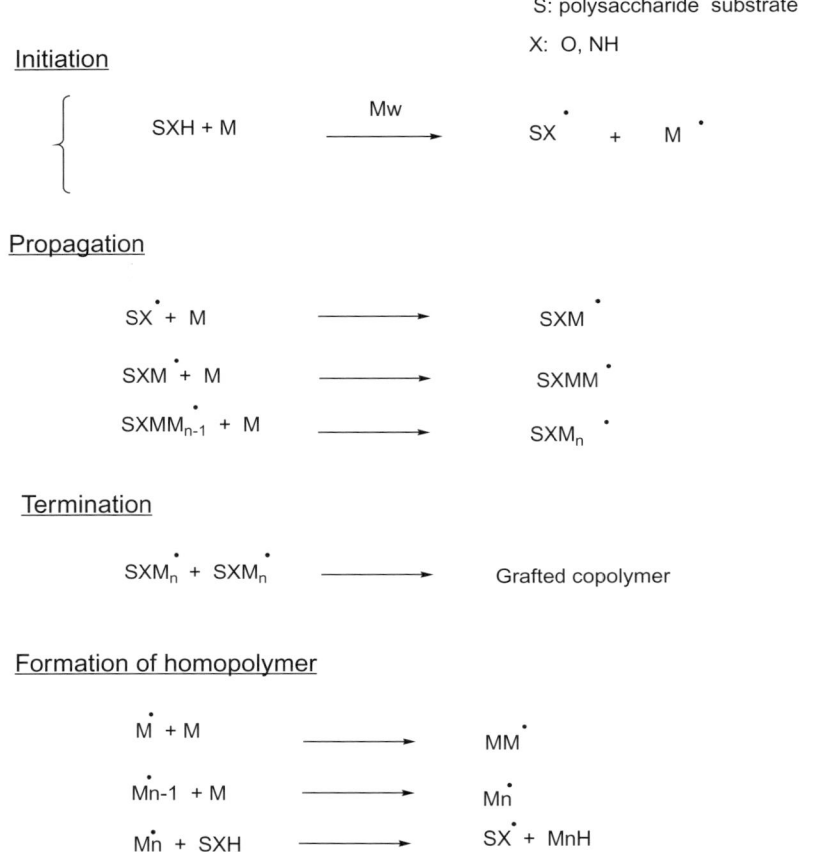

Scheme 6.1 Free-radical grafting polymerization of vinyl monomers onto polysaccharides.

The best results obtained were 72.2% grafting yield and 88.2% grafting efficiency. Thus, the reaction under microwave irradiation was proven to be a rapid and efficient method for modifying the chemical structures of these natural fibers. As a consequence, their properties have been changed. Crystallinity and moisture absorbance decreased, and acid and base stability as well as the thermal resistance were improved. Other common redox systems such as cerium ammonium nitrate (CAN)–KPS initiator were also used. KPS promotes the oxidation of Ce^{3+} in Ce^{4+}, which is considered as the active species in the grafting reaction. Matahwa and co-workers[6] have successfully grafted N-isopropylacrylamide and methyl acrylate onto cellulose using this redox system. Reaction temperatures lower than 60 °C were required to obtain efficient copolymer grafting. Indeed, at higher temperatures (80 °C) no grafting was observed but an undesired crosslinking reaction of the cellulose occurred. On the other hand, Kaith et al.[7] have investigated the synthesis of Flax-graft copolymers of binary vinyl monomeric mixtures in the presence of FAS–H_2O_2 initiator and the reinforcement effect of these copolymers on the phenol-formaldehyde matrices. Depending on the vinyl monomers, various grafting yields were reached. According to the authors, one possible explanation is a miscibility difference of monomers with water. The resulting modified fibers show better strength than the parent fibers.

Using a difunctionalized monomer, crosslinked chitosan-*graft*-acrylic acid copolymers have been synthesized from N,N'-methylenebisacrylamide (MBA), in the presence of ammonium cerous sulfate (ACS) as initiator.[8] Different experimental parameters were studied to optimize the grafting degree (89.6%) and grafting efficiency (86.5%) of the resin. Whereas similar results were obtained by the conventional heating method, the microwave-assisted synthesis was achieved at a much faster rate. Moreover, the resulting material was proven to be a very effective superabsorbent material.

Different studies have described the efficient grafting of vinyl monomers onto polysaccharides with only very low concentrations of potassium persulfate, under microwave irradiation. For instance, porous hybrid materials based on PMMA-grafted sulfated polysaccharide of the green seaweed *Chaetomorpha antennina* (CMsps) have been produced in the presence of $0.00037\,\mathrm{mol\,L^{-1}}$ potassium persulfate.[9]

A polyacrylic acid grafted *Artemisia* seed gum has also been synthesized under microwaves, in the presence of both potassium persulfate and partially neutralized acrylic acid.[10] Among different experimental parameters, the effect of the microwave irradiation time on the final copolymer water absorbency properties was investigated. From optimized reaction conditions, a highly superabsorbent graft copolymer (absorbing 400 times its own dry weight) was obtained.

Potato starch-*graft*-poly(acrylonitrile) has been synthesized in the presence of $0.0014\,\mathrm{mol\,L^{-1}}$ of ammonium peroxodisulfate, under microwave irradiation.[11] Interestingly, the presence of atmospheric oxygen (which is a very potent inhibitor for the polymerization of vinyl monomers) was not detrimental to the grafting reaction. A maximum of 225% grafting ratio and 98% grafting efficiency were reached. Notably, under similar experimental conditions, no graft copolymerization was observed using conventional heating, which required a higher peroxodisulfate concentration ($0.24\,\mathrm{mol\,L^{-1}}$) and inert atmosphere to produce only 10% grafting yield. Similar results have reported by the same research group for the synthesis of potato starch-*graft*-poly(acrylamide).[12] By microwave irradiation (at 720 W, 60 s reaction time), 160% grafting and 89% efficiency were reached using very low persulfate concentrations ($0.0025\,\mathrm{mol\,L^{-1}}$), in the presence of oxygen. Much higher initiator concentrations ($0.24\,\mathrm{mol\,L^{-1}}$) and an inert atmosphere were required with a heating process to only get 10% grafting and 5% efficiency. κ-Carrageenan-*graft*-poly(acrylamide)[13] or κ-carrageenan-*graft*-PMMA[14] have been synthesized using similar microwave irradiation conditions.

By first making a pretreatment of chitosan with ^{60}Co γ-rays, Duan et al.[15] were able to graft poly[rosin-(2-acryloyloxy)ethyl ester] (PRAEE) as a side

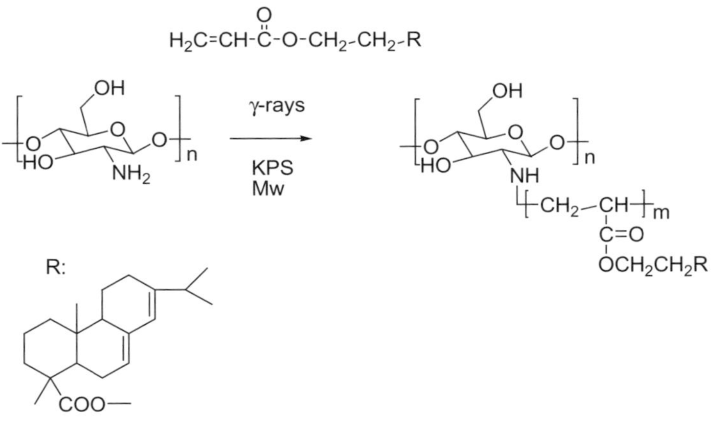

Scheme 6.2 Synthesis of chitosan-*graft*-poly[rosin-(2-acryloyloxy)ethyl ester].

chain on the polymer backbone, using potassium persulfate as initiator and under microwave irradiation (Scheme 6.2).

A grafting yield of 72.3% was obtained. The resulting materials showed interesting properties for the controlled release of fenoprofen calcium compared to the parent chitosan.

Microwave-assisted graft polymerization without any initiator or catalyst has attracted great interest recently. In this respect, Singh and co-workers have synthesized chitosan-*graft*-poly(methyl methacrylate).[16] The influence of different parameters such as microwave power, microwave irradiation time, and monomer concentration on the grafting reaction was studied. Similar works have been reported by the same authors concerning the synthesis of chitosan-*graft*-poly(acrylamide),[4] chitosan-*graft*-poly(acrylonitrile),[17] guar-*graft*-poly(acrylamide),[18] poly(acrylonitrile) grafted *Cassia simea* seed polysaccharide,[19] and guar-*graft*-poly(acrylonitrile).[20] As a general trend, an efficient grafting reaction could be achieved in very short times whereas conventional reaction procedures require longer time and result in lower efficiency.

Analogous observations have been reported for the synthesis of new polymeric flocculants [polyacrylamide (PAM)-*graft*-carboxymethyl starch (CMS)].[21] These polymer materials showed high flocculation efficacy, which could find application in the treatment of sewage wastewater. The properties of the graft copolymers were also investigated for lower gastrointestinal tract targeted drug delivery.[22] Interest in microwave-assisted reaction for the grafting of poly(acrylonitrile) (PAN) onto xyloglucan has also been reported.[23] Here, a grafting ratio of 92% was obtained in only 120 s of reaction time, at 60 °C and in the absence of any initiator, while not much than a 76% grafting ratio was reached with a conventional heating process after 24 h of reaction at 40 °C and in the presence of ceric ammonium nitrate–nitric acid redox system. Similarly, xanthan-*graft*-poly(acrylamide) has been synthesized by Kumar and co-workers.[24]

6.2.2 Radical Polymerization in Dispersed Media

Conventional polymerizations in aqueous dispersed media usually involve a fine dispersion of monomers in a water phase. Such polymerizations afford polymer colloids (latexes). Whereas polar solvents couple efficiently with microwave irradiation, unpolar phases behave differently. Depending on the presence of surfactant, co-surfactant, the polymerization kinetics, and the particle nucleation process, such a polymerization is considered as emulsion, miniemulsion, suspension or microemulsion polymerization. As well as a few modeling works of microwave-assisted emulsion polymerization,[25,26] many experimental investigations have recently been reported in this field.

A prime example concerns the synthesis of monodisperse polystyrene microspheres by dispersion polymerization.[27] The dispersant phase used was a water–ethanol mixture. While typical particle sizes produced by conventional heating reaction are in the range 0.5–10 microns, substantially smaller particles were obtained under microwave irradiation, namely 200–500 nm. The authors ascribe this effect to a higher rate of polymerization and a shorter nucleation period. The effect of monomer, initiator, and surfactant (polyvinylpyrrolidone) on both the particle size and particle size distribution was studied. The authors also reported a much higher stability over time of the microspheres obtained under microwave irradiation.

In contrast, Holtze and co-workers[28] have reported the synthesis of polystyrene particles through microwave-assisted miniemulsion polymerization. While carefully controlling the reaction temperature of a conventionally heated and a microwave-assisted reaction, the authors observed a comparable reaction rate. However, under pulsed microwave irradiations, extremely high molecular weight polymers (10^7 g mol^{-1}) were obtained even after one short 8-s pulse. The authors explain such an effect by the fundamental rules of heterophase polymerization. At the end of the microwave pulse, no very high molecular weight polymer or very high monomer conversion could be evidenced. The authors argue that during this pulse the radicals produced react or recombine until only one radical is left in each droplet. This radical is then able to grow efficiently during the cooling period (during which no other radical is produced), leading to very high molecular weights.[29] This so-called "post-effect" characterized by non-steady state free radical polymerization kinetics is supported by a compartmentalization effect and the high viscosity inside the polymer particles.[30] In such systems, medium hydrophobic initiators (AIBN, PEGA 200) were proven to be the most efficient, compared with more hydrophilic KPS. After only four microwave irradiation pulses (irradiation for 10 s at 1000 W power for each pulse) a monomer conversion as high as 90% was reached.

In emulsion polymerization, the initiator is generally water soluble and so the initiation step occurs in the water phase. As regards the monomer, a major portion forms 1–10 micron droplets, and to a very small extent it is solubilized in micelles (when the concentration of surfactant exceeds the critical micellar concentration – CMC) or dissolved in water. After a homogeneous and/or micellar nucleation, the polymerization proceeds in the polymer particles (with diffusion of the monomers from the droplets to the particles). One of the first articles reporting the synthesis of microwave-assisted colloids concerns the emulsion polymerization of styrene in the presence of sodium dodecyl sulfate as surfactant.[31] The authors used a short pulsed microwave irradiation, which leads to a 93% monomer conversion. Compared to conventional thermal conditions, a highly beneficial effect of such a microwave treatment both on the polymerization time (reduced by a factor 70) and on the molecular weights (increased by a factor 1.2) was established. More recently, different studies have shown that even higher activation effects can be observed with more polar monomers. In this respect, Sierra and co-workers[32] have compared the polymerization rates under both microwave (Rp_{Mw}) and conventional heating ($Rp_{Conventional}$) of methyl methacrylate (MMA) and styrene monomers. Comparing the behavior of MMA and butyl acrylate (BuA), Costa et al.[33] observed a 93% conversion of MMA in 6 min under microwave irradiation, whereas 30 min were required to reach 91% conversion of BuA. Moreover, the particle diameters were significantly reduced for the PMMA colloid produced with microwave heating (40% reduction compared with conventional lattices) whereas only an 8% reduction in size was observed for PBuA colloids. The authors ascribed these results to the very different water solubility of the two monomers (MMA: 1.5×10^{-1} mol L^{-1} at 50 °C, BuA: 6.2×10^{-3} mol L^{-1} at 50 °C). On the other hand, the observed particle size reduction was related to a higher number of nucleated particles, due to a very high decomposition rate of the initiator under microwave irradiation. Notably, the authors also have reported higher efficiencies when pulsed microwave irradiation was used. Consequently, six pulses, each of 27 s, were enough to reach 97% conversion, while 6 min of constant microwave irradiation were necessary to obtain 93% conversion. Examining at the polymerization of styrene and butyl acrylate, Jung and co-workers[34] have reported a high conversion rate for microwave-assisted polymerization of BuA, and also larger conversion rates for the copolymer synthesis with increasing BuA molar fraction. As an explanation, the authors have suggested a heating effect dependence upon the monomer polarity (respective dipole moments for BuA and styrene: 1.72 and 0.36 D)

Because "clean surface" latex particles are obtained, soapless emulsion polymerization is considered as an attractive method. Here, polymerization starts in the aqueous phase and when the polymer chains reach a critical size they collapse, forming small primary particles that are stabilized by the ionic groups generated from the initiator. Depending on the colloidal stability, the particles may then agglomerate into larger ones. By way of illustration, Murray et al.[35] have described the first successful microwave-assisted synthesis of a colloidal poly(N-isopropylacrylamide) microgel system. In a dedicated microwave oven equipped with a fiber-optic probe, N-isopropylacrylamide was polymerized at 70 °C, with a microwave power of 400 W, without any surfactant. Under such conditions, a reaction time of 1 h was sufficient whereas 6 h was necessary by conventional heating. In both cases, monodisperse particles with quite similar size were obtained. As shown by TEM, DSC, and turbidity analyses, the structure of both microgels were similar. Zhang and co-workers[36] have described the synthesis of stable polystyrene nanoparticles *via* a microwave approach. Interestingly, a narrow distribution of particle size was observed as compared with the broad one obtained by the conventional heating process.

Other studies have highlighted the beneficial effect of microwave irradiation in synthesizing such colloids. Bao *et al.*[37] have studied the synthesis of PMMA by surfactant-free emulsion polymerization. Considering that methyl methacrylate is relatively hydrophilic and water soluble (0.1 mol L^{-1} at 70 °C), the nucleation step was believed to occur by a homogeneous process. In addition, because the final particles size of the latex is closely related to the nucleation step, the high decomposition rate of the initiator under microwave irradiation should contribute to a higher number of particles and a better uniformity of the particle size. The authors have considered two different strategies to reduce the PMMA particle size. The first consisted of adding some polar solvent to the polymerization medium with the purpose of increasing the number of stable nuclei during the nucleation stage. Acetone was chosen as it enhances the solubility of MMA in water and reduces the surface tension of the polymerization system, thus increasing the latex stability. As a result, 99% conversion was reached after a 2 h reaction. A stable latex with particles around 45 nm was obtained. The second strategy relied on the use of a KPS/Na$_2$SO$_3$/CuSO$_4$ redox initiating system. In this case, a stabilizing effect of the copper ions absorbed by the latex particles was believed to be beneficial for reducing both the size (67 nm) and size distribution (PDI = 5.1%) of particles. In both cases, very stable colloids were obtained (270 days compared with 90 days for a latex synthesized by conventional surfactant-free emulsion polymerization).

Similarly, Ngai *et al.*[38] have studied the polymerization of styrene and observed a significant particle size decrease (from 280 to 70 nm) upon adding acetone (up to 50 wt%). The polymer particle size was further reduced (down to 20 nm) by the introduction of a small amount of MMA. Taking into account the higher water-solubility of MMA, they suggested that the free radical produced by the decomposition of KPS should first react with MMA rather than with styrene.

Monodisperse thermoresponsive poly(styrene-*co*-N-isopropylacrylamide) particles have been synthesized (100–130 nm) by Yi *et al.*[39] As NiPAM is more hydrophilic and reactive than styrene, the persulfate radicals most probably initiate first the formation of PNIPAM oligomers. Once these hydrophilic oligomers reach a critical length they precipitate and form primary particles. The styrene is then supposed to diffuse from the monomer droplets and further polymerize. The stability of the resulting particles is due to the hydrophilic SO$_4$ end groups as well as the hydrophilic PNIPAM chains, which are mostly concentrated at the particle–water interface. The authors showed that the size of the poly(styrene-*co*-NIPAM) particles decreases with high NiPAM : styrene ratio. Stable colloids were obtained after reaction for 1 h under microwave irradiation. Upon increasing the KPS concentration from 3.14 to 14.6 mmol L^{-1}, the particle size decreased from 220.7 to 142.3 nm.

An et al.[40,41] have reported the synthesis of crosslinked nanoparticles with a narrow polydispersity, without any surfactant, under microwave irradiation. Stable crosslinked nanoparticles bearing hydroxyl functional groups were successfully obtained with unprecedented solid contents. Notably, monomer conversions up to 96% were reached.

Another soapless emulsion polymerization concerning butyl methacrylate has been described by He et al.[42] The influence of monomer concentration, initiator (potassium persulfate) amount and the addition of ethanol on the monomer conversion, particle size, and particle size distribution were studied. Comparing the results with those obtained by a conventional thermal procedure, the authors showed that microwave treatment induces higher polymerization rate and gives smaller polymer particles with, in this case, also a narrower particle size distribution. Moreover, the addition of ethanol increased the polymer particle diameter.

Whereas most microwave-assisted reactions are performed at 2.45 GHz (either mono-modal or multi-modal) Zhu and co-workers[43] have investigated the emulsion polymerization of MMA under high-power pulsed microwave irradiation (PMI) at 1.25 GHz. A significant effect of irradiation power on the conversion rate of MMA was observed. In the same way, it was shown that the degree of conversion depended on the irradiation energy. In contrast, the pulse width had no effect on the polymerization rate. The same authors[44] have also reported the synthesis of polystyrene using the PMI method. A small effect of pulse width on the conversion rate was, this time, observed.

Palacios and Valverde[45] have studied the effect of microwave irradiation on styrene emulsion polymerization kinetics. Experiments were performed at 50 °C with an irradiation power of 389 W. Without added initiator, microwave irradiation could not initiate the polymerization of styrene (5% yield after 33 min). In contrast, the combined effects of microwave irradiation and potassium persulfate (12.4 mmol L^{-1}) were very efficient as a 91% yield was reached after 33 min reaction; only 69% yield, in the presence of 19.26 mmol L^{-1} of potassium persulfate, was obtained through a conventional heating reaction. The authors pointed out that under microwave irradiation the polymerization rate was 26 times higher than with conventional heating conditions.

Luo and co-workers[46] have reported the preparation of TiO_2/poly(styrene) core–shell nanospheres by combining a sol–gel procedure reaction and an emulsion polymerization under microwave irradiation. First, a TiO_2 colloid is formed by microwave-assisted hydrolysis of tetra-n-butyl titanate in a basic aqueous medium. Then, this colloid is used as a seed in the emulsion polymerization of styrene, still under microwave heating. TEM analyses revealed the formation of spherical and smooth particles surfaces with a TiO_2 core and a polystyrene shell. The diameter and distribution of the nanosphere were affected by the styrene concentration. For a styrene concentration ranging from 2% to 3.5%, monodisperse particles (123–161 nm) were obtained.

Based on a PNIPAM-g-PEO seed emulsion, Zhao and co-workers[47] have reported the syntheses of PMMA coated sub-micron PNIPAM particles, under microwave irradiation. Because of their potential application in medicine or biology, the design of magnetic nanoparticles has attracted great interest. Recently, Huang et al.[48] have synthesized magnetic Fe_3O_4/poly(styrene-co-acrylamide) composite nanoparticles by microwave-assisted emulsion polymerization. The experimental procedure involved magnetite nanoparticles as ferrofluid and a pre-emulsification of the system by sonication. The resulting particles were smaller (60 nm) and more uniform than those obtained with conventional heating (150–250 nm).

In terms of organic–inorganic nanocomposite material manufacturing, the microwave-assisted synthesis of poly(acrylamide)-calcium phosphate nanocomposites has been used by Tang et al.[49] From calcium salts, phosphate, and acrylamide aqueous medium solution, homogeneously dispersed inorganic nanoparticles in the poly(acrylamide) matrix were obtained in one reaction. Depending on the pH of the aqueous solution, different calcium phosphate phases were obtained. Small (<10 nm) nanoparticles of amorphous calcium phosphate were formed in weak acidic medium, whereas crystalline hydroxyapatite nanorods (10–20 nm in diameter and around 25–49 nm long) were obtained in basic medium. For the authors, this method not only avoids the commonly reported drawbacks of organic–inorganic nanocomposites prepared with a conventional heating treatment (such as an inhomogeneous dispersion and aggregation of the inorganic particles) but also appears to be efficient, simple, economic, and environment friendly.

The synthesis of polymer-grafted carbon nanotubes is expected to avoid problems associated with the partial incompatibility between carbon nanotubes and polymer matrices, which are taken to be responsible for the inhomogeneous dispersions or phase segregation of carbon nanotubes in composite materials. In this respect, Wu and co-workers[50] have reported the synthesis of polystyrene and PMMA-wrapped carbon nanotubes by *in situ* emulsion polymerization activated by microwave irradiation. The authors mentioned that microwave irradiation induces a very rapid decomposition of the initiator into radicals, which leads to both the polymerization of monomer and the grafting of oligomer radicals onto the carbon nanotubes. As a result, high yields of carbon nanotube surface functionalization (20–40 wt% of polymer grafted) were achieved in short reaction times (25–30 min).

Hu et al.[51] have also synthesized, by soapless microwave-assisted emulsion polymerization, submicron PMMA particles doped with Eu^{3+}. Based on XPS analysis, europium ions were found mainly at the particle surface. To account for this observation, the authors suggested that the hydrophobic alkyl chain of the $Eu(C_7H_{15}COO)_3$ precursor may enter the inner of the particles whereas the more hydrophilic Eu^{3+} may remain at the particle surface.

6.2.3 Controlled/Living Radical Polymerization in Dispersed Media

In the last decade, controlled/living radical polymerization in dispersed media has been of considerable interest for the design and synthesis of controlled molecular weight polymers with well-defined macromolecular architectures. Different polymerization methods have been investigated (nitroxide-mediated polymerization, atom transfer radical polymerization, reversible addition fragmentation transfer). In all cases, both the efficiency and the rapidity of the initiation process are critical points of the living polymerization. In this respect, the microwave-assisted technique appears particularly well adapted.

6.2.3.1 Nitroxide-mediated Polymerization (NMP)

Research related to the nitroxide-mediated miniemulsion polymerization of styrene under microwave irradiation has been reported.[52] Stable latexes were synthesized in the presence of sodium dodecyl sulfate (surfactant) and hexadecane (co-surfactant). The particle sizes were similar to those observed for particles prepared by a conventional miniemulsion polymerization. Under microwave conditions, M_n increases linearly with monomer conversion, confirming the controlled/living character of the polymerization. Moreover, good agreement between the theoretical and microwave-assisted M_n–conversion profiles was noted.

6.2.3.2 Atom Transfer Radical Polymerization (ATRP)

Monodisperse polymer particles have been produced by atom transfer radical polymerization performed in emulsion and under microwave irradiation.[53] A halogen-containing poly(ethylene glycol) (PEG-Cl) was used as macro-initiator for the synthesis of poly(ethylene glycol)-*block*-poly(styrene) copolymer. The PEG-*b*-poly(styrene) particles were spherical, monodispersed and smaller than those prepared by a conventional thermal procedure. To explain such a result, it was suggested that microwave irradiation should have an enhancing effect on the initiation efficiency of the PEG-Cl macro-initiator. As a large amount of active species is generated in a very short time, a high number of particles are formed.

6.2.3.3 Reversible Addition–Fragmentation Chain Transfer (RAFT)

An and co-workers[54] have realized successfully the polymerization of *N*-isopropylacrylamide (NIPAm) and *N*-isopropylmethacrylamide (NIPMAm) by RAFT in aqueous medium under microwave irradiation. For this purpose, water-soluble macromolecular chain transfer agents (Macro-CTA) were considered as both macro-initiators and latex stabilizers. Poly(NIPAm) particles were prepared and then used as seeds for the synthesis of core–shell copolymers poly(NIPAm)/poly(NIPMAm). The excellent control of molecular weight afforded by the microwave-assisted RAFT polymerization was related both to the water-soluble nature of the macro chain transfer agent and to the very fast microwave initiation process. With microwave irradiation, the conversion of the monomer was very efficient (up to 90–100%) within 5 min, whereas many hours of reaction were required under conventional thermal conditions to obtain similar results.

6.3 Step-growth Polymerization in Aqueous Media

One of the key points required for the synthesis of high molecular weight polymers by step-growth polymerization is the high conversion degree of the reaction involved in the polymerization process. In this respect, and taking into account that many organic reactions proceed in quantitative fashion under microwave irradiation, the use of microwaves for the step-growth polymer synthesis makes sense and offers the possibility of developing polymers that are not accessible by conventional heating process. On the other hand, there is increasing interest in developing environmentally friendly polymer synthesis processes. In this perspective, microwave-assisted step-growth polymerization in aqueous media appears as a challenging issue.

6.3.1 Synthesis of Poly(ether)s

Surpateanu and co-workers[55,56] have synthesized poly(ether)s from a dihalogenated oxetane monomer and different bisphenols *via* a nucleophilic substitution reaction (Scheme 6.3).

The reaction proceeds by phase-transfer catalysis in a nitrobenzene–water (1:1) mixture, with tetrabutylammonium bromide (TBAB) as transfer agent. Whereas 4–6 h were required with a conventional thermal procedure to obtain the targeted oligomers, 20 min reaction under microwave irradiation was sufficient. To explain such a result, the authors have suggested a better transfer of the catalytic complex from aqueous to organic phase in the case of the microwave-assisted reaction. Better yields were also obtained.

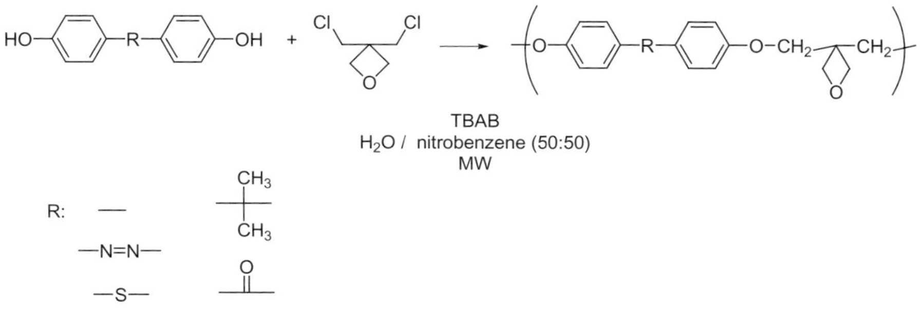

Scheme 6.3 Synthesis of poly(ether)s.

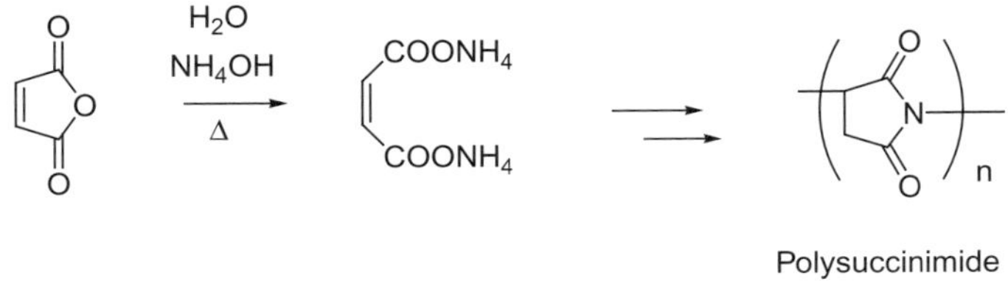

Scheme 6.4 Synthesis of polysuccinimide from maleic anhydride and ammonia.

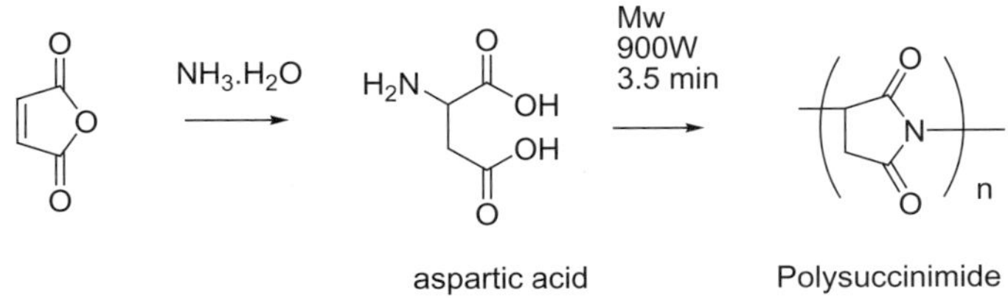

Scheme 6.5 Synthesis of polysuccinimide from *in situ* generated aspartic acid.

6.3.2 Synthesis of Poly(amide)s and Poly(imide)s

Polysuccinimide (PSI) is commonly synthesized in two steps, from maleic anhydride and ammonia (Scheme 6.4).

However, conventional polymerization procedures require a long reaction time. Recently, using microwave irradiation, Yu and co-workers[57] were able to obtain a polysuccinimide in only 10 min. The polymerization initially occurs in the aqueous medium and is finished in the solid state. The average molecular weight (M_w) of the resulting polymer was 1009 g mol^{-1} (PDI = 1.24). This new synthetic procedure is very promising as polysuccinimide is a precursor of polyaspartic acid (obtained from a basic hydrolysis), which is known for its excellent biodegradability and biocompatibility. Other authors[58] have described the synthesis of polysuccinimide under microwave irradiation from aspartic acid generated *in situ* by the reaction of ammonia with maleic anhydride (Scheme 6.5).

The influence of the maleic anhydride–ammonia molar ratio, microwave output power, and irradiation time on the polymer yield has been studied. A molar ratio of 1.2, an output power of 900 W and an irradiation time of 3.5 min were found to be optimum conditions to obtain a very high yield.

Imai *et al.*[59] have investigated the polymerization of 12-aminododecanoic acid in different solvents under microwave irradiation. In the case of water as reaction medium, a poly(amide) with 0.35 dL g^{-1} of inherent viscosity was obtained after only 5 min irradiation (Scheme 6.6).

In general, the poly(imide) synthesis is conducted in two stages: first to prepare a poly(amic acid), which is then converted into poly(imide). The reaction solvents usually used are polar aprotic solvents such as NMP, which must contain a very low quantity of water to obtain high molecular weight polymers. Long reaction times are also required. Recently, it was shown that

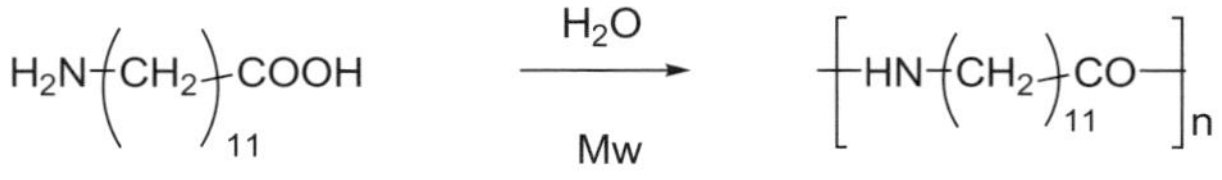

Scheme 6.6 Polyamide synthesis from polycondensation of 12-aminododecanoic acid.

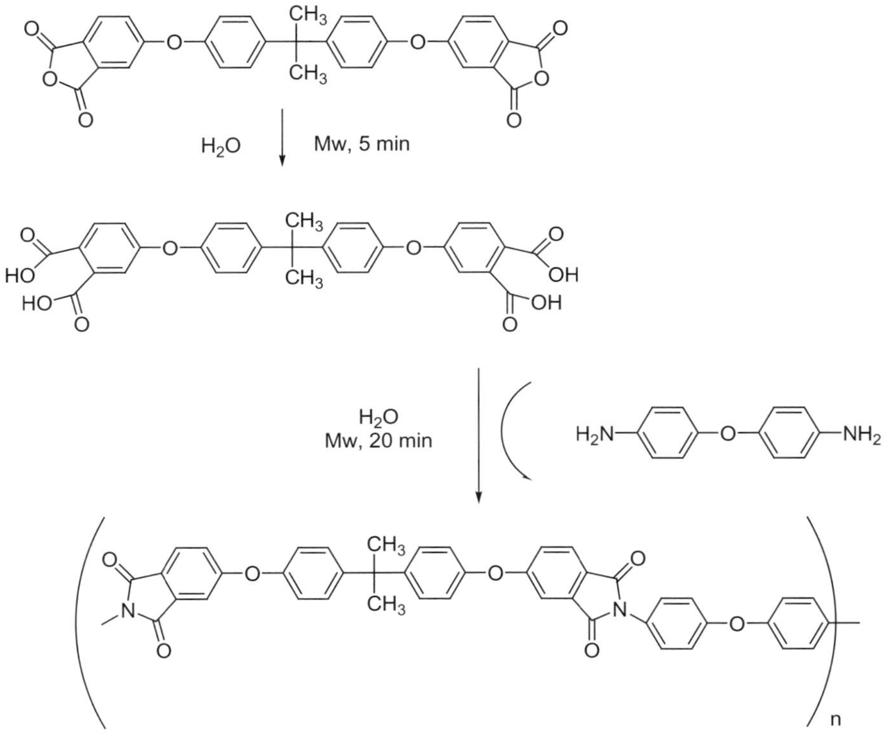

Scheme 6.7 Poly(imide) synthesis.

poly(imide)s can be prepared in a very short time, under microwave irradiation and using water as reaction medium. A series of high molecular weight poly(imide)s based on 4,4′-oxydianiline and 4,4′-(4,4′-isopropylidenediphenoxy)bis(phthalic anhydride) were synthesized (Scheme 6.7).

Different experimental parameters (monomer concentration, reaction time, microwave irradiation power, preliminary dianhydride hydrolysis) were investigated to optimize the polymerization conditions and to obtain high molecular weight polymers.[60] The best results were obtained with a one-pot, two-step reaction, consisting of first the hydrolysis of the dianhydride before adding the diamine. Compared to classical conventional procedures, extremely short polymerization reaction times were required under microwave irradiation. Moreover, the properties (T_g, solubility) of the resulting polymers were comparable to the analogous polymers produced by conventional thermal polymerization.

In contrast, Dao and co-workers[61] have used a high-throughput methodology to synthesize a series of copolyimides in water under microwave irradiation. The obtained polymers have high molecular weights and high glass transition temperatures (T_gs), while maintaining good solution processability.

6.3.3 Synthesis of Polytriazoles

The synthesis of a novel cyclodextrin-containing polymer by microwave-assisted reaction of 1,3-bis(azidomethyl)benzene with propargyl β-cyclodextrin has been described recently.[62] This cycloaddition reaction, achieved in presence of copper catalyst, is commonly referred to as "click chemistry." In the present case, the reaction was performed under microwave irradiation in a tBuOH–H_2O (1:1) mixture (Scheme 6.8).

6.3.4 C–C Coupling Polymerizations

Transition metal catalyzed coupling polymerizations have become powerful tools for the synthesis of semiconducting polymers such as poly(para-phenylene)s, poly(thiophene)s, and related polymers. The Suzuki cross-coupling reaction is one of the reactions used extensively. Recently, Alesi[63] *et al.* have investigated the synthesis of oligothiophenes *via* the Suzuki reaction under microwave irradiation in aqueous ethanol (1:1) medium. From a model reaction involving 5-bromo-2-thiophenecarboxaldehyde and 2-thienylboronic acid they have

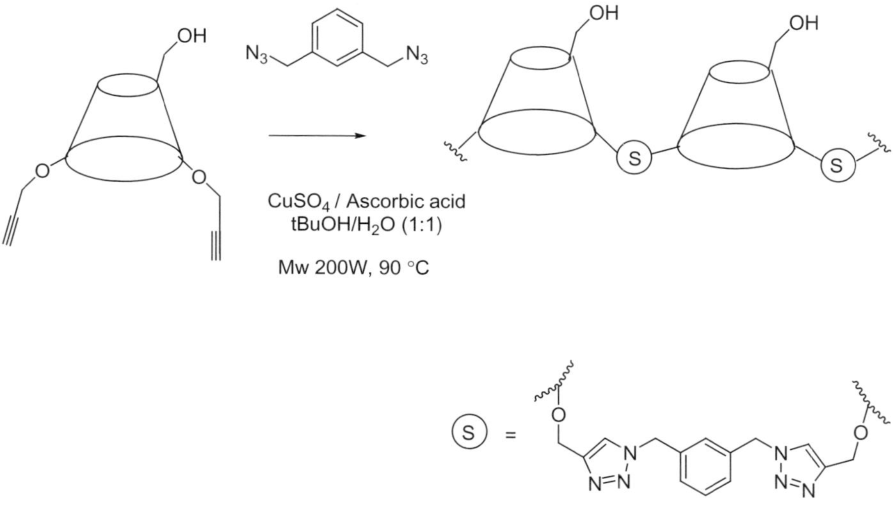

Scheme 6.8 Synthesis of a new cyclodextrin-containing polymer *via* the click chemistry reaction.

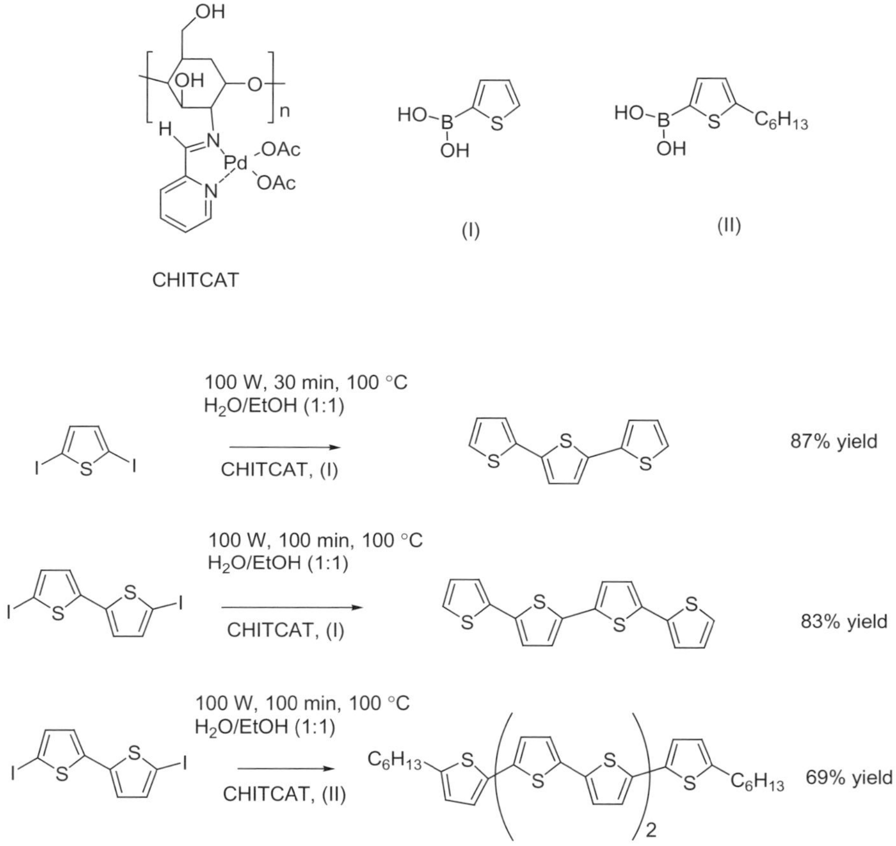

Scheme 6.9 Synthesis of oligothiophene semiconductors by Suzuki coupling.

shown that the use of chitosan palladium complex (CHITCAT) and potassium fluoride (KF) as a base affords a complete conversion of reagents into the expected product after only 2-min microwave irradiation, while only 55% conversion is obtained after 48 h reaction under a conventional heating process. Performing the reaction in aqueous ethanol solution was found to be more efficient as the KF solubility is better than in toluene, which is commonly used in Suzuki reactions. A series of oligothiophenes containing up to six units were synthesized with a high purity (Scheme 6.9).

Recently, Hitachi has developed and patented a very convenient procedure for the synthesis of conjugated polymers through a Suzuki coupling reaction.[64] The reaction was performed in a toluene–water mixture, in the presence of tricaprylmethylammonium chloride (Aliquat), potassium carbonate (K_2CO_3), and palladium catalyst (Scheme 6.10). Whereas very long reaction times (>10 h) were necessary with a conventional thermal procedure, polymers were

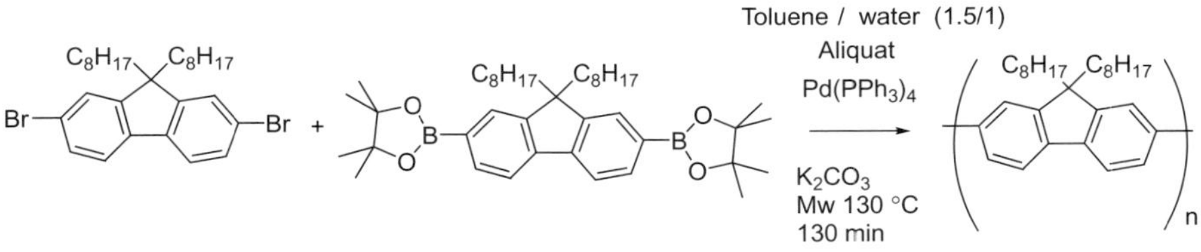

Scheme 6.10 Synthesis of conjugated polymers through Suzuki coupling.

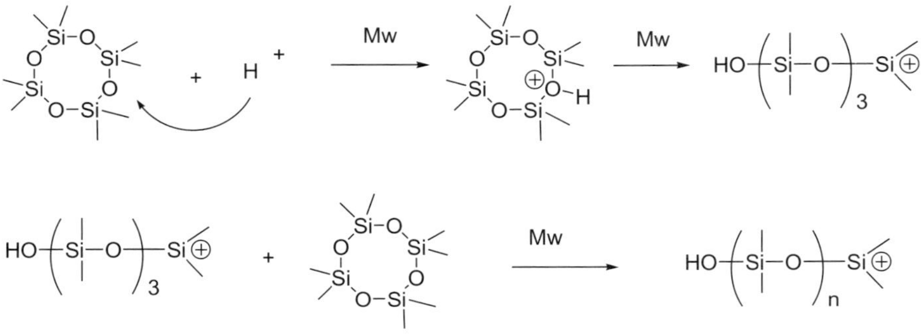

Scheme 6.11 Cationic polymerization of octamethylcyclotetrasiloxane.

isolated after 2 h 10 min reaction under microwave irradiation. This high-throughput technique appears very promising for the development of new conjugated polymers as electroluminescent materials

6.4 Miscellaneous Polymer Synthesis in Aqueous Media

6.4.1 Cationic Polymerization

Recently, poly(siloxane) (PSI) nanoparticles have been synthesized by cationic ring-opening polymerization under emulsion conditions with microwave heating (Scheme 6.11).[65]

The particle sizes of the latex produced under microwave irradiation are substantially smaller than those obtained by a conventional heating. Moreover, the authors observed a significant enhancement of the polymerization rate under microwave irradiation.

6.4.2 Polymer Modifications

The synthesis of new macromolecular structures from the chemical modification of polymer precursors provides an easy way to either tailor the properties of existing polymers or to access new polymer structures. In this connection,

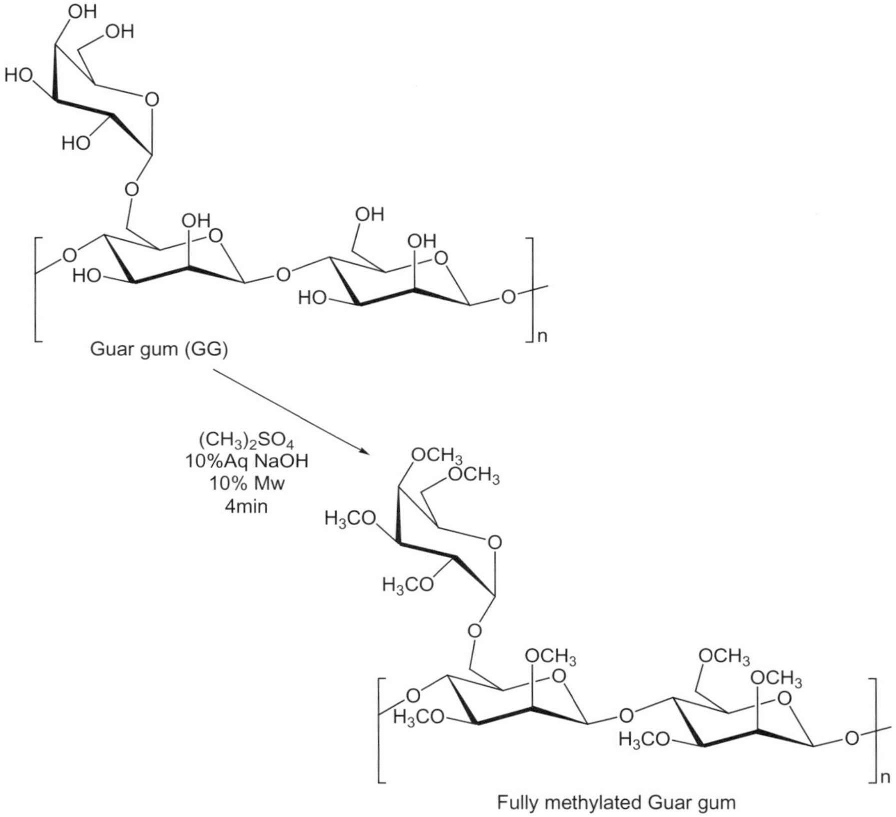

Scheme 6.12 Full methylation of guar gum.

polysaccharides have been extensively used as scaffolds for different post-modification reactions.

6.4.2.1 Complete O-Methylation of Polysaccharides

The O-methylation reaction is of great importance for obtaining structural information on various polysaccharides. However, conventional heating process cannot allow a complete modification. Singh and co-workers[66] have reported the full methylation of guar gum with dimethyl sulfate in basic aqueous medium, within 4 min, using microwave heating (Scheme 6.12).

This new microwave-assisted method opens up new perspectives for the structural identification of unknown polysaccharides.

6.4.2.2 Cationization of Polysaccharides

The cationization of polysaccharides can be performed with different reagents. Depending on the reactant used [3-chloro-2-hydroxypropyl) trimethylammonium chloride (CHPTMAC) or 2-(diethylamino)ethyl chloride hydrochloride (DEAEC)], exclusive quaternary ammonium or a mixture of tertiary and quaternary ammonium groups can be obtained (Scheme 6.13).

The synthesis of cationic polysaccharides using CHPTMAC or DEAEC usually requires long reaction times, in alkaline medium. Koroskenyi and co-workers[67] have achieved the amination of starch in aqueous medium under microwaves within 3 min reaction. The interest in using microwave heating was proven by the higher degree of substitution (DS) reached (Table 6.1).

Concerning chitosan, the amination reaction involves the amine groups of the polymer chain (Scheme 6.14).

In this case, the microwave heating process was, however, less effective. This result was attributed to the different swelling behavior of starch and chitosan in alkaline medium.

6.4.2.3 Post-modification by Suzuki Coupling

Li and co-workers[68] have prepared biaryl α-ketophosphonic acids (Scheme 6.15) by aqueous microwave-assisted Suzuki coupling on a solid support. A polymer having an aryl boronic acid group was used.

In the same manner, Blettner and co-workers[69] have realized the cross-coupling of arylboronic acids with an ester triflate-*para*-substituted benzoate end group of a poly(ethylene glycol). The reaction was performed in water, in the presence of palladium acetate. In this reaction, poly(ethylene glycol) was considered both as a soluble solid support and phase-transfer catalyst (Scheme 6.16).

Certainly as a consequence of the shorter reaction time under microwave-irradiation, no side reaction regarding the hydrolysis of the ester triflate (which can attain up to 45% yield under thermal reaction conditions) was observed.

6.4.2.4 Post-modification by Click Chemistry

The 1,3-Huisgen's dipolar cycloaddition ("click reaction") is a well-known method enabling the synthesis of complex architectures in quantitative yields. Further interest in this cycloaddition reaction is the possibility of using aqueous media. In this regard, a microwave-assisted click reaction opens up wide perspectives for the rapid synthesis of new functional polymers.

Yoon *et al.*[70] have investigated the monofunctionalization of dendrimers using click chemistry (Scheme 6.17). The reaction was performed in the presence of sodium ascorbate and copper sulfate hydrate ($CuSO_4 \cdot 5H_2O$) in a *t*-butanol–water mixture (1:1). At 100 W irradiation power and with a shut-off temperature fixed at 100 °C, the reaction proceeds in 10 min with a very high yield. Similarly, Rijkers *et al.*[71] have efficiently attached azido peptides onto dendritic alkynes to obtain multivalent dendritic peptides. By performing the reaction at 100 °C in a THF–H_2O (1:1) mixture or in aqueous DMF, quasi-quantitative grafting yields were obtained, after 10 min microwave irradiation.

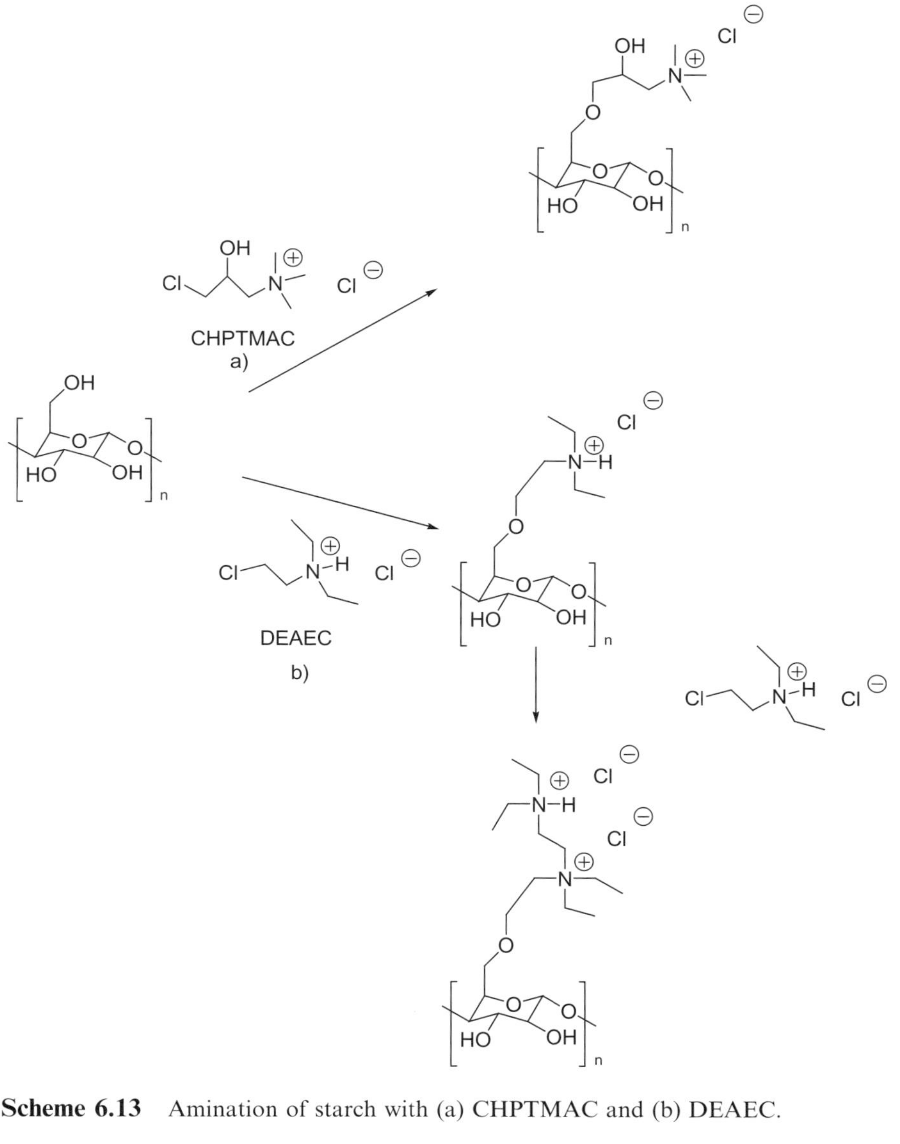

Scheme 6.13 Amination of starch with (a) CHPTMAC and (b) DEAEC.

Table 6.1 Degree of substitution (DS) for starch amination with conventional and microwave heating procedures.

Amination agent	DS for conventional heating (2 h, 70 °C)	DS for microwave-assisted procedure (3 min, 260 W)
CHPTMAC	0.83	1.54
DEAEC	0.64	1.24

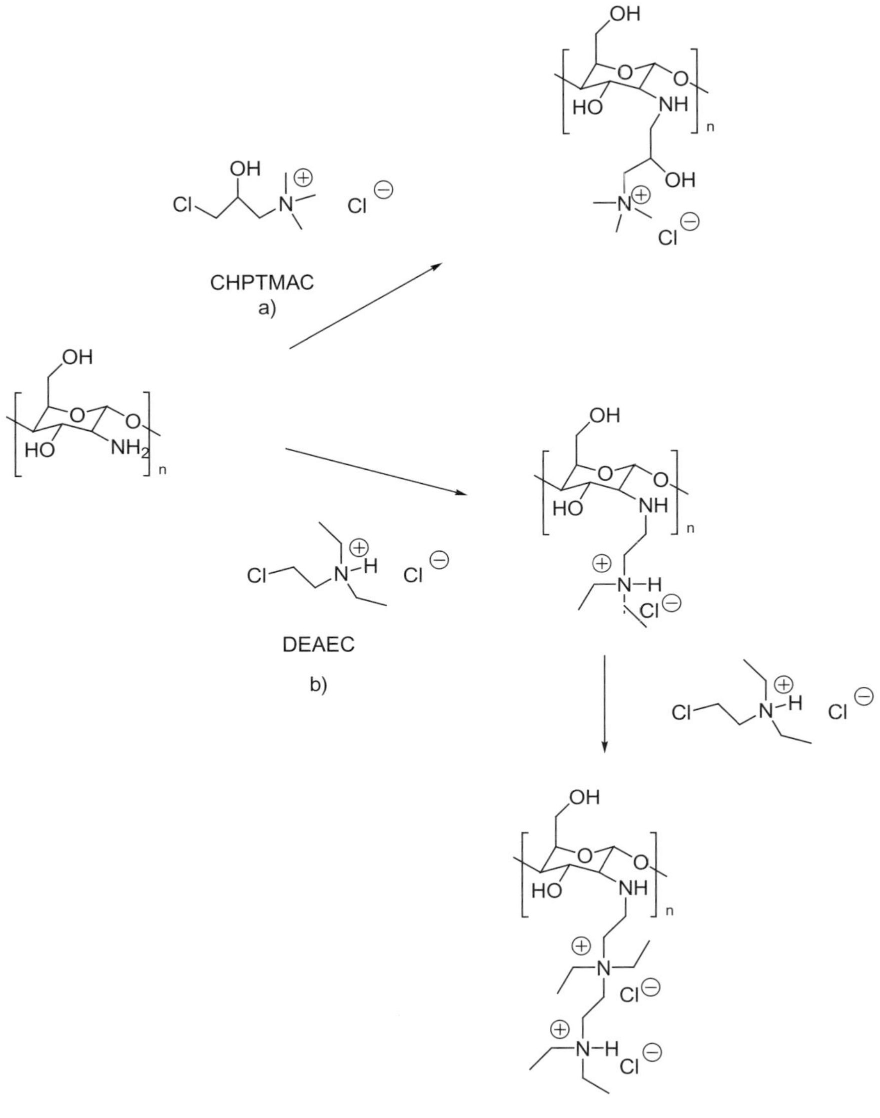

Scheme 6.14 Amination of chitosan with (a) CHPTMAC and (b) DEAEC.

Such results prove the interest of microwave heating given that only 43–56% yields were obtained after 16 h reaction under conventional conditions.

Miller and co-workers[72] have described an efficient procedure to covalently graft an alkynylated sugar onto an azide functionalized peptide (Scheme 6.18).

When performed under microwave irradiation, a full conversion was reached in less than 10 min reaction. Multiple labeled carbohydrate oligonucleotides

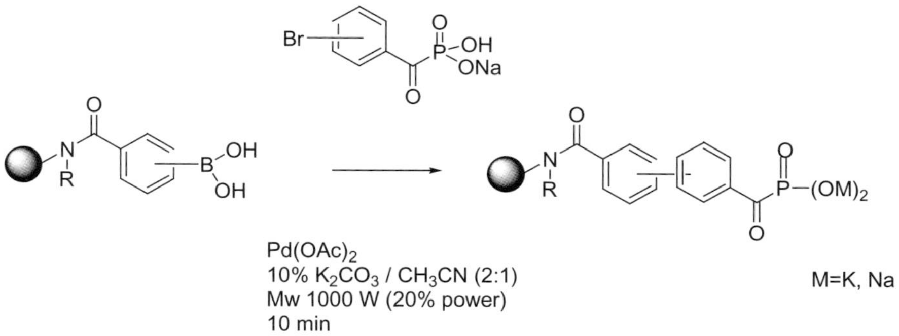

Scheme 6.15 Suzuki coupling reaction on a polymer-bound solid support.

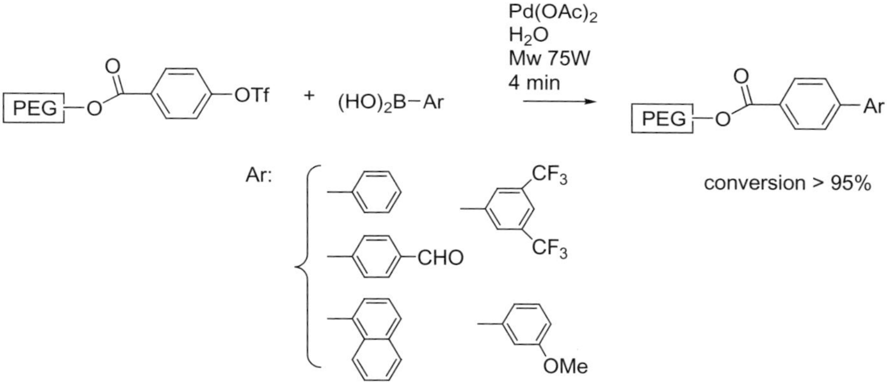

Scheme 6.16 End-functionalization of poly(ethylene glycol) precursors.

have been synthesized following a similar approach.[73] The reaction was performed between a trivalent alkyne oligonucleotide and a monovalent azide, in a methanol–water (1:1) mixture.

6.4.2.5 Post-modification by Thioesterification

A novel thioesterification reaction involving 3-mercaptopropionic acid (MPA) has been used to synthesize a glycosylated peptide thioester.[74] The method was validated on a model experiment: in 40% aqueous MPA, 30 min irradiation (microwave, 150 W, 80 °C) was sufficient for an almost complete reaction. If only 20% and 10% MPA were used, 1 h was necessary. In contrast, a 70% yield was reached after 7 days (168 h) by a conventional thermal procedure (at room

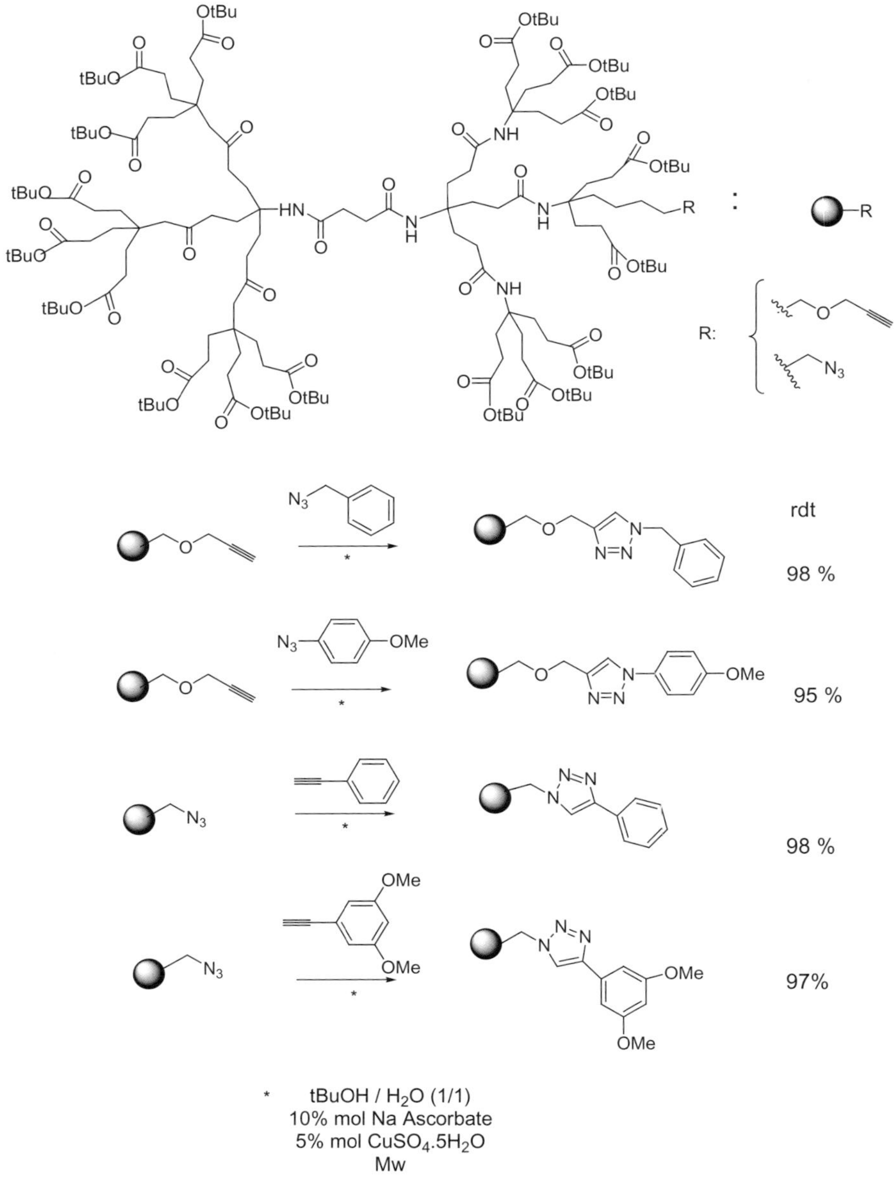

Scheme 6.17 Monofunctionalization of dendrimers by click chemistry.

temperature). Based on these results, an *N*-acetylglucosaminylated peptide thioester has been prepared (Scheme 6.19). In this case, isolated yields of the desired product were only 4.2% and 3.7% (with, respectively, 20% and 10% MPA). However, these results are twice the values reported in the literature.

Scheme 6.18 Functionalization of an azide-containing peptide by click chemistry.

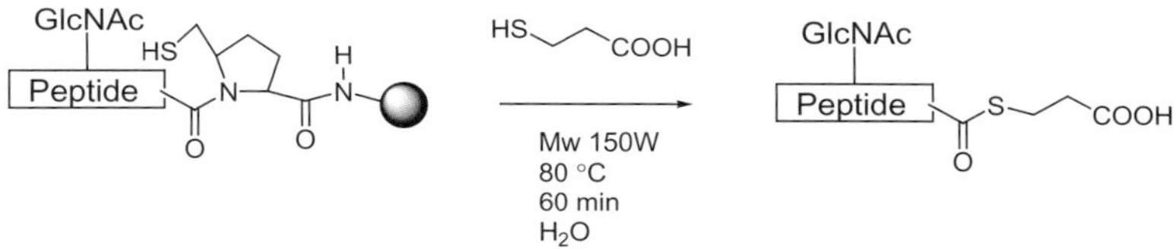

Scheme 6.19 Preparation of a glycosylated peptide thioester.

6.4.2.6 Polyvinylpyrrolidone Grafting

Agar-*graft*-PVP and κ-carrageenan-*graft*-PVP have been synthesized by direct grafting of polyvinylpyrrolidone (PVP) onto the respective polysaccharides.[75] The grafting reaction was performed under microwave irradiation, in the presence of potassium persulfate, leading to the formation of hydrogels.

6.4.2.7 Grafting Polyaniline onto Gum Acacia

Tiwari *et al.*[76] have reported the synthesis of gum acacia-*graft*-polyaniline in the presence of catalytic amounts of ammonium peroxodisulfate, (Scheme 6.20).

In this process the grafted polymer formation resulted from the reaction of the growing polyaniline chains with the gum acacia macroradicals. Depending on their grafting ratio, the graft copolymers present pH switching behavior.

6.4.3 Solid-phase Peptide Synthesis in Water

Galanis and co-workers have reported recently a solid-phase peptide synthesis, in water, under microwave irradiation.[77] Leu-enkephalin was synthesized successfully in high yields and purity, without racemization.

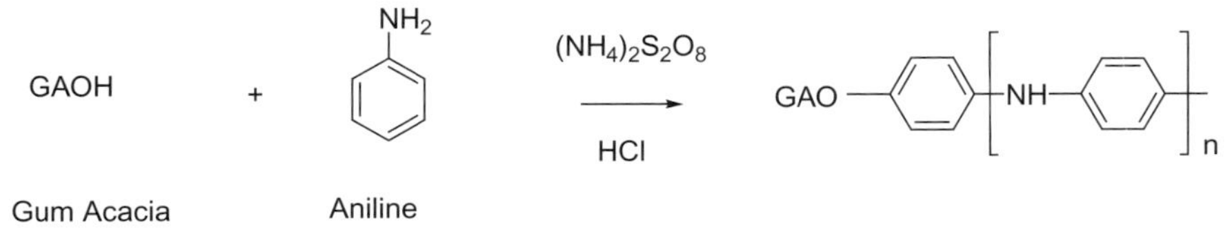

GAOH + [Aniline structure with NH₂] → (NH₄)₂S₂O₈ / HCl → GAO—[⌬—NH—⌬]ₙ

Gum Acacia   Aniline

Scheme 6.20 Grafting polyaniline onto gum acacia.

6.5 Experimental

This section describes a few experimental procedures for the microwave-assisted synthesis of polymers in aqueous medium. Specific attention has been given to choosing examples that use dedicated microwave devices and to representing different types of polymerization.

6.5.1 PMMA Grafting onto a Polysaccharide[9]

This reaction was performed in a Milestone Start S apparatus. In an open vessel, polysaccharide extracted from *Chaetomorpha antennina* (CMsps) (1 g) was dissolved in distilled water (100 mL). MMA [CMsps:MMA (1 : 6.2 molar ratio)] and potassium persulfate (0.00037 mol L^{-1}) were added. The reaction mixture was then submitted to microwave irradiation. The temperature was set at 100 °C for 2.5 min (the set temperature was reached within 1 min and maintained for an additional 1.5 min). After cooling, the reaction mixture was precipitated in isopropyl alcohol and the copolymer isolated by centrifugation. The homopolymer formed was obtained by Soxhlet extraction with toluene, leading to a 91% yield of graft copolymer.

6.5.2 PMMA Synthesis in Emulsion Polymerization[33]

This reaction was performed in an Anton Paar device, equipped with two temperature probes. Some 16 mL of a mixture composed of 80 wt% of water, 0.08 wt% of KPS, 0.14 wt% of Disponil FES32 [(CH$_3$-(CH$_2$)$_{11}$-(OCH$_2$-CH$_2$)$_4$-OSO$_3$Na] and 0.5 wt% of Disponil A3065 (CH$_3$-(CH$_2$)$_{11}$-(OCH$_2$-CH$_2$)$_{19}$-OH) was placed in a vial and purged with nitrogen. MMA (19.04 wt%) was then added. The vials were closed with screw caps and submitted to microwave irradiation. Two different procedures were used:

1. Microwave-assisted polymerization at constant temperature: the vials were heated from room temperature to 80 °C and kept at this temperature. The sample is then quenched in an ice bath.
2. Microwave-assisted polymerization with pulsed irradiations at constant power: the sample was heated from room temperature to 80 °C through a 27 s microwave irradiation at 1400 W and then immediately cooled in an ice bath for 4 min. This cycle is repeated one to six times.

In both cases, polymerization was stopped by adding four drops of a hydroquinone solution.

6.5.3 Synthesis of PMMA Nanoparticles[40]

This reaction was performed with a Biotage Initiator Eight apparatus. Potassium persulfate (0.01 g, 37 µmol), deionized water nitrogen pre-purged (4 mL), and MMA (0.05 g, 0.5 mmol) were introduced in a vial that was then sealed. The reaction mixture was irradiated with microwaves (23 ± 2 W) at 70 °C for 1 h.

6.5.4 Synthesis of Polyacrylamide–Calcium Phosphate Nanocomposites[49]

This reaction was performed in a CEM Discover apparatus. $CaCl_2 \cdot 2H_2O$ (0.458 g), $(NH_4)_2S_2O_8$ (0.025 g), and acrylamide (2.708 g) were dissolved in deionized water (40 mL) and maintained under magnetic stirring. A 0.192 M $(NH_4)_2HPO_4$ aqueous solution (10 mL) was then added. To maintain the solution as pH 10, $NH_3 \cdot H_2O$ was added. The reaction mixture was heated to 80 °C and maintained at this temperature for 45 min by microwave irradiation.

6.5.5 Nitroxide-assisted Synthesis of Styrene in Miniemulsion[52]

This reaction was performed in a CEM Discover apparatus. In a 10-mL pressure reaction vessel, styrene (0.5 g, 4.8 mmol), hexadecane (0.09 g, 18 wt% with respect to styrene), and hydroxyTEMPO (0.002 g, 0.0116 mmol) were mixed in water (2 g) with sodium dodecyl sulfate (0.0172 g), and KPS (1.87 mg, 0.0069 mmol). The reaction mixture was stirred vigorously for 5 min and subjected to ultrasonication (3 min, 400 W, Xinzhi JY98-3D). The reaction was degassed under vacuum/argon cycles and sealed. The reaction mixture was then submitted to microwaves (130 W) in conjunction with air-cooling to keep the temperature at 135 °C. After the reaction, the polymerization was quenched by rapidly cooling the reaction vessel in a cold bath.

6.5.6 Synthesis of Poly(ether)s[56]

This reaction was performed in a PROLABO Synthewave apparatus. An aromatic diphenol (4.5 mmol) and 3,3'-bis(chloromethyl)oxetane (4.5 mmol) were mixed in a 100 mL dedicated reactor with 40% aqueous NaOH (25 mL). Tetrabutylammonium bromide (1 mmol) and nitrobenzene (25 mL) were then added. The reaction mixture was submitted to microwave irradiation, at 100 °C and for 20 min. After cooling, the reaction mixture was poured into methanol. The oligomers were collected by filtration.

6.5.7 Synthesis of Poly(imide)s[61]

This reaction was performed in a BIOTAGE apparatus. A 0.01 M solution of hydrolyzed 4,4'-(hexafluoroisopropylidene) diphthalic anhydride (6FDA)

(10 mL) and an equimolar amount of 4,4′-oxydianiline (4,4′-ODA) were added to a 20-mL reactor equipped with a magnetic stirrer bar. The vial was sealed and submitted to microwave irradiation at 200 °C and for 60 min. A similar procedure was used to synthesize a 4,4′-ODA/6FDA/PMDA (2:1:1) copolymer (45 min microwave irradiation, 190 °C).

6.5.8 Polymerization by C–C Coupling[64]

This reaction was performed in a MILESTONE Microsynth apparatus. 2,7-Dibromo-9,9-dioctylfluorene (0.4 mmol), a 3 vol.% anisole solution of tricaprylmethylammonium chloride (Aliquat) (8 mL), an anisole solution of $Pd(PPh_3)_4$ (0.4 mmol), and a 2 M aqueous solution of K_2CO_3 (5.3 mL) were added to a dedicated high-pressure Teflon. The reaction mixture was submitted to microwave irradiation (300 W power), with the following temperature program: 10 min from room temperature to 90 °C and 120 min at 90 °C. The reaction mixture was then poured into methanol–water (9:1) (150 mL) and the polymer isolated by filtration.

6.5.9 Cationic Polymerization[65]

This reaction was performed in a MAS-I model equipped with an online IR temperature sensor and compressed air for cooling. Octamethylcyclotetrasiloxane (30 g, 0.101 mol), water (150 g, 8.33 mol), sodium dodecyl sulfate (2 g, 5.74 mmol), mono-octylphenyl ether (2 g, 3.10 mmol), hexamethyldisiloxane (0.2 g, 1.23 mmol), and dodecylbenzene sulfonic acid (3 g, 9.19 mmol) were added to a round-bottom flask equipped with a nitrogen inlet, a condenser, and a mechanical stirrer. The reaction mixture was pre-emulsified under vigorous stirring before polymerization and irradiated with microwaves (700 W) at 85 °C. After cooling, the polymerization was quenched by adding a 20 wt% NaOH solution.

6.5.10 Polymer Modification by Click Chemistry[70]

The reaction was performed in a CEM Discover apparatus. Acetylenyl dendrimer (20 mg, 0.0067 mmol) and benzyl azide (0.89 mg, 0.0067 mmol) were suspended in t-BuOH–H_2O (1:1) in the presence of 10 mol% sodium ascorbate and 5% of $CuSO_4 \tilde{n} 5H_2O$. The reactor was then sealed and the reaction mixture submitted to microwave irradiation. The experimental parameters are as follows: power-time control method, 100 W, 10 min, $P_{max} = 65$ psi, $T_{max} = 100$ °C. The desired compound was obtained as a yellowish solid (20.2 mg).

References

1. M. T. Radoiu, D. I. Martin, I. Calinescu and H. Iovu, *J. Hazard. Mater.*, 2004, **106**, 19.

2. K. Xu, W. D. Zhang, Y. M. Yue and P. X. Wang, *J. Appl. Polym. Sci.*, 2005, **98**, 1050.
3. T. Zheng, P. Wang, Z. Zhang and B. Zhao, *J. Appl. Polym. Sci.*, 2005, **95**, 264.
4. V. Singh, A. Tiwari, D. N. Tripathi and R. Sanghi, *Polymer*, 2006, **47**, 254.
5. B. S. Kaith, R. Jindal, A. K. Jana and M. Maiti, *Carbohydr. Polym.*, 2009, **78**, 987.
6. H. Matahwa, V. Ramiah, W. L. Jarrett, J. B. McLeary and R. D. Sanderson, *Macromol. Symp.*, 2007, **255**, 50.
7. B. S. Kaith and S. Kalia, *Polym. Compos.*, 2008, **29**, 791.
8. H. Ge, P. Wan and D. Luo, *Carbohydr. Polym.*, 2006, **66**, 372.
9. G. Prasad, K. Prasad, R. Meena and A. K. Siddhanta, *J. Mater. Sci.*, 2009, **44**, 4062.
10. J. Zhang, S. Zhang, K. Yuan and Y. Wang, *J. Macromol. Sci., Part A: Pure Appl. Chem.*, 2007, **44**, 881.
11. V. Singh, A. Tiwari, S. Pandey and S. K. Singh, *eXPRESS Polym. Lett.*, 2007, **1**, 51.
12. V. Singh, A. Tiwari, S. Pandey and S. K. Singh, *Starch/Staerke*, 2006, **58**, 536.
13. R. Meena, K. Prasad, G. Mehta and A. K. Siddhanta, *J. Appl. Polym. Sci.*, 2006, **102**, 5144.
14. K. Prasad, R. Meena and A. K. Siddhanta, *J. Appl. Polym. Sci.*, 2006, **101**, 161.
15. W. Duan, C. Chen, L. Jiang and G. H. Li, *Carbohydr. Polym.*, 2008, **73**, 582.
16. V. Singh, D. N. Tripathi, A. Tiwari and R. Sanghi, *Carbohydr. Polym.*, 2006, **65**, 35.
17. V. Singh, D. N. Tripathi, A. Tiwari and R. Sanghi, *J. Appl. Polym. Sci.*, 2005, **95**, 820.
18. V. Singh, A. Tiwari, D. N. Tripathi and R. Sanghi, *Carbohydr. Polym.*, 2004, **58**, 1.
19. V. Singh and D. N. Tripathi, *J. Appl. Polym. Sci.*, 2006, **101**, 2384.
20. V. Singh, A. Tiwari, D. N. Tripathi and R. Sanghi, *J. Appl. Polym. Sci.*, 2004, **92**, 1569.
21. G. Sen, R. Kumar, S. Ghosh and S. Pal, *Carbohydr. Polym.*, 2009, **77**, 822.
22. G. Sen and S. Pal, *Int. J. Biol. Macromol.*, 2009, **45**, 48.
23. A. Mishra, J. H. Clark, A. Vij and S. Daswal, *Polym. Adv. Technol.*, 2008, **19**, 99.
24. A. Kumar, K. Singh and M. Ahuja, *Carbohydr. Polym.*, 2009, **76**, 261.
25. M. Aldana-Garcia, J. Palacios and E. Vivaldo-Lima, *J. Macromol. Sci., Part A: Pure Appl. Chem.*, 2005, **42**, 1207.
26. J. Gao and C. Wu, *Langmuir*, 2005, **21**, 782.
27. Z.-S. Xu, Z.-W. Deng, X.-X. Hu, L. Li and C.-F. Yi, *J. Polym. Sci., Part A: Polym. Chem.*, 2005, **43**, 2368.
28. C. Holtze, M. Antonietti and K. Tauer, *Macromolecules*, 2006, **39**, 5720.
29. C. Holtze and K. Tauer, *Macromol. Rapid Commun.*, 2007, **28**, 428.

30. K. Tauer, M. Mukhamedjanova, C. Holtze, P. Nazaran and J. Lee, *Macromol. Symp.*, 2007, **248**, 227.
31. R. Correa, G. Gonzalez and V. Dougar, *Polymer*, 1998, **39**, 1471.
32. J. Sierra, J. Palacios and E. Vivaldo-Lima, *J. Macromol. Sci., Part A: Pure Appl. Chem.*, 2006, **43**, 589.
33. C. Costa, A. F. Santos, M. Fortuny, P. H. H. Araujo and C. Sayer, *Mater. Sci. Eng., C*, 2009, **29**, 415.
34. H. M. Jung, Y. Yoo, Y. S. Kim and J. H. Lee, *Macromol. Symp.*, 2007, **249/250**, 521.
35. M. Murray, D. Charlesworth, L. Swires, P. Riby, J. Cook, B. Z. Chowdhry and M. J. Snowden, *J. Chem. Soc., Faraday Trans.*, 1994, **90**, 1999.
36. W. Zhang, J. Gao and C. Wu, *Macromolecules*, 1997, **30**, 6388.
37. J. Bao and A. Zhang, *J. Appl. Polym. Sci.*, 2004, **93**, 2815.
38. T. Ngai and C. Wu, *Langmuir*, 2005, **21**, 8520.
39. C. Yi, Z. Deng and Z. Xu, *Colloid Polym. Sci.*, 2005, **283**, 1259.
40. Z. An, W. Tang, C. J. Hawker and G. D. Stucky, *J. Am. Chem. Soc.*, 2006, **128**, 15054.
41. C. J. Hawker, G. D. Stucky and Z. An, One-step microwave preparation of well-defined and functionalized polymeric nanoparticles, US Patent 2008009558A1, 2008.
42. W.-D. He, C.-Y. Pan and T. Lu, *J. Appl. Polym. Sci.*, 2001, **80**, 2455.
43. X. Zhu, J. Chen, N. Zhou, Z. Cheng and J. Lu, *Eur. Polym. J.*, 2003, **39**, 1187.
44. X. Zhu, J. Chen, Z. Cheng, J. Lu and J. Zhu, *J. Appl. Polym. Sci.*, 2003, **89**, 28.
45. J. Palacios and C. Valverde, *New Polym. Mater.*, 1996, **5**, 93.
46. H. L. Luo, J. Sheng and Y. Z. Wan, *Mater. Lett.*, 2008, **62**, 37.
47. H. Zhao, H. Chen, Z. Li, W. Su and Q. Zhang, *Eur. Polym. J.*, 2006, **42**, 2192.
48. J. Huang, P. Hui, Z. Xu and C. Yi, *React. Funct. Polym.*, 2008, **68**, 332.
49. Q.-L. Tang, K.-W. Wang, Y.-J. Zhu and F. Chen, *Mater. Lett.*, 2009, **63**, 1332.
50. H.-X. Wu, X.-Q. Qiu, W.-M. Cao, Y.-H. Lin, R.-F. Cai and S.-X. Qian, *Carbon*, 2007, **45**, 2866.
51. J. Hu, H. Zhao, Q. Zhang and W. He, *J. Appl. Polym. Sci.*, 2003, **89**, 1124.
52. J. Li, X. Zhu, J. Zhu and Z. Cheng, *Radiat. Phys. Chem.*, 2006, **76**, 23.
53. Z. Xu, X. Hu, X. Li and C. Yi, *J. Polym. Sci., Part A: Polym. Chem.*, 2007, **46**, 481.
54. Z. An, Q. Shi, W. Tang, C.-K. Tsung, C. J. Hawker and G. D. Stucky, *J. Am. Chem. Soc.*, 2007, **129**, 14493.
55. N. Hurduc, D. Abdelylah, J.-M. Buisine, P. Decock and G. Surpateanu, *Eur. Polym. J.*, 1997, **33**, 187.
56. V. Baudel, F. Cazier, P. Woisel and G. Surpateanu, *Eur. Polym. J.*, 2002, **38**, 615.
57. Y.-q. Yu, Z.-z. Li, H.-j. Tian, S.-s. Zhang and P.-k. Ouyang, *Colloid Polym. Sci.*, 2007, **285**, 1553.

58. J.-l. Huang, Y.-l. Zhang, Z.-h. Cheng and H.-c. Tao, *J. Appl. Polym. Sci.*, 2007, **103**, 358.
59. Y. Imai, *A new facile and rapid synthesis of polyamides and polyimides by microwave-assisted polycondensation*, in Step-Growth Polymers for High-Performance Materials: New Synthetic Materials., ACS Symposium Series, ed. J. L. Hedrick and J. W. Labadie, Vol. **624**, American Chemical Society, Washington DC, 1996, ch. 27, pp. 421–430.
60. R. Brunel, C. Marestin, V. Martin and R. Mercier, *High Performance Polymers*, 2010, **22**, 1, 82–94.
61. B. N. Dao, A. M. Groth and J. H. Hodgkin, *Macromol. Rapid Commun.*, 2007, **28**, 604.
62. A. Binello, B. Robaldo, A. Barge, R. Cavalli and G. Cravotto, *J. Appl. Polym. Sci.*, 2008, **107**, 2549.
63. S. Alesi, F. Di Maria, M. Melucci, D. J. Macquarrie, R. Luque and G. Barbarella, *Green Chem.*, 2008, **10**, 517.
64. Y. Tsuda, Y. Morishita, S. Nomura, Y. Hoshi and S. Funyuu, Process for producing conjugated polymer, US Patent, 2009036623A1, 2009.
65. Y. Bo, H. Hui, C. Riping and J. Demin, e-Polym., 2008, no. 57. http://www.e-polymers.org/journal/papers/hhui_150408.pdf.
66. V. Singh, A. Tiwari, D. N. Tripathi and T. Malviya, *Tetrahedron Lett.*, 2003, **44**, 7295.
67. B. Koroskenyi and S. P. McCarthy, *J. Polym. Environ.*, 2002, **10**, 93.
68. X. Li, A. K. Szardenings, C. P. Holmes, L. Wang, A. Bhandari, L. Shi, M. Navre, L. Jang and J. R. Grove, *Tetrahedron Lett.*, 2005, **47**, 19.
69. C. G. Blettner, W. A. Koenig, W. Stenzel and T. Schotten, *J. Org. Chem.*, 1999, **64**, 3885.
70. K. Yoon, P. Goyal and M. Weck, *Org. Lett.*, 2007, **9**, 2051.
71. D. T. S. Rijkers, G. W. van Esse, R. Merkx, A. J. Brouwer, H. J. F. Jacobs, R. J. Pieters and R. M. J. Liskamp, *Chem. Commun.*, 2005, 4581.
72. N. Miller, G. M. Williams and M. A. Brimble, *Org. Lett.*, 2009, **11**, 2409.
73. C. Bouillon, A. Meyer, S. Vidal, A. Jochum, Y. Chevolot, J.-P. Cloarec, J.-P. Praly, J.-J. Vasseur and F. Morvan, *J. Org. Chem.*, 2006, **71**, 4700.
74. F. Nagaike, Y. Onuma, C. Kanazawa, H. Hojo, A. Ueki, Y. Nakahara and Y. Nakahara, *Org. Lett.*, 2006, **8**, 4465.
75. K. Prasad, G. Mehta, R. Meena and A. K. Siddhanta, *J. Appl. Polym. Sci.*, 2006, **102**, 3654.
76. A. Tiwari and V. Singh, *Carbohydr. Polym.*, 2008, **74**, 427.
77. A. S. Galanis, F. Albericio and M. Grotli, *Org. Lett.*, 2009, **11**, 4488.

CHAPTER 7
Microwave-assisted Synthesis of Nanomaterials in Aqueous Media

BABITA BARUWATI, VIVEK POLSHETTIWAR[a] AND RAJENDER S. VARMA

Sustainable Technology Division, National Risk Management Research Laboratory, U. S. Environmental Protection Agency, 26 W. Martin Luther King Dr., MS 443, Cincinnati, Ohio 45268, USA

7.1 Introduction

Whether it is termed a revolution or simply a continuous evolution, it is clear that development of new materials on a smaller and smaller length scale and the understanding of these materials are at the root of progress in many areas of materials science.[1] This is particularly true in the development of existing bulk materials in the nanometer (10^{-9} m) region for various important applications, including electronic, magnetic, catalytic, optical, environmental remediation, and many more. Nanomaterials are simply defined by their size, which imparts an interdisciplinary character to this field of research. Structures having at least one dimension of nanometer length are defined as nanomaterials. They might be one-dimensional like rods or tubes, two-dimensional like sheets, or three-dimensional like spheres. The interest and effort to develop existing materials in

---

[a]Current address: KAUST Catalysis Center (KCC), King Abdullah University of Science and Technology, Thuwal 23955, Kingdom of Saudi Arabia

---

RSC Green Chemistry No. 7
Aqueous Microwave Assisted Chemistry: Synthesis and Catalysis
Edited by Vivek Polshettiwar and Rajender S. Varma
© Royal Society of Chemistry 2010
Published by the Royal Society of Chemistry, www.rsc.org

the nanometer length scale come from the extraordinary physical properties of these materials compared to their bulk counterparts. These extraordinary properties arise due to the enormous increase in the surface-to-volume (S/V) ratio and the predominance of quantum mechanical phenomena exhibited by materials in this particular length scale. There is a growing curiosity to understand the behavior of materials in the nanometer length scale and to exploit the new properties exhibited by these materials purely as a consequence of the smallness of size. The increase of S/V ratio in nanomaterials can be illustrated by a simple example. Assuming a cube of side 1 cm, the corresponding S/V ratio can be calculated as follows:

$$\text{Area} = 6 \times 1\,\text{cm}^2 = 6 \times 10^{-4}\,\text{m}^2, \text{volume} = 1\,\text{cm}^3 = 10^{-6}\,\text{m}^3, \text{S/V} = 6 \times 100\,\text{m}^{-1}$$

Now, if the sides of the above cube were reduced to 1 mm, the corresponding S/V ratio would be 10^4 times the above number; the S/V ratio was increased by 10 000 times by a simple decrease of size by one order. Similar calculations for a cube with 1 nm sides would result in an exponential increase in the S/V ratio. This provides nanomaterials with an exceptionally increased number of active sites and, as a consequence, improved surface-related properties.

Upon realization that the properties of the materials are based solely on the size and shape, control over the synthetic methodologies becomes an issue. This is because the growth of the materials at the nanoscale is largely dependent on the thermodynamic and kinetic barriers in the reaction, as defined by the reaction trajectory; growth is influenced by vacancies, defects, and surface reconstruction events. Most synthetic methods for producing nanomaterials use conventional heating due to the need for high-temperature-initiated nucleation followed by controlled precursor addition to the reaction. Conventional thermal techniques rely on conduction of blackbody radiation to drive the reaction. The reaction vessel acts as an intermediary for energy transfer from the heating mantle to the solvent and finally to the reactant molecules. This can cause sharp thermal gradients throughout the bulk solution and inefficient, non-uniform reaction conditions. This is a common problem encountered in chemical scale-up and one that is more problematic for nanomaterials, where uniform nucleation and growth rates are critical to material quality.

Microwave (MW) heating methods can address the problem of heating inhomogeneity. It is well known that the interaction of dielectric materials, liquids or solids, with microwaves leads to what is generally known as dielectric heating.[2] The use of microwave irradiation provides increased reaction kinetics and rapid initial heating, which results in enhanced reaction rates. This also provides energy efficiency by reducing the reaction times from hours to minutes when compared with the conventional heating methods, which culminates in clean reaction products and higher yields. By judiciously choosing the solvents, passivating ligands, and reactants, the precursors can be selectively heated preferentially with regard to the solvent or passivating ligand to produce specific nanostructures. As a result, the use of microwave irradiation as an efficient, environmentally friendly, and economically viable heating method for the production of nanomaterials has increased significantly. Even household microwave ovens have been used to form nanoparticles,[3–7] although the crystallinity and optical properties of the resulting materials appear to be lower in these systems.

The dielectric heating effect in microwaves is created by the interaction of the dipole moment of molecules with the high-frequency electromagnetic radiation. Water has a high dipole moment, which makes it one of the best solvents for microwave heating.[8] Based on its high dipole moment, completely benign nature, and cost effectiveness, water is the best solvent for the synthesis of nanoparticles using microwaves.

The present chapter discusses the synthesis of various nanomaterials under microwave irradiation conditions using water as a solvent. Application of these nanomaterials in catalysis is also reviewed.

7.2 Synthesis of Metal Nanoparticles using Water under Microwave Irradiation

In recent years, intensive research has been devoted to the synthesis of metallic nanoparticles[9] because of their potential applications in catalysis,[10,11] photonics, plasmonics,[12,13] information storage,[14] surface-enhanced Raman scattering (SERS),[15–17] and biological labeling, imaging, and sensing.[18–21] Many studies have demonstrated that the intrinsic properties of metal nanostructures can be effectively tailored by controlling their size, shape, composition, crystallinity, and structure.[22–24] Various methods have been developed for the synthesis of these property-tailored nanoparticles, including sodium borohydride ($NaBH_4$) reduction,[25] a polyol process,[26–28] use of plant extracts,[29,30] photoreduction,[31] etc. As most of these methods use either toxic chemicals and solvents or longer periods of conventional heating, microwave-mediated aqueous synthesis of the nanoparticles is a desirable option that provides the opportunity to synthesize the nanoparticles within minutes without creating environmental as well as economic hazards. The following sections discuss the synthesis of different metal nanoparticles using microwave irradiation in aqueous media.

7.2.1 Gold (Au) Nanoparticles

Gold nanoparticles are the most studied metal nanoparticles because of their numerous applications in catalysis, optoelectronics, drug delivery, *etc*. Among the various methods developed to synthesize these metal nanoparticles, microwave-mediated synthesis using water as a solvent is growing as an important method from both economical and environmental points of view. Different morphologies of gold nanoparticles have been reported using this method.

Kundu[32] et al. have reported the synthesis of spherical, polygonal, rod-shaped, and prism-shaped Au nanoparticles by using microwave heating in water (Scheme 7.1). The reduction of Au salt was performed in cetyl (trimethyl)ammonium bromide (CTAB) micellar media in the presence of alkaline 2,7-dihydroxynaphthalene (2,7-DHN) as a new reducing agent. The method generated exclusively nanoparticles with specific morphologies; the size and shape of the nanoparticles could be tailored by changing the ratio of metal salt to surfactant molar ratio, concentration of 2,7-DHN, and the microwave heating time. The authors propose that the radicals or the solvated electrons formed during the microwave heating of a solution mixture containing 2,7-DHN enabled the reduction of Au(III) to Au(0). When the nucleation of Au(0) started, growth occurred in multiple steps and produced different anisotropic particles. The reduction of Au(III) to Au(0) furnished small spherical nuclei. The nuclei subsequently formed Au atoms and assembled into a crystalline particle. The surfactant CTAB molecules slowed down the growth of the crystal faces by becoming attached to the surfaces of the crystals. When the surfactant concentration was lower (10^{-2} M), the interaction of the cationic parts with the specific surface was weak. As a result, the growth of particles occurred in all directions, leading to the formation of spherical particles. With a high surfactant concentration (10^{-1} M or higher), the cationic CTAB in aqueous solution formed polygonal micellar shapes (rod-like and others). This rod-like micelle promoted the formation of rod-shaped nanoparticles. The formation of triangular and prism structures were explained on the basis of formation of trimeric clusters and subsequent nucleation and growth.

Wang et al.[33] have also reported the synthesis of Au nanoplates by microwave irradiation of a mixture of sodium tris-citrate and hydrogen tetrachloroaurate ($HAuCl_4$) (Scheme 7.2). The formation of the nanoplates was discussed

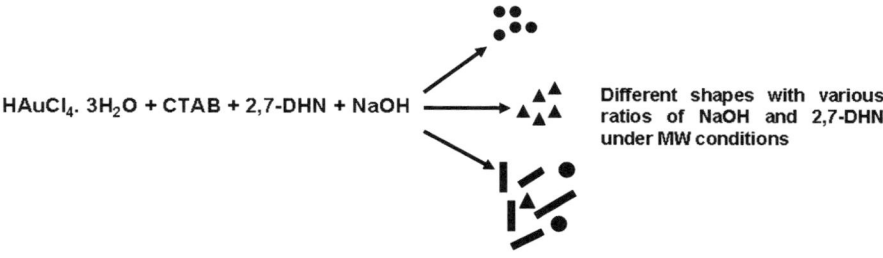

Scheme 7.1

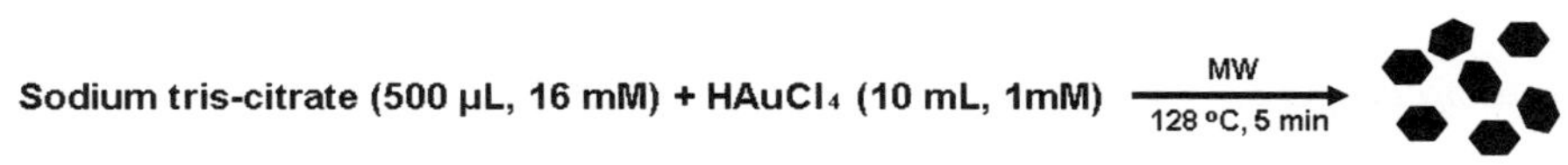

Scheme 7.2

on the basis of the strong kinetic control over the growth of the various faces provided by microwave irradiation, which produced uniform-sized gold nanoplates in the size range of 100 nm. The authors showed that conventional heating was not able to provide the particles with the same morphology even though the same reaction conditions were used.

Liu et al.[34] have synthesized Au nanoparticles and rods using the cationic surfactant tetrabutylammonium bromide (TBAB) in water under microwave irradiation conditions (Scheme 7.3). The study showed the influence of concentration of TBAB and temperature on the formation of Au nanoparticles. With a higher rate of temperature increase, Au nanorods of higher aspect ratios were formed. This may be due to particle nucleation, which is an extremely sensitive function of temperature. It might be possible that any increase in temperature within the particle has an important effect on the dynamics of particle growth, which is responsible for the formation of the novel shape of the particular high aspect ratio Au nanorods. Although the method predominantly formed Au nanorods, formation of Au nanoparticles was also observed, which diminished its value for certain applications.

An expeditious synthetic method for Au nanoparticles has been reported by Fan et al.[35] in which chitosan is used as both a reducing agent and stabilizing agent for the production of Au nanoparticles under microwave heating conditions (Scheme 7.4). The method produced mostly spherical nanoparticles with uniform size but a very small percentage of triangular as well as rod-shaped particles was also seen in the TEM micrographs. Although the protocol appeared very simple and convenient, it required longer exposure time (25 min) to microwave than other methods.

The synthesis of different morphologies of Au nanoparticles using sugar as reducing agent under microwave heating has been reported by Nadagouda and Varma.[36] The authors reported the synthesis of nanoparticles (Figure 7.1), prisms, hexagons, and rods using a wide variety of sugars (D-glucose, sucrose, mannose, etc.) as reducing agent in the presence of poly(vinylpyrrolidone) as a capping agent and microwaves as the rapid heating technique. The

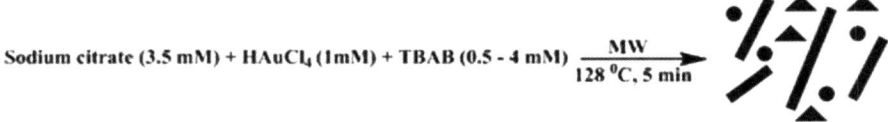

Sodium citrate (3.5 mM) + HAuCl$_4$ (1mM) + TBAB (0.5 - 4 mM) $\xrightarrow[128\ ^0C,\ 5\ min]{MW}$

Scheme 7.3

HAuCl$_4$ (350 μL, 2%) + Chitosan (50 mL, 1 mg/mL, 1% acetic acid) $\xrightarrow[25\ min]{MW}$

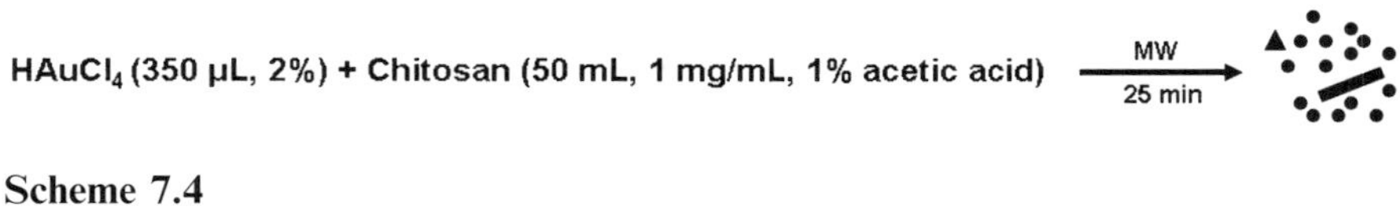

Scheme 7.4

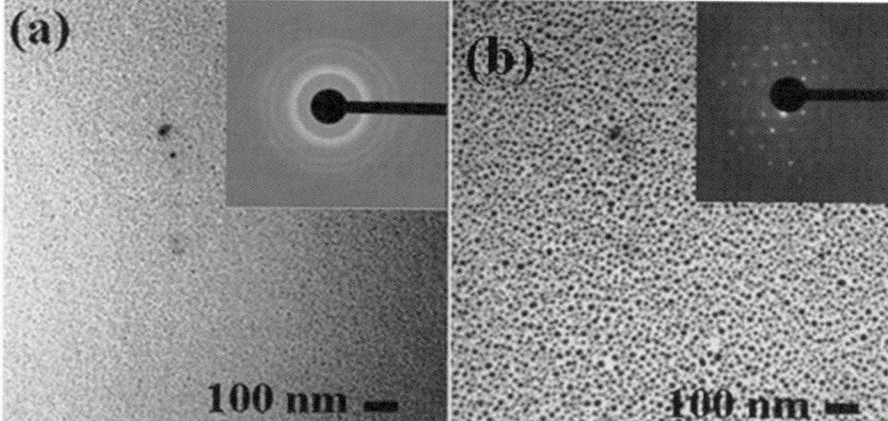

Figure 7.1 Gold nanoparticles synthesized with (a) glucose and (b) sucrose under microwave irradiation conditions.

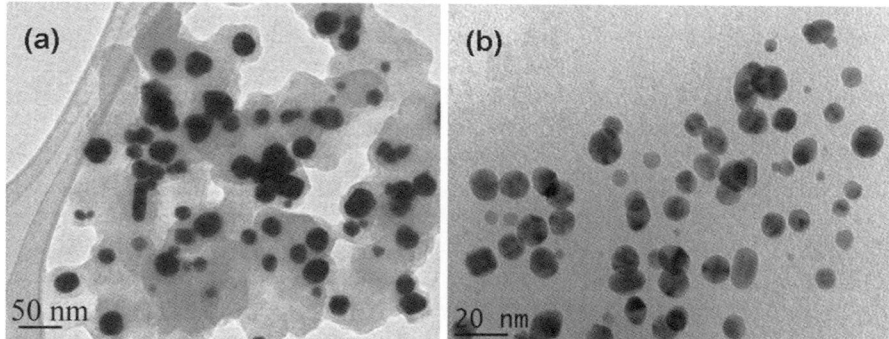

Figure 7.2 Au nanoparticles synthesized using (a) red wine and (b) pomace with a microwave power level of 50 W for 60 s.

nanoparticles produced were in the size range 10–20 nm and were highly dispersed. The reaction time required was less than a minute. The method described the effect of sugar concentration on the size of the ensuing nanoparticles. When low concentrations of sugar were used, hexagonal, triangular, and rod-shaped particles with submicron sizes were obtained.

Baruwati and Varma[37] have reported the synthesis of Au nanoparticles along with other noble metal nanoparticles using red wine and red grape pomace extract as a single source of reducing and capping agent under microwave irradiation conditions. The nanoparticles produced by this method were highly dispersed and in the size range of 5–20 nm (Figure 7.2). The reaction time required in this synthesis procedure was only 45–60 s. It was shown that the particles produced by red grape pomace extract were better dispersed than particles produced by red wine itself. In contrast, the white wine produced highly agglomerated nanoparticles. This variation in dispersity was explained on the basis of the amounts of various polyphenolics present in pomace and red wine that are not in white wine. This method also produced other noble metals like silver (Ag), palladium (Pd), and platinum (Pt) as well as iron (Fe) in the particle size range 5–20 nm.

7.2.2 Silver (Ag) Nanoparticles

The synthesis of Ag nanoparticles has been of much interest to the scientific community because of their numerous applications in catalysis, electronics, photonics, optoelectronics, sensing, and pharmaceuticals. The antibacterial activity of Ag nanoparticles has been proven by various studies and consequently they have been incorporated into many commercial products.

Most importantly, Ag nanomaterials are the most suitable choice for surface-enhanced Raman spectroscopy (SERS) studies.[38,39] Enlightening results have been produced since SERS spectroscopy was discovered in 1974.[40–43] With ever increasing demand, both in basic research and industry, the development of green and sustainable methods for producing these nanomaterials has become an immediate need. Many researchers have developed various methodologies, including the use of microwave as a source of heating and water as the solvent of choice, for rapid and greener production of the Ag nanoparticles.

Hu et al.[44] have realized the production of uniformly-sized Ag nanoparticles in aqueous medium under microwave irradiation using L-arginine or L-lysine as the reducing agent and soluble starch as the stabilizing agent (Scheme 7.5). The particles so-produced were highly dispersed and in the size range 20–30 nm. Experimentation with different amino acids showed that the presence of basic amino acids having two amino groups in each molecule, such as L-lysine or L-arginine, was indispensable for the synthesis of uniformly-sized Ag nanoparticles. The study also showed the formation of uniformly-sized nanoparticles under microwave irradiation conditions compared to the agglomerated products obtained under conventional hydrothermal heating.

Kundu et al.[45] have reported the synthesis of spherical Ag nanoparticles in the size range 20–30 nm with very high dispersity using 2,7-dihydroxynaphthalene (2,7-DHN) in a non-ionic surfactant (TX-100) medium (Scheme 7.6). The method produced nanoparticles and nanochains within a minute by the application of microwaves. Conventional heating produces highly agglomerated particles within 20–30 min.

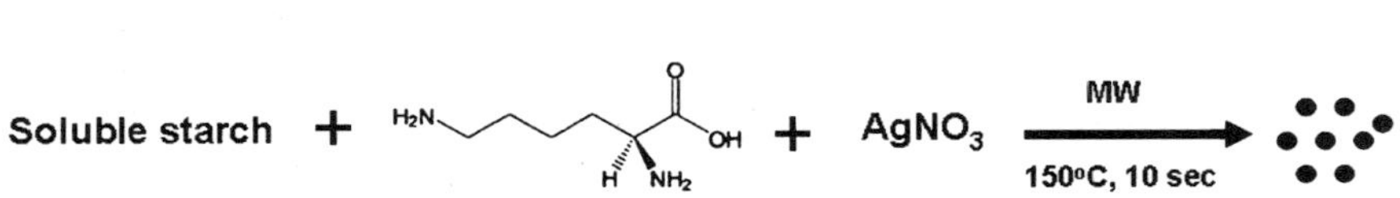

Scheme 7.5

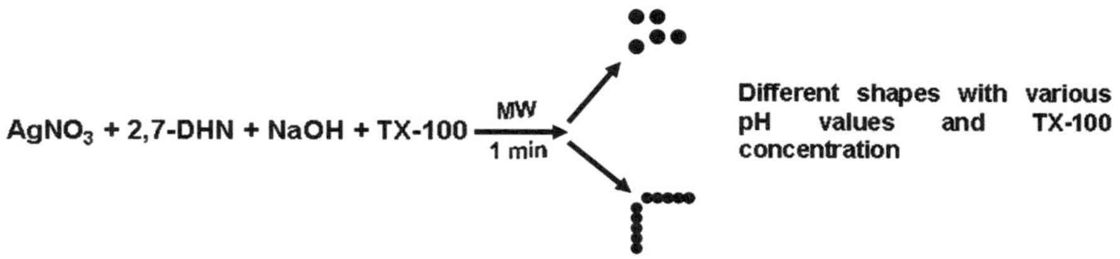

Scheme 7.6

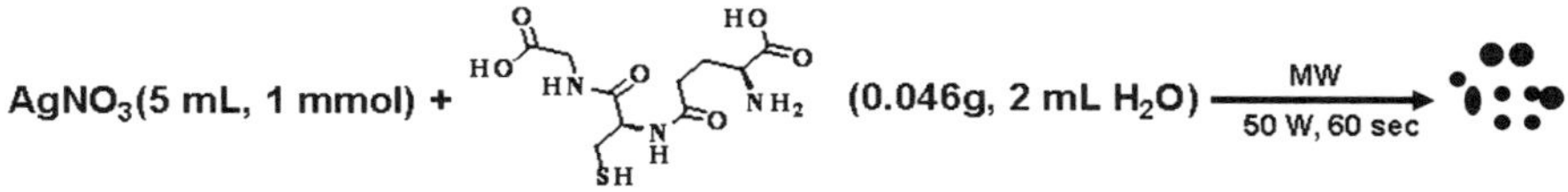

Scheme 7.7

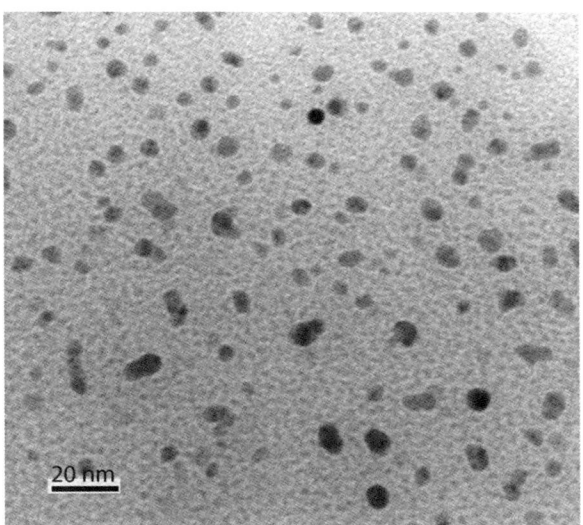

Figure 7.3 TEM micrograph of Ag nanoparticles synthesized at a microwave power level of 50 W for 60 s; the AgNO$_3$ to glutathione mole ratio is 1.0:0.15.

Baruwati et al.[46a] have reported the synthesis of highly dispersed Ag nanoparticles using microwave in water using a completely benign tripeptide, glutathione (Scheme 7.7). The method produced highly crystalline Ag nanoparticles in the size range 5–10 nm within 1 min (Figure 7.3). The study concentrated on the variation of microwave exposure time as well as the ratio of concentration of the starting materials to produce nanoparticles of varying sizes. It was observed that particle size and agglomeration increased with the increase in exposure time. The change in particle morphology was also observed by the variation of precursor concentration.

Baruwati and Varma[37] have reported the synthesis of Ag nanoparticles using red grape pomace extract under microwave irradiation conditions (Scheme 7.8). No other reducing or stabilizing agents were necessary in the method, which produced highly crystalline and nearly monodispersed Ag nanoparticles with a size range of 7–15 nm (Figure 7.4). The reaction time was less than a minute. This method is much simpler and greener than many other methods for the synthesis of Ag nanoparticles, as no harmful or hazardous chemical was used as a reducing or stabilizing agent; pomace extract, which is very rich in polyphenolics, served both functions. Interestingly, iron[46b] and silver nanoparticles[46c] generated using green tea polyphenols do not exhibit the usual toxicity.

Another environmentally friendly method for the synthesis of Ag nanoparticles, using the sodium salt of carboxymethyl cellulose (CMC) as the reducing and stabilizing agent under microwave irradiation conditions, has been reported by Chen *et al.*[47] The nanoparticles were synthesized simply by mixing the starting reagents and exposing them to the microwaves. Microwave heating promoted partial hydrolysis of CMC, producing aldose as the final hydrolysate that, in turn, reduced Ag^+ to Ag. With continuing hydrolyzation,

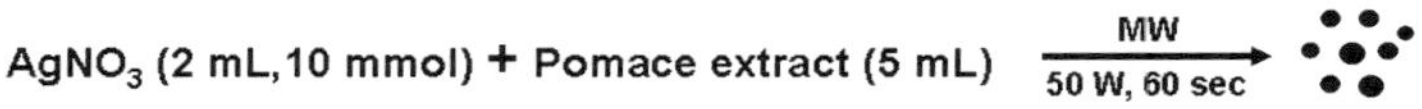

$$AgNO_3 \text{ (2 mL, 10 mmol)} + \text{Pomace extract (5 mL)} \xrightarrow[\text{50 W, 60 sec}]{\text{MW}} \bullet\bullet\bullet\bullet\bullet\bullet\bullet$$

Scheme 7.8

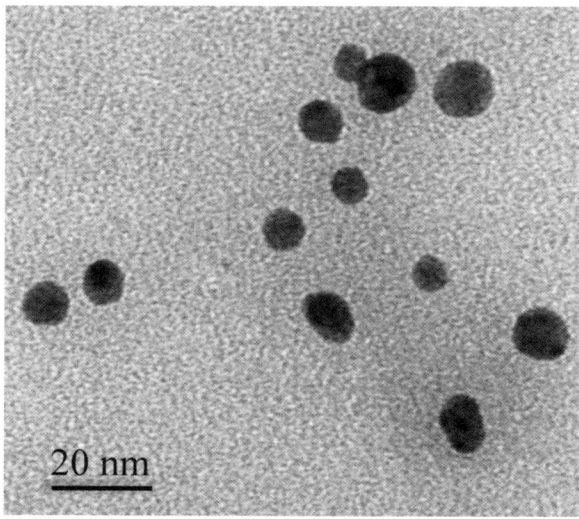

Figure 7.4 TEM micrograph of Ag nanoparticles synthesized using wine pomace at 50 W microwave power and 60-s exposure time.

Ag nanoparticles grew gradually, and CMC further formed a stable protecting layer on the surface of the Ag nanoparticles. Concentrations of CMC higher than 0.1% generated Ag_2O, creating unfavorable conditions for the formation of Ag nanoparticles. No observable agglomeration of the particles was found even after 58 days at room temperature. The method is simple and produced particles that were quite well dispersed for many applications.

7.2.3 Palladium (Pd) and Platinum (Pt) Nanoparticles

In addition to Au and Ag nanoparticles, Pt and Pd nanoparticles are also of much importance because of their tremendous use in the field of catalysis, fuel cells, and microelectrodes.[48]

Synthesis of Pt colloids in a single-step process, involving the direct microwave-based heat-treatment of an aqueous solution containing H_2PtCl_6 and 3-thiophenemalonic acid, has been reported by Luo.[48] The study showed that simply heating H_2PtCl_6 and 3-thiophenemalonic acid solutions in a domestic microwave oven operated 300 W for 8 min generated Pt nanoparticles (Scheme 7.9) in the size range 5–20 nm. The particle size depended on the ratio of 3-thiophenemalonic acid to H_2PtCl_6, decreasing with increasing 3-thiophenemalonic acid. Direct mixing of both components in the absence of microwave heating did not produce any particles even after several months, indicating that microwaves accelerated the chemical process.

Another method of synthesizing spherical Pt nanoparticles in aqueous medium under MW has been illustrated by Pal *et al.* (Scheme 7.10).[49] The procedure used hydrazine as a reducing agent and polyacrylamide as a stabilizing agent. The particles produced were in the size range 10–80 nm with a 3–5 nm acrylamide coating. Under identical reaction conditions with Pd, the method produced star-shaped Pd nanoparticles without any noticeable acrylamide coating. The authors suggested that the coating on the Pt nanoparticles

[H_2PtCl_6 (0.06M, 200 µL) +H_2O(10 mL)] + Thiophenemalonic acid $\xrightarrow[\text{300W, 8 min}]{\text{MW}}$ Nano Pt

Scheme 7.9

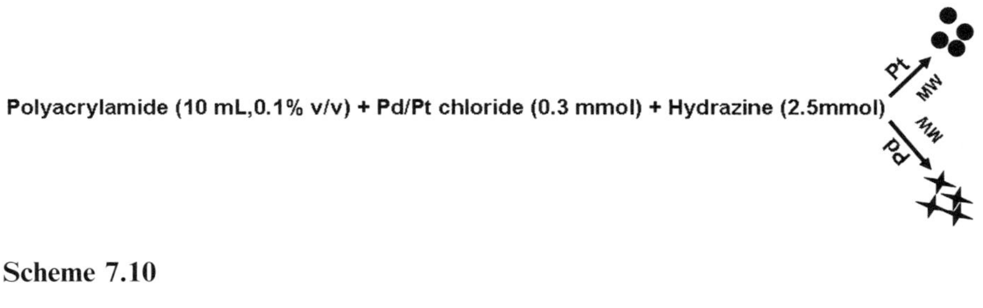

Polyacrylamide (10 mL, 0.1% v/v) + Pd/Pt chloride (0.3 mmol) + Hydrazine (2.5 mmol)

Scheme 7.10

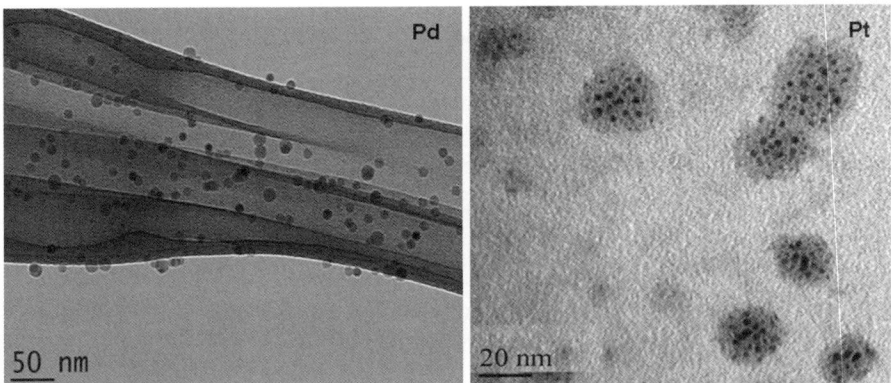

Figure 7.5 Pd and Pt nanoparticles synthesized with red grape pomace extract at 50 W MW power within 60 s.

might be responsible for the different growth pattern than that observed for the Pd nanoparticles.

Following the same method used for the production of Au and Ag nanoparticles, using red grape pomace extract, Baruwati and Varma[37] were able to produce both Pd and Pt nanoparticles in the size range 7–10 nm for Pd and 3–5 nm for Pt. The particles were highly dispersed and spherical in shape (Figure 7.5).

7.3 Synthesis of Metal Oxide Nanoparticles in Aqueous Medium under MW Irradiation Conditions

The strong dependence of structural, magnetic, optical, electronic, and catalytic properties of the nanoscale materials on morphology is a phenomenon of interest for both fundamental scientific understanding and for many practical and technological applications. To advance the basic understanding of the principles that determine the shape and to provide tailored building blocks for nanodevices, various methods have been developed to control the shape, dimensionality, and assembly of nanostructures.[50–53] With increasing concern about the environmental safety risks that could be created by the application of several environmentally hazardous chemicals and solvents, the design of synthetic methods that use environment friendly solvents, like water and microwave as a rapid and non-conventional method of heating, has also become a subject of intense research. This section describes some methods for the synthesis of metal oxide nanoparticles using water and microwave as the source of heating. Among the various metal-oxide nanoparticles, TiO_2, ZnO, and different ferrite nanoparticles will be discussed in detail because of their technological applications.

7.3.1 Synthesis of Titania (TiO_2) Nanoparticles

TiO_2 is one of the most studied metal oxides as it is clean, cheap, and versatile. It has been intensively used in catalysis,[54] photocatalysis,[55] chemical gas sensors,[56] self-cleaning and antiseptic coatings,[57] dye-sensitized solar cells (DSSCs),[58] and spintronic devices.[59] It has also been studied for its photo-electrochemical properties in water splitting for hydrogen production.[54] The physical and chemical properties of TiO_2 have been realized through various synthetic methodologies, including sol–gel, hydrothermal, template,[56] sonochemical,[58,59] etc. The use of microwaves as the source of heating has overcome the problem of energy efficiency along with its other advantages like (i) rapid heating, (ii) faster kinetics, (iii) phase purity, (iv) higher yield, and (v) better reliability and reproducibility.[60]

Wilson et al.[61] have reported a comparative study between the conventional hydrothermal and microwave hydrothermal procedure for the synthesis of TiO_2 nanoparticles. TiO_2 colloid was prepared by adding a mixture of titanium isopropoxide and isopropanol to water, with a subsequent addition of nitric acid, and heating at 80 °C. The colloids were than subjected to conventional heating and microwave heating in parts. The study showed that the microwave hydrothermal processing of colloidal TiO_2 solutions allowed rapid heating to the desired temperature (5–10 min compared to 15 h in conventional heating) and extremely rapid kinetics of crystallization. Highly nanocrystalline material with a narrow size distribution in the range 4–5 nm was obtained in the process. The authors suggested that the implications of this technique could provide faster processing time for obtaining a more crystalline material with a smaller size and a more regular shape.

The synthesis of TiO_2 nanowires in the anatase phase using microwave hydrothermal method was reported by Chung et al.[62] This method utilized commercial TiO_2 powder and 10 M NaOH solution under microwave irradiation conditions to synthesize the nanotubes. The temperature was varied from 110 to 210 °C for 2 h at 350 W to discern the effect of temperature on the morphology of the nanowires. Nanowires were obtained only at temperatures above 200 °C and the reaction time required was 2 h. The conventional method required more than 24 h to generate similar nanostructures. Thus microwaves made the process faster and more energy efficient.

Suprabha et al. have synthesized TiO_2 nanocubes, nanospheres, and nanotubes by hydrolyzing $TiCl_3$ in the presence of different hydrolyzing agents and pH under microwave irradiation conditions (Scheme 7.11).[63] The method utilized NH_4OH (basic), NaCl (neutral), and NH_4Cl (acidic) solutions as the different hydrolyzing media to produce nanocubes, nanospheres, and nanorods, respectively. The product obtained using NH_4OH was in anatase phase while the others were found to be in rutile phase. The formation of different morphologies was realized by varying the hydrolyzing speed and subsequent condensation during reaction, dehydration, and calcination steps.

Jia et al.[64] have described another method of synthesizing anatase TiO_2 nanorods with mesopores under microwave conditions. This method used

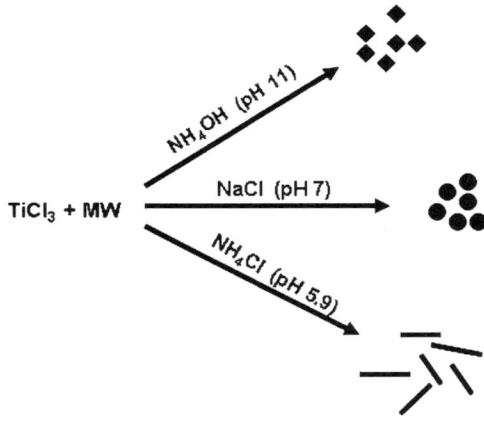

Scheme 7.11

TiCl$_4$ as the metal precursor and the tri-block copolymer poly(ethylene oxide)–poly(propylene oxide)–poly(ethylene oxide) as a surfactant and structure stabilization agent. The nanorods were formed by oriented attachment of the primary particles. Microwave treatment in the reaction procedure helped in maintaining the organic–inorganic hybrid nanostructures and in turn nanopores, which otherwise are disrupted during conventional heating methods. Another advantage of using microwaves was the perfect crystallization of nanorods before the calcination process that helped in preventing heavy aggregation and in maintenance of the nanostructures.

TiO$_2$ nanoparticles in the rutile phase have been synthesized under microwave irradiation conditions by Chen et al.[65] The method involved the addition of TiCl$_4$ diluted with HCl to water and successive microwave treatment at 120 °C for 1.5 h and at 180 °C for 1 h. X-Ray analysis showed that the resultant material was pure phase crystalline rutile instead of anatase. This extremely rapid crystallization was apparently due to the generation of localized high temperatures as a result of microwave heating.

TiO$_2$ has been used as a photocatalyst for several decades but mainly in the UV region of the solar spectrum due to its large band gap of 3.2 eV. Recently, doping TiO$_2$ with non-metal atoms like N, C, or S to obtain visible light activation by the redshift of the absorption spectrum has gained momentum.[66–70] Among the cationic dopants, N has attracted the most attention because its p-states contribute to the band gap narrowing by mixing with the O 2p states.[71]

Synthesis of N-doped TiO$_2$ nanoparticles *via* a microwave hydrothermal method has been reported by Zhang et al.[72] Hexamethylenetetramine (HMT) was used as the N source and TiCl$_3$ was used as the Ti source. The simple method required only the mixing of the precursors in water and exposure to microwave irradiation at 190 °C for 5–60 min; TiO$_2$ particles were produced after exposure of only 5 min, which is not possible by conventional heat treatment. Further, the product formed by microwave treatment was in the anatase phase while those formed by conventional treatment were in the rutile phase. Such phenomenon was caused by the long time used to heat up to the desired temperature by conventional heating, which would cause the phase transformation of TiO_2 into the more stable rutile phase. It was also observed that the metastable monoclinic phase of Ti was obtained under microwave irradiation conditions by variation of the reaction time.

7.3.2 Synthesis of ZnO Nanoparticles

Similar to TiO_2 nanoparticles, ZnO is a versatile smart semiconductor that has been garnering attention due to its wide range of technological applications in field-effect transistors,[73,74] chemical sensors,[75,76] field emitters,[77–79] transparent conductors,[80–82] ultraviolet light emitting devices,[83–86] *etc*. Its importance in technological applications comes from its wide direct band gap of 3.37 eV at room temperature and a large exciton binding energy of about 60 meV along with the electrical and optical properties of a II-VI semiconductor. ZnO has excellent thermal and chemical stability, a large piezoelectric constant, and an easily modified electric conductivity.

The physical properties of ZnO are highly dependent on its morphology, size, aspect ratio, orientation, *etc.*, which dictate its technological applications. Consequently, significant scientific efforts have been made in the synthesis of shape-controlled nanostructures of ZnO to explore the potential as a smart and functional material. Numerous nanostructured ZnOs with varying morphologies have been fabricated, such as nanowires,[85] nanosheets,[87] nanotubes,[88] and nanoflowers.[89] Controlled growth of hierarchical ZnO crystal structures has also received particular attention to support demands for more complex structures that impart greater control over material and device properties.[90–92]

Rapid synthesis of ZnO nanowires using microwave heating has been reported by Unalan *et al.*[93] and entailed the synthesis of vertically aligned ZnO nanowires on different substrates, including n-type silicon, glass, and poly (ethylene terephthalate) (PET). The process involved spin coating of a 10 mM solution of zinc acetate dehydrate and 1-propanol on the substrates at 2000 rpm for 30 s and annealing at 100 °C for 1 min after each spin coating to enhance adhesion. A uniform seed layer was obtained after three layers of spin coating. Vertical ZnO nanowires were then grown by dipping the substrates in a mixture of equimolar 25 mM zinc nitrate hexahydrate and hexamethylenetetramine solution in deionized water and subsequent heating in a commercially available microwave oven (2.45 GHz) at different power settings (120, 385, and 700 W) at atmospheric pressure. The exposure time was varied from 1 to 30 min. For comparison, several samples were grown at 90 °C using a furnace. The nanowires produced by the microwave method had fewer defects than those grown in the conventional heating. In addition, the method produced nanowires in a large area within a negligible amount of time.

Another method for the rapid production of different ZnO nanostructures has been reported by Zhang et al.,[94] wherein $Zn(CH_3COO)_2 \cdot 2H_2O$ was used as the reactant, a mixture of water and ethylene glycol as solvents, and no other additives were used. Different nanostructures, including rods, hexagonal prisms, peanuts, butterflies, and hierarchical nanospheres, were obtained within 30 min by changing only the mixing ratio of water to ethylene glycol under microwave irradiation conditions. When conventional heating was used, no nanostructures were obtained up to 1 h and only a small amount of nanorods after 3 h. Deployment of microwave heating, in contrast, caused rapid reaction and nucleation rate and the formation of massive seeds throughout the bulk solution, thereby increasing the efficiency.

Another morphology-controlled synthesis of ZnO nanostructures has been reported by Cho et al. (Scheme 7.12).[95] A choice of various basic ZnO structures such as nanorods, nanocandles, nanoneedles, nanodisks, nanonuts, microstars, microUFOs, and microballs were synthesized simply at low temperature (90 °C) with low power microwave-assisted heating (about 50 W) and a subsequent aging process. These varying morphologies were obtained by changing the precursor chemicals, the capping agents, and the aging times. Still more complex ZnO structures, including ZnO bulky stars, cakes, and jellyfishes, were constructed by microwave irradiation of a mixture of the as-prepared basic ZnO structures and some other precursor solutions. This fast, simple, and reproducible method did not require any template, catalyst, or surfactant but controlled the morphology of ZnO crystals from simple to a complex variety.

Huang et al.[96] have reported the synthesis of one-dimensional ZnO nanostructures. With an extremely simple procedure of mixing $Zn(NO_3)_2 \cdot 6H_2O$ and NaOH in particular amounts under microwave hydrothermal conditions, this method produced different three-dimensional morphologies, including nanorods, nanowires, nanodandelions, and radial nanospindles, within a 20 min reaction time. The products were found to be highly crystalline even in this short reaction time because microwaves provide rapid heating and fast supersaturation that leads to lower crystallization temperatures and shorter crystallization times. This study delineated that microwaves not only accelerated

Scheme 7.12

the reaction between the materials but also led to the growth and crystallization of ZnO with complex three-dimensional morphology. The reaction mechanism leading to the formation of different nanostructures was discussed on the basis of oriented attachment, wherein larger particles were grown at the expense of small primary nanoparticles *via* an orientated attachment process, in which the adjacent nanoparticles were self-assembled by sharing a common crystallographic orientation and docking of these particles at a planar interface. Small particles may aggregate in an oriented fashion to produce a larger single crystal, or they may aggregate randomly and reorient, recrystallize, or undergo phase transformations to produce larger single crystals. This type of growth could lead to the formation of faceted particles or anisotropic growth if there is sufficient difference in the surface energies of different crystallographic faces.

7.3.3 Synthesis of Ferrite Nanoparticles

Ferrite nanoparticles have long been of much scientific and technological research interest due to their applications in electronics,[97] magnetic recording media,[98,99] ferrofluids,[100] catalysis,[101] and in gas sensors.[102,103] Significant research efforts have been made towards using these nanomagnetic particles as MRI contrasting agents[104–106] and in drug delivery systems.[107,108] The physical and chemical properties of ferrite nanocrystals are greatly influenced by the synthesis route and for this reason various approaches, including hydrothermal, sol–gel, reverse and normal micelles, co-precipitation, sonochemical reactions, ball milling *etc.*,[109–113] have been adopted to produce ferrites with specific properties. The major problem associated with most of these methods is poor control of the particle size, distribution, and dispersity, longer reaction times, and the high temperature required for crystallization. Application of microwave as a heating technique could eliminate some of the difficulties, especially the problem of longer reaction time and the high crystallization temperature.

The synthesis of nanosized $MgFe_2O_4$ nanoparticles within 25 min at 150 °C using a microwave hydrothermal method has been reported by Verma *et al.*[114] The particles produced were in the size range 2–3 nm and were slightly agglomerated. This might be because of the magnetic attraction among the particles as well as high surface energy because of the extremely small sizes. The particles were superparamagnetic in nature at temperatures above 50 K. The blocking temperature for these particles was 38 K. This mild and simple method could be used for large-scale production of nano-ferrites provided the problem of agglomeration is controlled.

Another method of producing α-Fe_2O_3 nanorods under aqueous microwave conditions has been reported by Zhang *et al.*,[115] which used $FeCl_3 \cdot 6H_2O$ as the iron source and poly(vinylpyrrolidone) (PVP, MW = 300 000) as the structure-directing agent in an aqueous solution. The temperature used ranged from 120 to 180 °C and the rods formed were 60–80 nm in diameter and 300–500 nm long. Scheme 7.13 shows the formation mechanism.

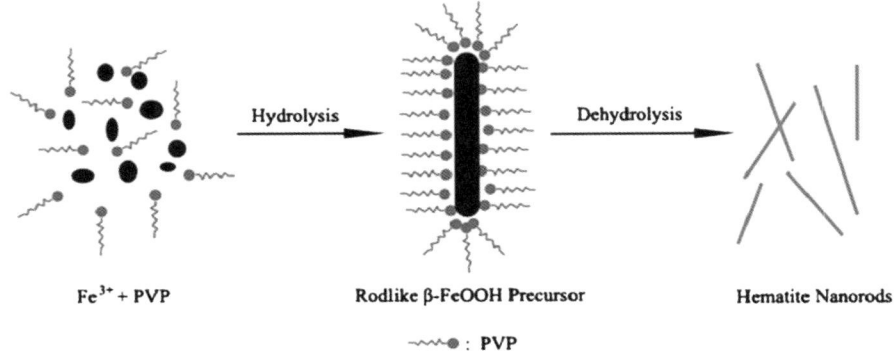

Fe^{3+} + PVP   Rodlike β-FeOOH Precursor   Hematite Nanorods

∼∼∼● : PVP

Scheme 7.13

The nanorods were found to be weakly ferromagnetic at low temperatures instead of the characteristically anti-ferromagnetic behavior that could be correlated with the one-dimensional nanostructures. The shape anisotropy of the rod-like nanostructures can affect the magnetic property and increase the coercivity, where the magnetic spins are preferentially aligned along the long axis of the one-dimensional nanostructures.[115]

Katsuki et al.[116] have reported the synthesis of monodispersed α-Fe$_2$O$_3$ nanoparticles simply by heating different molar solutions of Fe(NO$_3$)$_3$·9H$_2$O at various temperatures. This research used 0.02, 0.05, and 0.1 M solutions of Fe(NO$_3$)$_3$·9H$_2$O at 100, 120, and 140 °C, under both conventional and microwave hydrothermal conditions. The same morphology could be obtained under microwave hydrothermal conditions in a time much shorter than the conventional technique. Microwave heating facilitated the formation of smaller particles with a narrow size distribution because of the uniform nucleation.

The synthesis of monodispersed ferrite nanoparticles, including NiFe$_2$O$_4$, MnFe$_2$O$_4$, Co Fe$_2$O$_4$, and Fe$_2$O$_3$, under microwave conditions in water–organic mixed solvents has been reported by Baruwati et al.[117] The procedure utilized nitrate salts in water as precursor material and oleic acid in toluene as the capping agent. Under microwave hydrothermal conditions, the two different solvents diffused into each other, formed an oleate complex, and then decomposed to form the monodispersed particles in the organic phase that were highly dispersible in organic solvents such as hexane, toluene, etc. The microwave temperature used in the method was 160 °C and the time required was 1 h. The particles formed in the case of NiFe$_2$O$_4$ and other ferrites except Fe$_2$O$_3$ were in the size range 5–7 nm (Figure 7.6a). When the time period was increased to 2 h, particle sizes increased to 7–9 nm. For Fe$_2$O$_3,$ there were some 30-nm cubic particles (Figure 7.6b). The study also compared the conventional hydrothermal technique where the temperature applied was more than 225 °C for 2 h.

Polshettiwar et al.[118] have reported on the synthesis of Fe$_2$O$_3$ snowflake and micro-pine structures (Figure 7.7) simply by heating K$_4$[Fe(CN)$_6$] in water

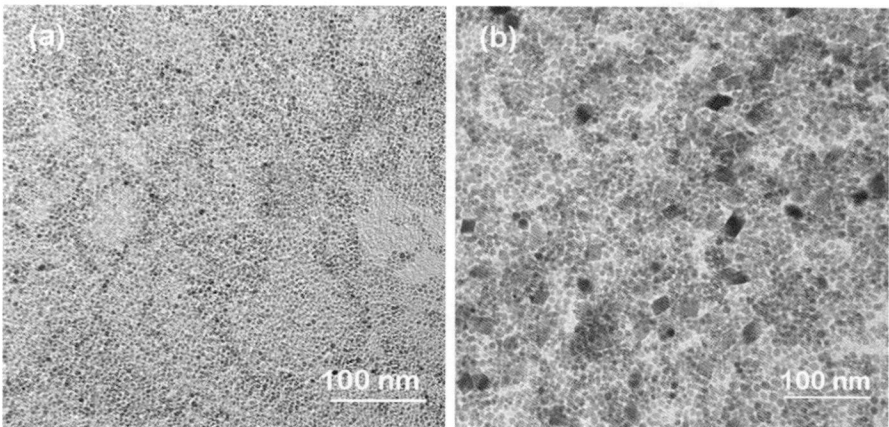

Figure 7.6 TEM micrograph of (a) NiFe$_2$O$_4$ and (b) Fe$_2$O$_3$ nanoparticles synthesized under microwave irradiation conditions at 160 °C for 1 h.

Figure 7.7 (a) Fe$_2$O$_3$ snowflake; (b) and (c) micro-pine structures.

under microwave irradiation conditions. The reaction temperature and time used in the process were 150 °C and 3 h, respectively, for the optimum yield. TEM images of the synthesized micro-pine α-Fe$_2$O$_3$ particles showed the single-crystal structure. Closer inspection of these particles revealed well-defined and highly-ordered branches (Figures 7.7b and c) distributed on both sides. The TEM observations also revealed that the micro-pine dendrites self-assembled to form a six-fold-symmetric structure (Figure 7.7a).

Figure 7.8 SEM micrographs of Fe_2O_3 snowflakes, CoO biprisms, Mn_2O_3 triangular nanorods, and Cr_2O_3 nanospheres.

This extremely simple method was generalized to produce various other metal oxide nanostructures, including CoO, Mn_2O_3, Cr_2O_3, and MoO_2, by heating their corresponding cyano- salts in water under microwave irradiation conditions (Figure 7.8). CoO and MoO_2 then were converted into Co_3O_4 and MoO_3 without any change in morphology by calcining at 700 °C.

7.3.4 Synthesis of Quantum Dots in Aqueous Medium under Microwave Conditions

Over the past decade, the synthesis of high quality colloidal semiconductor nanocrystals (referred to as quantum dots, QD) has been a subject of intense research because of their size-dependent properties and flexible processing chemistry. These colloidal semiconductor nanoparticles far superior to conventional dye molecules in terms of flexible photoexcitation, sharp photoemission, and superb resistance to photobleaching.[119] Their optical properties can be tailored by changing the size and composition to meet specific wavelength requirements.[99] They have been successfully used in optoelectronics, nonlinear optical devices, QD lasers, solar cells, and bio-tagging.[120–125] Great progress has been made in preparation of QDs, and numerous high-quality QDs, such as CdSe, CdTe, and some alloy nanocrystals, have been synthesized by the organometallic approach and the aqueous approach.[126–129]

The synthesis of water-soluble ZnSe nanocrystals in aqueous medium under microwave conditions has been reported by Qian et al.[130] ZnSe QDs prepared by microwave irradiation were water-soluble, had high crystallinity, and their photoluminescence (PL) quantum yield was up to 17%. These properties were remarkable improvements over ZnSe QDs prepared by conventional aqueous synthesis methods. The method eliminated the use of expensive, environmentally unfriendly reagents such as trioctylphosphine (TOP), tributylphosphine (TBP), and trioctylphosphine oxide (TOPO).

ZnS nanoballs have been synthesized in saturated water solutions under microwave by Zhao et al.[131] Zinc acetate and thioacetamide were used as Zn and sulfur (S) source, respectively. $Zn(Ac)_2$ (0.05 mol) was mixed with thioacetamide (TAA) (0.06 mmol). Water (40 mL) was added to the mixture and the suspension was placed in a microwave refluxing system and irradiated at 280 W for 15 min. The reactions were carried out under ambient air and the ensuing products were highly crystalline and were about 300 nm in diameter.

CdSe, PbSe, and Cu_2-xSe nanoparticles have been prepared under microwave irradiation conditions by Zhu et al.[132] where $CdSO_4$, $Pb(Ac)_2$, and $CuSO_4$ reacted with Na_2SeSO_3 in water in the presence of complexing agents [potassium nitrilotriacetate ($N(CH_2$-$COOK)_3$ – NTA) for CdSe and PbSe or triethanolamine for Cu_2-xSe] in a microwave refluxing system. Although the method is simple and good for producing CdSe nanoparticles in the 4–5 nm size range, both PbSe and Cu_2-xSe nanoparticles were larger (30–80 nm size range) and were agglomerated. The method also described how different phases of CdSe were obtainable by varying the microwave heating times. A 10-min irradiation afforded CdSe in the cubic (sphalerite) phase while after 30 min the CdSe was obtained in the hexagonal cadmoselite phase.

A one-pot method for synthesis of CdSe/ZnS core–shell QDs using microwave radiation has been reported by Schumacher et al.[133] and was based on the addition of a water-soluble Zn^{2+} complex, $Zn(NH_3)_4^{2+}$, to a solution containing CdSe initial nanocrystals and 3-mercaptopropionic acid (MPA). Subsequent microwave heating of these initial solutions for less than 2 h generated high-quality CdSe/ZnS-based QDs possessing good photoluminescent quantum yield (13%) and biocompatibility.

A seed-mediated approach for rapid synthesis of high quality alloyed quantum dots (CdSe–CdS) in the aqueous phase by microwave irradiation within an hour has been reported by Qian et al.[134] In the synthesis, CdSe seeds were first formed by the reaction of NaHSe and Cd^{2+}, and then alloyed quantum dots (CdSe–CdS) were rapidly produced by releasing of sulfide ions from 3-mercaptopropionic acid as sulfide source with microwave irradiation. The quantum yield was up to 25%.

7.4 Nanoparticles as Catalysts

Nanoparticles have emerged as a sustainable substitute to the usual materials, as high-surface-area heterogeneous catalysts[135] and supports.[136] The nano-size of the particles increases the uncovered surface area of the active part of catalyst thus enhances the contact among reactants and catalyst dramatically, mimicking homogeneous catalysts. The technical challenge is the synthesis of a nanosize catalyst, to permit facile movement of materials in the reacting phase, and control over morphology of nanostructures to modify the physical and chemical properties. The advancement of solution-based controlled synthesis of nanomaterials has made this feasible without complexity.[119,137]

A Pt metal core coated with a mesoporous silica nano-catalyst was discovered by Somorjai et al.[138] This porous silica coating helps the Pt cores to heat up to an elevated temperature of 750 °C in air but also allowed easy access of the catalytically active Pt. It is an excellent nano-catalyst system, with exceptional activity (due to the boost in contact of nano-Pt with reactants), which allows their usage in high-temperature catalytic reactions such as ethylene hydrogenation and CO oxidation (due to the thermal strength of the catalyst).

The size of catalyst particles as well as their shape and morphology also have a noteworthy effect on catalysis. Xie recently established that Co_3O_4 in the form of nanorods allowed favored exposure of catalytically active sites and amplified catalytic activity significantly for the oxidation of CO at very low temperature.[139] The catalyst was stable and active even in impure feed gases containing H_2O and CO_2. In addition to the reactivity and strength of the catalyst, selectivity was also influenced by controlling the shape of the catalytic nanoparticles. Lee et al. have achieved selective isomerization of trans-olefins to their less thermodynamically favorable cis isomers by means of (111) facets of the Pt-metal.[140]

Chandler and co-workers have synthesized a Pt/SiO_2 catalyst by using 4-hydroxyl terminated polyamidoamine dendrimer-encapsulated nanoparticles as a precursor. This dendrimer-encapsulated bimetallic nanoparticles showed superb catalytic activity for various reactions.[141] Pt nanoparticles were prepared and stabilized in polyamidoamine dendrimers, which were then deposited onto high surface area silica. Following thermal activation and treatment with hydrogen (Figure 7.9), the materials were found to be highly active catalysts for oxidation and hydrogenation reactions.[142]

In continuation of their work, the same authors also prepared Pt[143] and Au[144] supported on a titania catalyst using a similar procedure. Pt/TiO_2 was found to be active for CO oxidation, as well as toluene hydrogenation results; however, it was observed that catalyst activity reduced (in the case of toluene hydrogenation) with harsh reduction treatments.

Various metal-silica catalysts have been prepared by incorporation of surfactant-stabilized metal (Pd, Ir, Rh) nanoparticles in the pores of silica.[145] These silica catalysts ([Pd]x-MCM-41, [Ir]x-MCM-41 and [Rh]x-MCM-41) were found to be very active and selective catalysts for the hydrogenation of a range of cyclic olefins.

Monodispersed Ni–Pd (core–shell) bimetallic nanoclusters have been synthesized by Rothenberg et al. using combined electrochemical and "wet chemical" techniques.[146] The materials show good catalytic activity for the Hiyama reaction with excellent product yield for various aromatic halides (Scheme 7.14). Notably, minor amounts of homocoupling by-product were noticed.

Most of the hydrogenation catalysts are based on precious metals, such as Pd, Rh, ruthenium (Ru) and nickel (Ni), due to their good activity and applicability. However, the higher cost and toxicity of these regularly used metals has prompted a search to develop economic catalysts – that use more readily available metals – for such protocols. In this line, de Vries and his co-workers have developed a, low-cost, benign iron-based hydrogenation catalyst. The iron nanoparticles were synthesized by reduction of $FeCl_3$ using EtMgCl and were used effectively as a catalyst for the hydrogenation of various alkenes and alkynes.[147]

Visible-light-driven hydrogen production by water splitting is one of the elegant approaches to converting solar energy into a fuel and a key area in

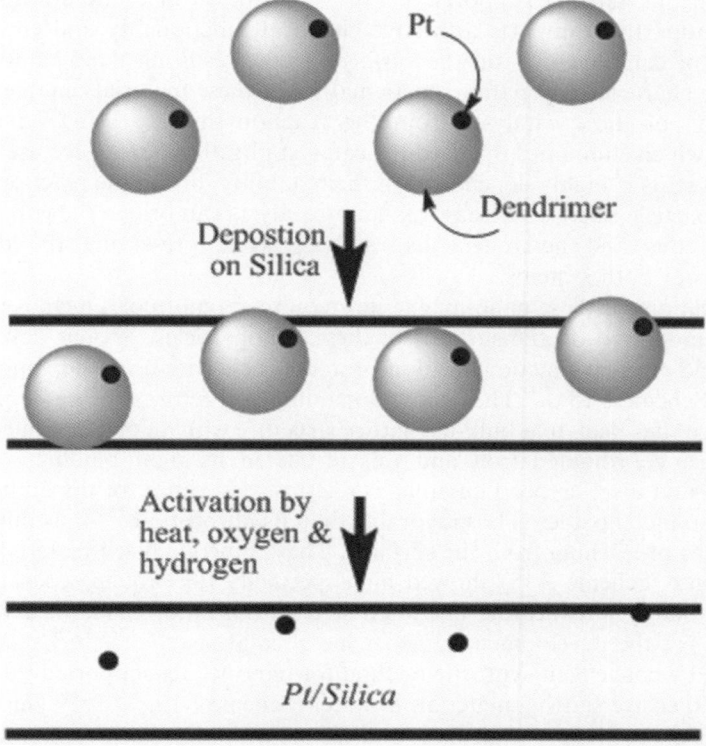

Figure 7.9 Synthesis of a Pt/silica catalyst.

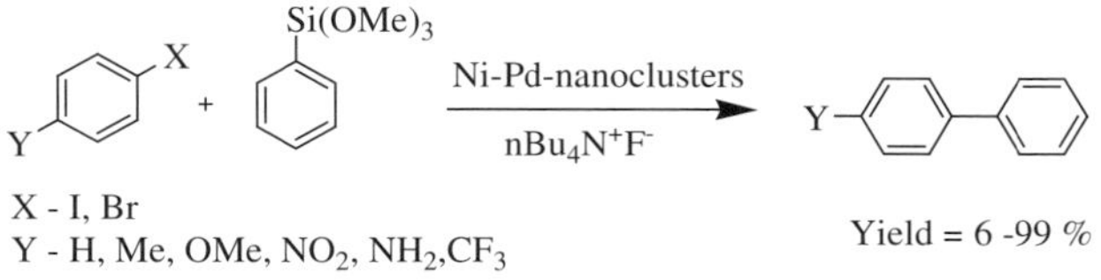

Scheme 7.14

photosynthesis. Nano-catalysis can be used to realize this goal and recently MoS_2 nanoparticles 10 nm in size were used as a competent catalyst for hydrogen evolution from a transition metal complex containing molecular systems in visible light.[148] MoS_2 nanoparticles catalyzed the reduction of protons more efficiently than the conventional MoS_2/Al_2O_3 catalyst. This confirmed the advantage of using nano-catalysts over conventional catalysts.

Magnetic nanoparticles have emerged recently as a viable alternative to conventional materials as robust, readily available, high-surface-area, heterogeneous catalyst supports.[149] Post-synthetic surface alteration of magnetic nanoparticles imparts attractive chemical functionality and enables the creation of catalytic sites on the surfaces of the resulting nano-catalyst. The insoluble character and paramagnetic nature of these materials enables simple separation of these catalysts from the reaction mixture using an external magnet, which eliminates the need for catalyst filtration. Extensive use of these nanomaterials is highly dependent on their stability during the reaction as well as their particle size.[136] These novel nano-catalysts can bridge the gap between homogeneous and heterogeneous catalysis while preserving the desirable attributes of both systems.

Applications of these nano-magnetites require the molecules/ligands involved to be immobilized on the surfaces of these nanoparticles. Xu has developed a simple and efficient functionalization protocol for ferrite using dopamine as an anchor (Scheme 7.15).[150] They chose dopamine as it returns the un-coordinated Fe surface sites back to a bulk-like lattice structure with an octahedral geometry for oxygen-coordinated iron and this results in its tight binding with iron oxide,[151] and also, as per Langmuir isotherms, desorption of dopamine molecules from metal oxides is less favorable than its absorption,[152] thus minimizing its chances of leaching from the surfaces. These functionalized materials with a Ni-complex (Scheme 7.15) showed high specificity for protein separation and excellent stability to heating and high salt concentration. This idea was then explored for the development of other metal catalysts.[136e,136g,137b,153] We have developed a convenient synthetic method for nano-ferrite-supported Pd catalyst from inexpensive starting materials in water (Scheme 7.16).[136e,136g] This catalyst catalyzed the oxidation of alcohols and olefins with high turnover numbers and excellent selectivity. Being magnetically separable it also eliminated the requirement of catalyst filtration after completion of the reaction, an additional sustainable attribute of this oxidation protocol.

Magnetite nanoparticle-supported chiral catalysts by Ru-complex formation with BINAP followed by functionalization on ferrite surfaces has also been discovered (Scheme 7.17). This material catalyzed the asymmetric hydrogenation of aromatic ketones under heterogeneous conditions with good activity and enantioselectivity. The catalyst was recovered using an external magnet and was re-used up to 14 times with no change in activity.[136a]

This concept of using a magnetically separable catalyst has also been explored by Jones for a one-pot multi-step reaction. A polymer resin-based acid catalyst and magnetic ferrite-based base catalyst for conversion of 1-(dimethoxymethyl)benzene into 2-benzylidenemalononitrile in a two-step, one-pot reaction was used (Scheme 7.18). Following completion of the reaction, resin-based catalyst was removed by decantation and ferrite-based catalyst with an external magnet, essentially in pure form.[136b]

Alper has developed a new method to homogenize these magnetically separable heterogeneous catalysts by growing polyaminoamido dendrons on silica-coated nano-ferrites up to three generations, followed by phosphination.[136c] The material was then complexed with Rh and used as a catalyst for hydroformylation reaction (Scheme 7.19). The as-synthesized catalyst was

Microwave-assisted Synthesis of Nanomaterials in Aqueous Media

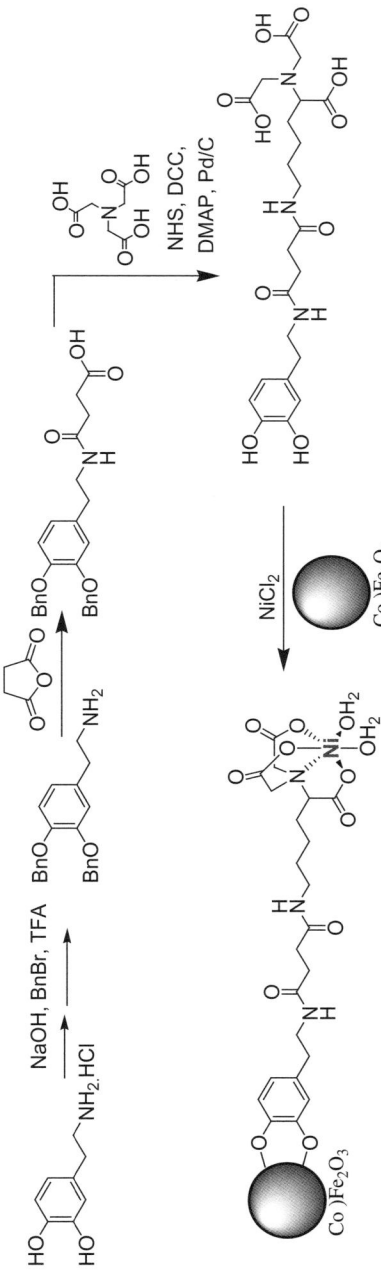

Scheme 7.15

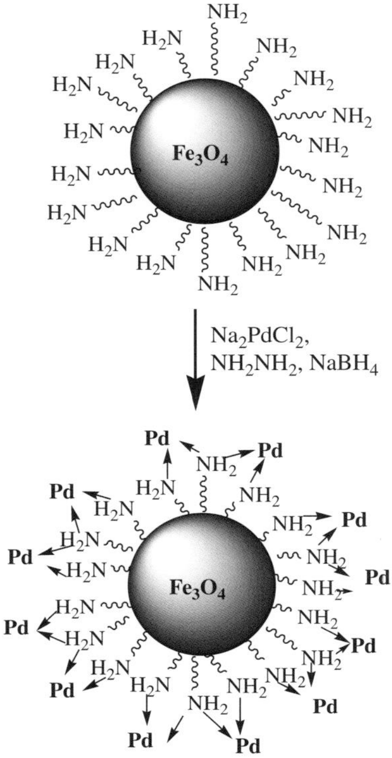

Scheme 7.16

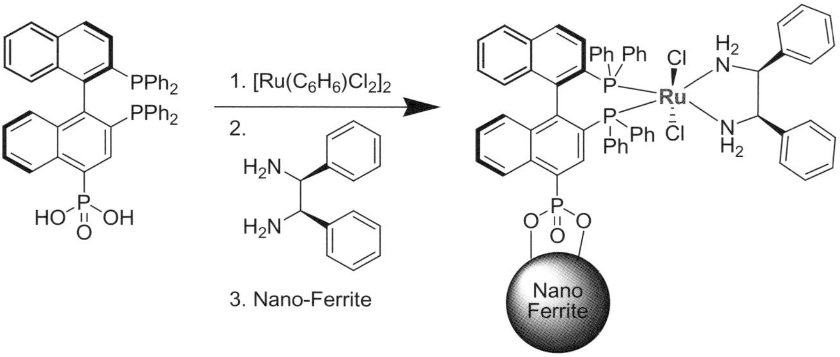

Scheme 7.17

stable and more soluble in organic solvent and catalyzed the reaction with excellent reactivity and selectivity.

Likewise, Kitamura has synthesized a catalyst for de-allylation by immobilization of [CpRu(η^3-C_3H_5)(2-pyridinecarboxylato)]PF_6 on silica-coated

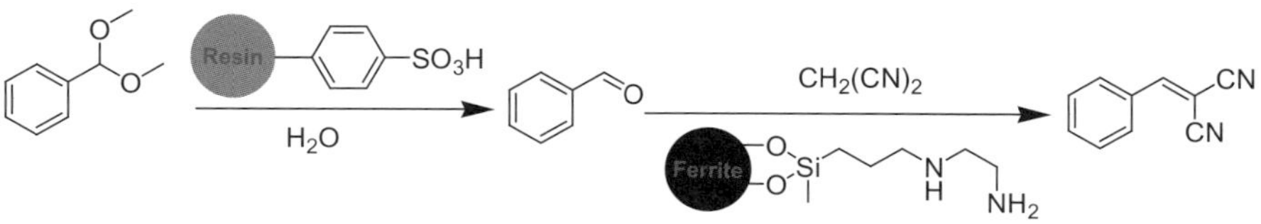

Scheme 7.18

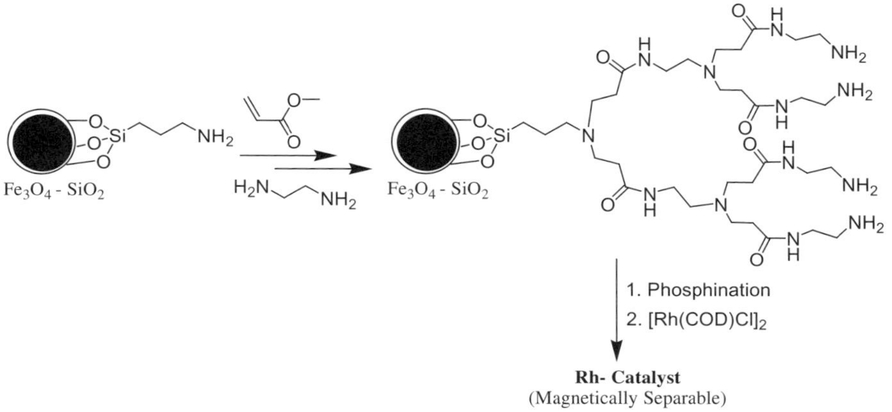

Scheme 7.19

nano-ferrite particles.[154] The materials showed high saturation magnetization and levels of dispersibility with weak coercive forces. The catalyst was very active for de-allylation reactions (Scheme 7.20) and no additional additives were needed to complete the reaction. Volatile allyl ethers were the only co-products of the reaction. Catalyst was removed using an external magnet and reused.

A pioneer in catalysis, Beller has used this magnetically separable nano-ferrite as a support for the development of a Ru-Fe_3O_4 catalyst[136f] and tested its catalytic prowess for synthesis of various sulfonamides successfully with excellent yield (Scheme 7.21). It was also observed that only one equivalent of alcohol was needed to complete the reaction and the catalyst was used several times, making the protocol green and sustainable.

The magnetic nanoparticle-supported 4-N,N-dimethylaminopyridine analogue has been prepared and used as a heterogeneous catalyst for a range of synthetic transformations with good activity and recyclability (Scheme 7.22). The ease of recovery as well as superb stability of this catalyst system made it easily recyclable using an external magnet for up to 30 times without any noticeable loss in activity.[136d] Similarly, a chiral version of this catalyst was also tested successfully for an enantioselective acylation.[155]

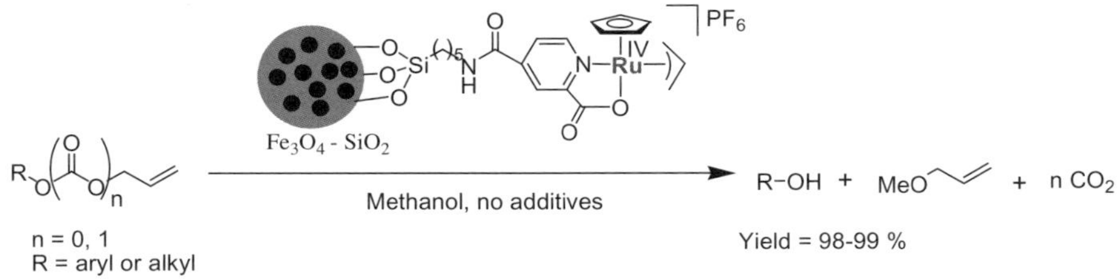

Scheme 7.20

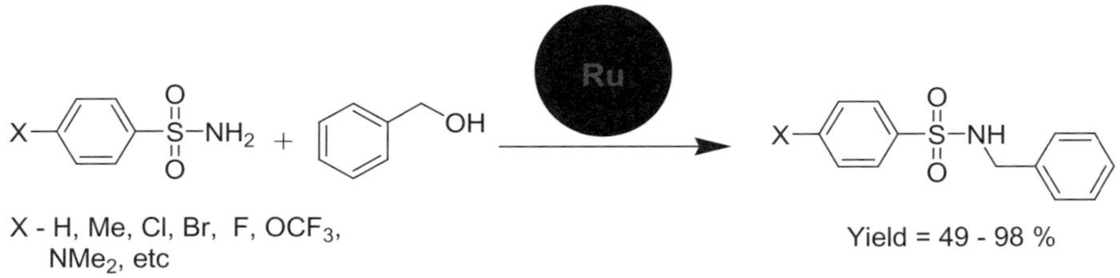

Scheme 7.21

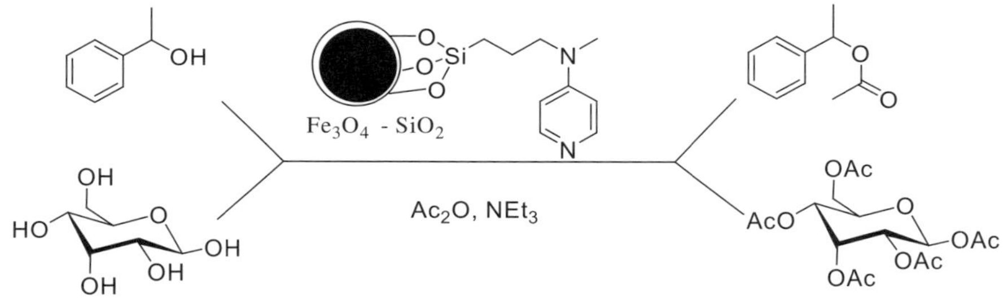

X – H, Me, Cl, Br, F, OCF$_3$, NMe$_2$, etc

Yield = 49 - 98 %

Scheme 7.22

7.5 Ruthenium Hydroxide Nano-catalyst for Microwave-assisted Hydration of Nitriles in Water

A recently developed aqueous hydration of nitrile protocol is good with respect to reaction conditions and product yield[156] but uses expensive Ru complexes as catalysts and needs traditional work-up using toxic organic solvents for isolation of the product. The same hydration protocol can be easily conducted in a green and sustainable way by using Ru hydroxide nano-catalyst under aqueous microwave conditions.[157] The nano-Ru(OH)$_x$ catalyst has been prepared in two steps: initially magnetic nanoparticles were functionalized by post-synthetic functionalization[137b] *via* sonication of nano-ferrites with dopamine in aqueous

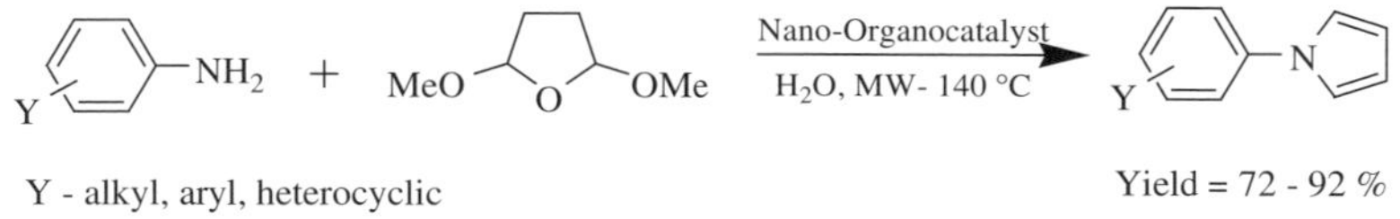

Scheme 7.23

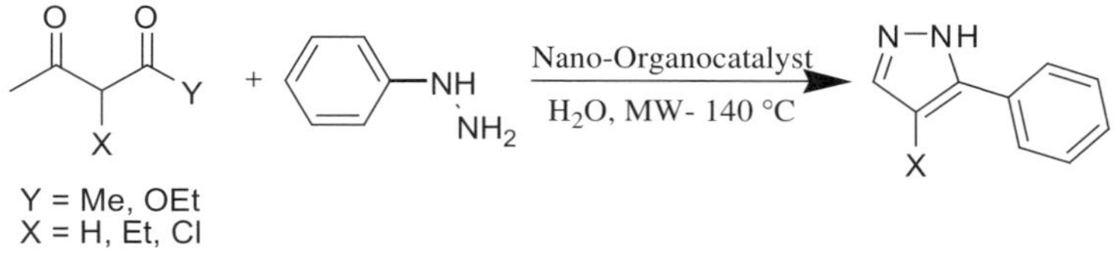

Y — alkyl, aryl, heterocyclic

Yield = 72 - 92 %

Scheme 7.24

Y = Me, OEt
X = H, Et, Cl

medium. The next step was the addition of Ru chloride, followed by hydrolysis using sodium hydroxide solution (Scheme 7.23). This protocol showed excellent chemoselectively; during hydration of the benzonitrile-containing dioxole ring, the reaction proceeded only at the cyano group to afford the corresponding amide, while keeping the ring untouched (Scheme 7.24). Metal leaching is an important criterion for selecting a heterogeneous catalyst because metal contamination is highly regulated by chemical industries.[158] This catalyst was successfully used several times and left no remnants of metal in the end product.

7.6 Glutathione-based Nano-organocatalyst for Microwave-assisted Synthesis of Heterocycles in Water

The outstanding nature of heterocyclic nuclei to act as biomimetics and pharmacophores has extensively contributed to their unique value as traditional key elements of numerous drugs.[159] Organocatalysis has become a very noteworthy area of research and this metal-free approach has attracted global interest. Although a wide range of reactions has been developed using this strategy, most of these transformations are generally conducted in organic solvents. In a recently developed aqueous protocol, it was observed that the addition of water often accelerates the organocatalyst-mediated reaction, making the overall protocol efficient and green.[160] However, most of these reports use small amounts of water as reaction medium and excessive amounts of hazardous organic solvents during the work-up, which unfortunately defeats the core intention of reducing the environmental burden of organic contaminants.[161] These drawbacks have been successfully circumvented in a green and sustainable manner using glutathione-based nano-organocatalyst under

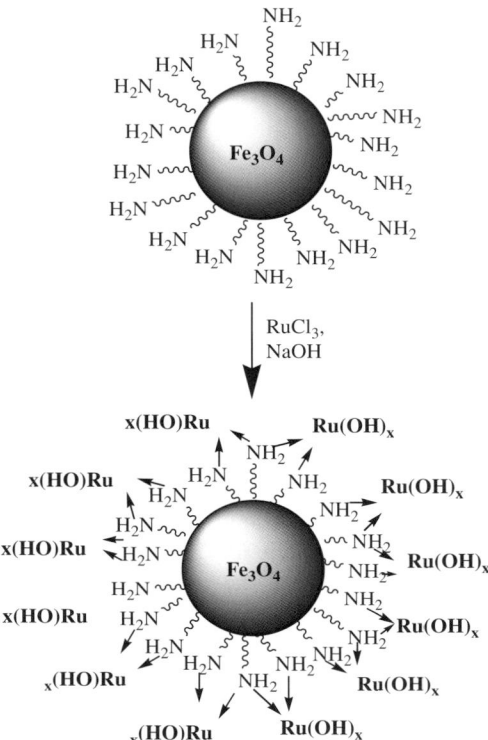

Scheme 7.25

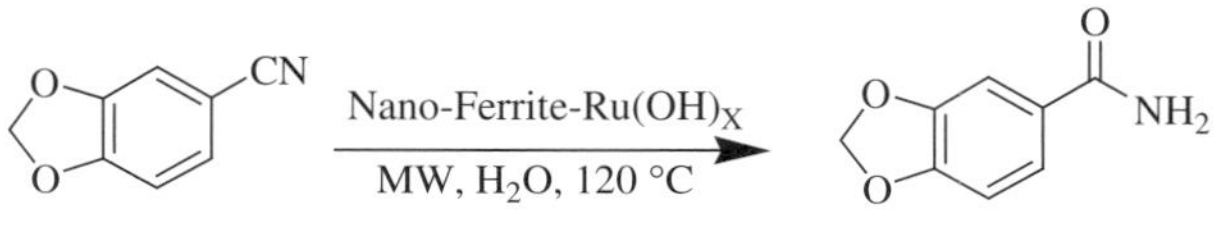

Scheme 7.26

aqueous microwave conditions.[162] The catalyst was prepared by sonochemical covalent anchoring of glutathione molecules through coupling of its thiol group and the free hydroxyl groups of ferrite surfaces (Scheme 7.25).

This glutathione-based nano-organocatalyst was used efficiently for the synthesis of a series of pyrrole heterocycles by the Paal–Knorr reaction under aqueous microwave conditions. It showed excellent catalytic activity and several amines reacted with tetrahydro-2,5-dimethoxyfuran to produce the respective pyrrole derivatives in good yields (Scheme 7.26). Using the above-developed strategy, various hydrazines and hydrazides reacted efficiently with 1,3-diketones to afford the desired pyrazoles in good yields (Scheme 7.27).

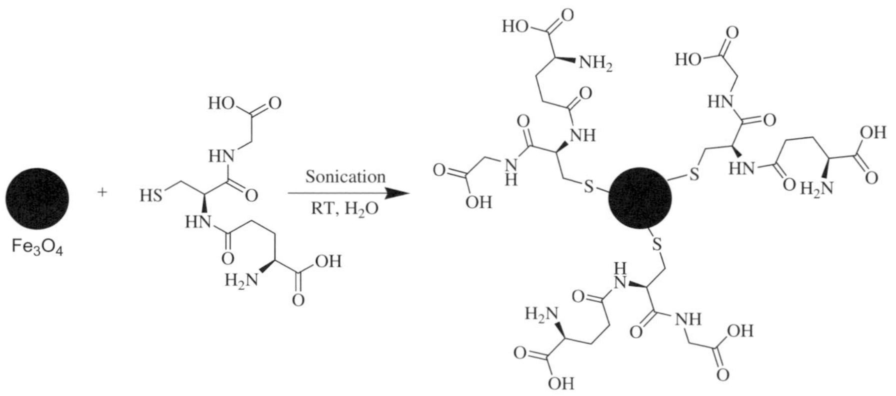

Scheme 7.27

7.7 Representative Experimental Procedures

7.7.1 Synthesis of Gold Nanoparticles

7.7.1.1 Method 1[37]

To obtain the pomace extract, pomace (100 g) was soaked in water (200 mL) for 30 min under magnetic stirring and then filtered through a sieve to remove the large solid particles. The wine-colored water was then used for the nanoparticle synthesis.

In a typical procedure 2 mL of 10 mmol solution of $HAuCl_4$ was placed in a 10 mL crimp-sealed thick-walled glass tube equipped with a pressure sensor and magnetic bar. Pomace extract (5 mL) was then added to it. The tube was then placed inside the cavity of a CEM Discover focused microwave synthesis system for 60 s operated at 50 W. Formation of the particles was observed from the change in color of the reaction mixture. The solution turned reddish brown at the end of the reaction. After completion of the reaction, the tube was cooled to room temperature and the particles were centrifuged and dispersed in water. These nanoparticles were then used for further characterizations.

7.7.1.2 Method 2[32]

Spherical Au nanoparticles were prepared by mixing 4 mL of a CTAB solution (10^{-2} M) with 200 μL of a aqueous Au(III) solution (10^{-2} M). An aqueous 2,7-DHN solution (2 mL) was then added to the reaction mixture such that the final concentration was 3.17×10^{-3} M. Finally, 100 μL of NaOH (1 M) was added, and the mixture was stirred for 30 s and then irradiated with microwaves for 90 s with an intermittent pause after every 10 s to cool the reaction vessel.

For the synthesis of anisotropic Au nanoparticles, the concentration of 2,7-DHN and NaOH was varied while keeping the other concentrations fixed. The final concentrations of 2,7-DHN and NaOH were 1.05×10^{-3} and 1.05×10^{-2} M, respectively.

For rod-shaped particles, the final concentrations of CTAB, Au(III), 2,7-DHN, and NaOH were 7.54×10^{-3}, 4.71×10^{-4}, 1.88×10^{-3}, and 9.43×10^{-3} M, respectively.

For nanoprism synthesis, a mixture was prepared with 4 mL of CTAB (0.1 M), 320 µL of a Au(III) solution (10^{-2} M), 280 µL of 2,7-DHN (10^{-2} M), and 10 µL of NaOH (1 M). The solution was stirred for 30 s and irradiated by microwave for 90 s. For all the above cases, Au particle formation started just after 10–20 s of microwave irradiation, as observed from the color change of the solution mixture and from the UV–visible spectrum. The resulting solution was centrifuged at 3000 rpm for 10 min to remove excess surfactants. The precipitate was re-dispersed in deionized water and centrifuged again at 2000 rpm for 5 min. This process was repeated twice. Finally, the precipitated Au nanoparticles were collected and re-dispersed in water for characterization. The solution became deep pink (for spherical), bluish pink (for mixed shapes), bluish purple (for rod shapes), and pinkish red (nanoprisms) after completion of the reaction. The color remained stable for at least 3 months in the dark under ambient environment without change in optical properties.

7.7.2 Synthesis of Silver Nanoparticles

7.7.2.1 Method 1[44]

Soluble starch (0.4 mmol) and L-Lysine or L-arginine (0.16 mmol) were added to deionized (4.0 mL) water in a 10 mL microwave sealed vessel. A 4 mL aliquot of an aqueous solution of $AgNO_3$ (20 mmol) was then added and the combined mixture further stirred. The mixture was heated to 150 °C by microwave under magnetic stirring and was maintained at this temperature for 10 s. The obtained solid product collected by centrifuging the mixture and was then washed several times with deionized water and finally dried in vacuum at 60 °C for characterization.

7.7.2.2 Method 2[46]

In a typical synthesis procedure, silver nitrate (1 mmol) was dissolved in distilled water (5 mL). To this, 0.046 g of glutathione (reduced) dissolved in 2 mL of water was added. The mixture was then transferred to a 10 mL crimp sealed thick-walled glass tube equipped with a pressure sensor and magnetic bar. The tube was then placed inside the cavity of a CEM Discover focused microwave synthesis system for 30–60 s operated at 50 W. The reaction was performed at different power levels to optimize the reaction conditions. Formation of the particles could be observed from the change of color of the reaction mixture.

The colorless solution turned reddish brown at the end of the reaction. After completion of the reaction, the tube was rapidly cooled to room temperature, centrifuged, and dispersed in water. The dispersion and centrifugation process were repeated twice to remove any unreacted silver nitrate or glutathione from the final product. These nanoparticles were then used for further characterizations.

7.7.3 Synthesis of Different TiO_2 Nanostructures[62]

To synthesize TiO_2 naocubes, $TiCl_3$ (20.0 mL) was added dropwise to 200 mL of NH_3 solution (1.5 M, pH 11) and microwave irradiation was performed for 20 min for complete precipitation. In the case of TiO_2 nanoparticles, $TiCl_3$ (5.0 mL) was added dropwise with continuous stirring to 200 mL of NaCl solution (5.0 M, pH 7) and the reaction mixture was irradiated for 60 min for complete precipitation. To synthesize nanorods, $TiCl_3$ (5.0 mL) was added dropwise to 200 mL of NH_4Cl solution (5.0 M, pH 5.9) and irradiated in a similar manner for 60 min.

7.7.4 Synthesis of ZnO Nanostructures[94]

ZnO nanorods, nanoneedles, nanocandles, nanodisks, and nanonuts were produced using zinc nitrate hexahydrate [$Zn(NO_3)_2 \cdot 6H_2O$, 98%, Aldrich] as the zinc cation precursor chemical, and hexamethylenetetramine [HMT, $(CH_2)_6N_4$, 99%, Junsei] as the hydroxide anion precursor chemical.

For the synthesis of ZnO rods, an aqueous solution (25 mL) composed of zinc nitrate hexahydrate (0.01 M) and HMT (0.01 M) was prepared at room temperature. The solution was irradiated by a temperature-controlled microwave synthesis system (2.45 GHz, single-mode, Greenmotif, IDX, Japan) at 90 °C for 15 min. The irradiated solution was then filtered with a polycarbonate membrane filter (ISOPORETM) that had pores 100 nm in diameter. The resultant white precipitate was washed several times with deionized water after filtration and dried in an oven.

In addition to zinc nitrate hexahydrate (0.01 M) and HMT (0.01 M), ethylenediamine (0.03 M, EDA, $NH_2CH_2CH_2NH_2$, 99%, Samchun) was used as a capping agent to synthesize ZnO needles.

For the production of ZnO candles, the irradiated ZnO nanorod-containing solution was cooled to room temperature and aged for 5 h at room temperature before filtration.

Disk and nutlike ZnO structures were obtained using triethyl citrate. For the synthesis of ZnO hexagonal disks, a transparent aqueous solution (25 mL) of zinc nitrate hexahydrate (0.01 M), HMT (0.01 M), and triethyl citrate (0.13 M) was prepared at room temperature. The procedure was the same as that for the nanorod synthesis. To produce ZnO nuts, the above irradiated ZnO nanodisk-containing solution was cooled to room temperature and aged for 100 min at room temperature before filtration.

ZnO stars, UFOs, and balls were synthesized using zinc acetate dihydrate [Zn(CH$_3$COO)$_2 \cdot$2H$_2$O, 99%, Samchun] and ammonia-water (28.0–30.0 wt%, Samchun) as zinc cation and hydroxide anion precursors, respectively. To obtain the ZnO stars, a transparent aqueous solution (25 mL) of zinc acetate dihydrate (0.01 M) and ammonia (about 0.16 M) was made at room temperature. To produce ZnO UFOs and balls, tripotassium citrate monohydrate [HOC(COOK)(CH$_2$COOK)$\cdot$2H$_2$O, 99%, Kanto] was added to the aqueous solution of zinc acetate dehydrate and ammonia. To prepare ZnO UFOs, zinc acetate dehydrate (0.01 M), ammonia (about 0.16 M), and tripotassium citrate monohydrate (0.01 M) were dissolved in deionized water (25 mL) under constant stirring. In the case of ZnO balls, the concentrations of zinc acetate dihydrate, ammonia, and tripotassium citrate monohydrate were 0.01, 0.32, and 0.01 M, respectively. The solution was heated by the microwave synthesis system at 90 °C for 15 min. Each microwave-irradiated solution was filtered, and the collected produce washed repeatedly with deionized water and dried in the oven.

7.7.5 Selected Methods for Production of Ferrite Nanoparticles

7.7.5.1 Method 1[117]

In a typical procedure, Fe(NO$_3$)$_3 \cdot$9H$_2$O (6.464 g, 16 mmol) and Ni(NO$_3$)$_2 \cdot$6H$_2$O (2.32 g, 8 mmol) were dissolved in distilled water (100.0 mL). The pH was adjusted to 9 by dropwise addition of ammonium hydroxide. The reaction mixture was then stirred magnetically at room temperature for 1 h. Oleic acid (27.4 mL) dissolved in toluene (100.0 mL) was added and the combined reaction mixture was then placed inside a microwave (power level 500 W). The temperature was then raised to 160 °C in 20 min and maintained at that temperature for 1 h. The resultant product was collected using the same procedure described above for conventional hydrothermal synthesis.

Similar reaction procedures were followed for the synthesis of both CoFe$_2$O$_4$ and MnFe$_2$O$_4$ nanoparticles, except that Co(NO$_3$)$_2 \cdot$6H$_2$O and Mn(NO$_3$)$_2 \cdot$4H$_2$O were used in place of Ni(NO$_3$)$_2 \cdot$6H$_2$O. For Fe$_2$O$_3$ nanoparticles, FeCl$_3$ (3.24 g, 20 mmol) and FeCl$_2 \cdot$4H$_2$O (1.98 g, 10 mmol) were used as precursors. If the temperature is maintained above 120 °C, hematite nanoparticles were formed.

7.7.5.2 Method 2[118]

In a typical synthetic procedure, 0.001 mole of the precursor [K$_4$Fe(CN)$_6$] was dissolved in distilled water (50 mL) and the solution was placed in a Teflon-sealed microwave-reactor. The reaction mixture was exposed to microwave irradiation at 180 °C for 3 h. After completion of the reaction, the mixture was allowed to cool to room temperature naturally and the metal oxides formed were isolated by centrifugation, washed with distilled water and methanol, and dried under a vacuum at 60 °C for 3–4 h.

Other metal oxide nanostructures could be obtained by replacing $K_4Fe(CN)_6$ with $K_3Co(CN)_6$, $K_3Mn(CN)_6$, $K_3Mo(CN)_8$, and $K_3Cr(CN)_6$.

7.8 Conclusion

Recent studies on the synthesis of various nanomaterials, including metals, metal oxides, and quantum dots, under relatively benign conditions and using microwave irradiation have been summarized in this chapter. The compilation could be very useful for identifying some of the simplest techniques for expedient assembly of various nanomaterials with well-defined morphologies for both laboratory and industrial applications. Rapid growth and exploration in this area is expected to continue in the coming years – uniformly smaller-sized nanoparticles could be easily synthesized using the selective and preferential heating technique of microwave irradiation, especially when a polar reaction media and metals that couple strongly to microwaves are involved.

References

1. H. Gleiter, *Prog. Mater. Sci.*, 1989, **33**, 223–315.
2. K. J. Rao, B. Vaidhyanathan, M. Ganguli and P. A. Ramakrishnan, *Chem. Mater.*, 1999, **11**, 882–895.
3. S. Verma, P. A. Joy, Y. B. Khollam, H. S. Potdar and S. B. Deshpande, *Mater. Lett.*, 2004, **58**, 1092–1095.
4. F. Bensebaa, F. Zavaliche, P. L'Ecuyer, R. W. Cochrane and T. Veres, *J. Colloid Interface Sci.*, 2004, **277**, 104–110.
5. A. L. Rogach, D. Nagesha, J. W. Ostrander, M. Giersig and N. A. Kotov, *Chem. Mater.*, 2000, **12**, 2676–2685.
6. J. J. Zhu, O. Palchik, S. G. Chen and A. Gedanken, *J. Phys. Chem. B*, 2000, **104**, 7344–7347.
7. A. V. Murugan, R. S. Sonawane, B. B. Kale, S. K. Apte and A. V. Kulkarni, *Mater. Chem. Phys.*, 2001, **71**, 98–102.
8. F.-K. Liua, Y.-C. Changb, F.-H. Koa and T.-C. Chu, *Mater. Lett.*, 2004, **58**, 373–377.
9. A. R. Tao, S. Habas and P. D. Yang, *Small*, 2008, **4**, 310–325.
10. L. N. Lewis, *Chem. Rev.*, 1993, **93**, 2693–2730.
11. R. M. Crooks, M. Q. Zhao, L. Sun, V. Chechik and L. K. Yeung, *Acc. Chem. Res.*, 2001, **34**, 181–190.
12. S. A. Maier, M. L. Brongersma, P. G. Kik, S. Meltzer, A. A. G. Requicha and H. A. Atwater, *Adv. Mater.*, 2001, **13**, 1501–1505.
13. P. V. Kamat, *J. Phys. Chem. B*, 2002, **106**, 7729–7744.
14. L. A. Peyser, A. E. Vinson, A. P. Bartko and R. M. Dickson, *Science*, 2001, **291**, 103–106.
15. M. Moskovits, *J. Raman Spectrosc.*, 2005, **36**, 485–496.
16. Z. Q. Tian, *J. Raman Spectrosc.*, 2005, **36**, 466–470.

17. H. X. Xu, E. J. Bjerneld, M. Käll and L. Börjesson, *Phys. Rev. Lett.*, 1999, **83**, 4357–4360.
18. T. A. Taton, C. A. Mirkin and R. L. Letsinger, *Science*, 2000, **289**, 1757–1760.
19. S. R. Nicewarner-Peña, R. G. Freeman, B. D. Reiss, L. He, D. J. Peña, I. D. Walton, R. Cromer, C. D. Keating and M. J. Natan, *Science*, 2001, **294**, 137–141.
20. A. J. Haes and R. P. Van Duyne, *J. Am. Chem. Soc.*, 2002, **124**, 10596.
21. X. H. Huang, I. H. El-Sayed, W. Qian and M. A. El-Sayed, *J. Am. Chem. Soc.*, 2006, **128**, 2115–2120.
22. Z. L. Wang, *Adv. Mater.*, 1998, **10**, 13.
23. S. J. Oldenburg, R. D. Averitt, S. L. Westcott and N. J. Halas, *Chem. Phys. Lett.*, 1998, **288**, 243–247.
24. C. J. Orendorff, T. K. Sau and C. J. Murphy, *Small*, 2006, **2**, 636–639.
25. Y. Qu, R. Porter, F. Shan, J. D. Carter and T. Guo, *Langmuir*, 2006, **22**, 6367–6374.
26. F. Gao, Q. Lu and S. Komarneni, *Chem. Mater.*, 2005, **17**, 856–860.
27. P. V. Silvert, R. H. Urbina and K. T. Elhsissen, *J. Mater. Chem.*, 1997, **7**, 293–299.
28. J. H. Yang, L. H. Lu, H. S. Wang, W. D. Shi and H. J. Zhang, *Cryst. Growth Des.*, 2006, **6**, 2155–2158.
29. S. S. Shankar, A. Rai, A. Ahmad and M. Shastry, *J. Colloid Interface Sci.*, 2004, **275**, 496–501.
30. S. Kundu, M. Mandal, S. K. Ghosh and T. Pal, *J. Colloid Interface Sci.*, 2004, **272**, 134–144.
31. J. L. Marignier, J. Belloni, M. O. Delcourt and J. P. Chevalier, *Nature*, 1985, **317**, 344–345.
32. S. Kundu, L. Peng and H. Liang, *Inorg. Chem.*, 2008, **47**, 6344–6352.
33. J. Wang and Z. Wang, *Mater. Lett.*, 2007, **61**, 4149–4151.
34. F.-K. Liu, Y.-C. Chang, F.-H. Ko and T.-C. Chu, *Mater. Lett.*, 2004, **58**, 373–377.
35. C. Fan, W. Li, S. Zhao, J. Chen and X. Li, *Mater. Lett.*, 2008, **62**, 3518–3520.
36. M. N. Nadagouda and R. S. Varma, *Crystal Growth Des.*, 2007, **7**, 686–689.
37. B. Baruwati and R. S. Varma, *ChemSusChem*, 2009, **2**, 1041–1044.
38. Y. Cui, B. Ren, J. L. Yao, R. A. Gu and Z. Q. Tian, *J. Phys. Chem. B.*, 2006, **110**, 4002–4006.
39. Z. Y. Li, J. Yuan, Y. Chen, R. E. Palmer and J. P. Wilcoxon, *Appl. Phys. Lett.*, 2005, **87**, 243103–243106.
40. B. J. Wiley, Y. C. Chen, J. M. McLellan, Y. J. Xiong, Z. Y. Li, D. Ginger and Y. N. Xia, *Nano. Lett.*, 2007, **7**, 1032–1036.
41. B. J. Wiley, S. H. Im, Z. Y. Li, J. McLellan, A. Siekkinen and Y. A. Xia, *J. Phys. Chem. B*, 2006, **110**, 15666–15675.
42. J. T. Zhang, X. L. Li, X. M. Sun and Y. D. Li, *J. Phys. Chem. B*, 2005, **109**, 12544–12548.

43. S. Pande, S. K. Ghosh, S. Praharaj, S. Panigrahi, S. Basu, S. Jana, A. Pal, T. Tsukuda and T. Pal, *J. Phys. Chem. C*, 2007, **111**, 10806–10813.
44. B. Hu, S.-B. Wang, K. Wang, M. Zhang and S.-H. Yu, *J. Phys. Chem. C*, 2008, **112**, 11169–11174.
45. S. Kundu, K. Wang and H. Liang, *J. Phys. Chem. C*, 2009, **113**, 134–141.
46. (a) B. Baruwati, V. Polshettiwar and R. S. Varma, *Green. Chem.*, 2009, **11**, 926–930; (b) M. N. Nadagouda, A. Castle, R. C. Murdock, S. M. Hussain and R. S. Varma, *Green Chem.*, 2010, **12**, 114–122; (c) M. C. Moulton, L. K. Braydich-Stolle, M. N. Nadagouda, S. Kunzelman, S. M. Hussain and R. S. Varma, *Nanoscale*, 2010, **2**, DOI: 10.1039/c0nr00046a.
47. J. Chen, J. Wang, X. Zhang and Y. Jin, *Mater. Chem. Phys.*, 2008, **108**, 421–424.
48. Y. Luo, *Mater. Lett.*, 2007, **61**, 1873–1875.
49. A. Pal, S. Shah, D. Chakraborty and S. Devi, *Aust. J. Chem.*, 2008, **61**, 833–836.
50. G. Schmid, *Nanoparticles: From Theory to Application*, Wiley-VCH Verlag GmbH, Weinheim, 2004.
51. G. A. Ozin and A. C. Arsenault, *Nanochemistry*, RSC Publishing, Cambridge, UK, 2005.
52. J. F. Wang, M. S. Gudiksen, X. F. Duan, Y. Cui and C. M. Lieber, *Science*, 2001, **293**, 1455–1457.
53. J. Hu, L. Li, W. Yang, L. Manna, L. Wang and A. P. Alivisatos, *Science*, 2001, **292**, 2060–2063.
54. A. Fujishima and K. Honda, *Nature*, 1972, **238**, 37–38.
55. M. Anpo and M. Takeuchi, *J. Catal.*, 2003, **216**, 505–516.
56. S. Akbar and S. Yoo, *Chem. Sens.*, 2004, **20**, 30–31.
57. Fujishima, K. Hashimoto and T. Watanabe, *TiO$_2$ Photocatalysis. Fundamentals and Applications*, BKC Inc, Tokyo, 1999, pp. 14–176.
58. G. K. Mor, K. Shanka, M. Paulose, O. K. Varghese and C. A. Grimes, *Nano Lett.*, 2006, **6**, 215–218.
59. G. Y. Gao, K. L. Yao, Z. L. Liu, J. Zhang, X. L. Li, J. Q. Zhang and N. Liu, *J. Magn. Magn. Mater.*, 2007, **313**, 210.
60. Y. B. Khollam, S. R. Dhage, H. S. Potdar, S. B. Deshpande, P. P. Bakare, S. D. Kulkarni and S. K. Date, *Mater. Lett.*, 2002, **56**, 571–577.
61. G. J. Wilson, G. D. Will, R. L. Frost and S. A. Montgomery, *J. Mater. Chem.*, 2002, **12**, 1787–1791.
62. C.-C. Chung, T.-W. Chung and T. C.-K. Yang, *Ind. Eng. Chem. Res.*, 2008, **47**, 2301–2307.
63. T. Suprabha, H. G. Roy, J. Thomas, K. P. Kumar and S. Mathew, *Nanoscale Res Lett.*, 2009, **4**, 144–152.
64. X. Jia, W. He, X. Zhang, H. Zhao, Z. Li and Y. Feng, *Nanotechnology*, 2007, **18**, 075602–075609.
65. Z. Chen, W. Li, W. Zeng, M. Li, J. Xiang, Z. Zhou and J. Huang, *Mater. Lett.*, 2008, **62**, 4343–4344.
66. S. In, A. Orlov, R. Berg, F. Garcia, S. P. Jilmenez, S. M. Tikhov, S. D. Wright and M. R. Lambert, *J. Am. Chem. Soc.*, 2007, **129**, 13790–13791.

67. W. Zhao, W. Ma, C. Chen, J. Zhao and Z. Shuai, *J. Am. Chem. Soc.*, 2004, **126**, 4728–4729.
68. R. Asahi, T. Morikawa, T. Ohwaki, K. Aoki and Y. Taga, *Science*, 2001, **293**, 269–271.
69. S. U. M. Khan, M. Al-Shahry and W. B. Ingler Jr, *Science*, 2002, **297**, 2243–2245.
70. S. Sakthivel and H. Kisch, *Angew. Chem. Int. Ed.*, 2003, **42**, 4908–4911.
71. R. Asahi, T. Morikawa, T. Ohwaki, K. Aoki and Y. Taga, *Science*, 2001, **293**, 269–271.
72. P. Zhang, B. Liua, S. Yina, Y. Wang, V. Petrykina, M. Kakihanaa and T. Satoa, *Mater. Chem. Phys.*, 2009, **116**, 269–272.
73. M. S. Arnold, P. Avouris, Z. W. Pan and Z. L. Wang, *J. Phys. Chem. B*, 2003, **107**, 659–663.
74. W. I. Park, J. S. Kim, G. C. Yi, M. H. Bae and H. J. Lee, *Appl. Phys. Lett.*, 2004, **85**, 5052.
75. G. S. T. Rao and D. T. Rao, *Sens. Actuators B*, 1999, **55**, 166–169.
76. X. Wang, C. J. Summers and Z. L. Wang, *Nano Lett.*, 2004, **4**, 423–426.
77. C. J. Lee, T. J. Lee, S. C. Lyu, Y. Zhang, H. Ruh and H. J. Lee, *Appl. Phys. Lett.*, 2003, **81**, 3648–3650.
78. C. X. Xu, X. W. Sun and B. J. Chen, *Appl. Phys. Lett.*, 2004, **84**, 1540–1542.
79. B. Banerjee, S. H. Jo and Z. F. Ren, *Adv. Mater.*, 2004, **16**, 2028–2032.
80. K. Tominaga, N. Umezu, I. Mori, T. Ushiro, T. Moriga and I. Nakabayashi, *Thin Solid Films*, 1998, **334**, 35–39.
81. Z. Zhou, Y. Li, L. Liu, Y. Chen, S. B. Zhang and Z. Chen, *J. Phys. Chem. C*, 2008, **112**, 13926–13931.
82. T. Minami, S. Ida, T. Miyata and Y. Minamino, *Thin Solid Films*, 2003, **445**, 268–273.
83. D. M. Bagnall, Y. F. Chen, Z. Zhu, T. Yao, S. Koyama, M. Y. Shen and T. Goto, *Appl. Phys. Lett.*, 1997, **70**, 2230–2232.
84. H. Cao, J. Y. Xu, D. Z. Zhang, S. H. Chang, S. T. Ho, E. W. Liu, X. Seelig and R. P. H. Chang, *Phys. Rev. Lett.*, 2000, **84**, 5584–5587.
85. M. H. Huang, S. Mao, H. Feick, H. Yan, Y. Wu, H. Kind, E. Weber, R. Russo and P. Yang, *Science*, 2001, **292**, 1897–1899.
86. X. Duan, Y. Huang, R. Agarwal and C. M. Lieber, *Nature*, 2003, **421**, 241–245.
87. Y. Yan, P. Liu, J. G. Wen, B. To and M. M. Al-Jassim, *J. Phys. Chem. B*, 2003, **107**, 9701–9704.
88. H. Yu, Z. Zhang, M. Han, X. Hao and F. Zhu, *J. Am. Chem. Soc.*, 2005, **127**, 2378–2379.
89. Y. Peng, A.-W. Xu, B. Deng, M. Antonietti and H. Cölfen, *J. Phys. Chem. B*, 2006, **110**, 2988–2993.
90. X. P. Gao, Z. F. Zheng, H. Y. Zhu, G. L. Pan, J. L. Bao, F. Wu and D. Y. Song, *Chem. Commun.*, 2004, 1428–1429.
91. T. L. Sounart, J. Liu, J. A. Voigt, J. Hsu, E. D. Spoerke, Z. Tian and Y. Jiang, *Adv. Funct. Mater.*, 2006, **16**, 335–344.

92. J. Y. Lao, J. G. Wen and Z. F. Ren, *Nano Lett.*, 2002, **2**, 1287–1291.
93. H. E. Unalan, P. H. N. Rupesinghe, S. Dalal, W. I. Milne and G. A. J. Amaratunga, *Nanotechnology*, 2008, **19**, 255608–2556012.
94. L. Zhang and Y.-J. Zhu, *Appl. Phys. A*, 2009, **97**, 847–852.
95. S. Cho, S.-H. Jung and Kun-Hong Lee, *J. Phys. Chem. C*, 2008, **112**, 12769–12776.
96. J. Huang, C. Xia, L. Cao and X. Zeng, *Mater. Sci. Eng. B*, 2008, **150**, 187–193.
97. R. F. Ziolo, E. P. Giannelis, B. A. Weinstein, M. P. O'Horo, B. N. Ganguli and V. Mehrotra, *Science*, 1992, **257**, 219–223.
98. F. Jorgensen, *The Complete Handbook of Magnetic Recording*, McGraw-Hill, New York, 1995.
99. P. Alivisatos, *Science*, 1996, **271**, 933–937.
100. M. V. Berkovsky, F. V. Medvedev and S. M. Krakov, *Magnetic Fluids Engineering Applications*, Oxford University Press, Oxford, UK, 1993.
101. E. Manova, T. Tsoncheva, D. Paneva, I. Mitov, K. Tenchev and L. Petrov, *Appl. Catal. A*, 2004, **277**, 119–127.
102. L. Satyanarayana, K. M. Reddy and S. V. Manorama, *Mater. Chem. Phys.*, 2003, **82**, 21–26.
103. K. M. Reddy, L. Satyanarayana, S. V. Manorama and D. K. Misra, *Mater. Res. Bull.*, 2004, **39**, 1491–1498.
104. L. X. Tiefenauer, A. Tscgirky, G. Kuhne and R. Y. Andres, *Magn. Reson. Imag.*, 1996, **14**, 391–402.
105. Z. X. Tang, C. M. Sorensen, K. J. Klabunde and G. C. Hadjipanayis, *Phys. Rev. Lett.*, 1991, **67**, 3602–3605.
106. G. Kulkarni, U. Kannan, K. R. T. Arunarkavalli and C. N. R. Rao, *Phys. Rev. B*, 1994, **49**, 724–727.
107. E. Viroonchatapan, M. Ueno, H. Sato, I. Adachi, H. Nagae, K. Tazawa and I. Horikoshi, *Pharm. Res.*, 1995, **12**, 1176–1183.
108. P. K. Gupta and C. T. Hung, *Life Sci.*, 1989, **44**, 175–186.
109. U. Lüders, A. Barthélémy, M. Bibes, K. Bouzehouane, S. Fufil, E. Jacquet, J.-P. Contour, J.-F. Bobo, J. Fontcuberta and A. Fert, *Adv. Mater.*, 2006, **18**, 1733–1736.
110. T. Sugimoto, Y. Shimotsuma and H. Itoh, *Powder Technol.*, 1998, **96**, 85–89.
111. T. Pannaparayil, R. Marande and S. Komarneni, *J. Appl. Phys.*, 1991, **69**, 5349–5358.
112. A. J. Lopez Perez, A. M. Quintela, J. Mira, J. Rivas and W. S. Charles, *J. Phys. Chem. B.*, 1997, **101**, 8045–8047.
113. M. P. V. K. Shafi, A. Gedanken, R. Prozorov and J. Balogh, *Chem. Mater.*, 1998, **10**, 3445–3450.
114. S. Verma, P. A. Joy, Y. B. Khollam, H. S. Potdar and S. B. Deshpande, *Mater. Lett.*, 2004, **58**, 1092–1095.
115. X. Zhang and Q. Li, *Mater. Lett.*, 2008, **62**, 988–990.
116. H. Katsuki, A. Shiraishi, S. Komernani, W. Moon, S. Toh and K. Kaneko, *J. Ceram. Soc. Jpn.*, 2004, **112**, 384–387.

117. B. Baruwati, M. N. Nadagouda and R. S. Varma, *J. Phys. Chem. C*, 2008, **112**, 18399–18404.
118. V. Polshettiwar, B. Baruwati and R. S. Varma, *ACS Nano*, 2009, **3**, 728–736.
119. W. J. Parak, T. Pellegrino and C. Plank, *Nanotechnology*, 2005, **16**, R9.
120. A. A. Mamedov, A. Belov, M. Giersig, N. N. Mamedova and N. A. Kotov, *J. Am. Chem. Soc.*, 2001, **123**, 7738–7739.
121. V. I. Klimov, A. A. Mikhailovsky, S. Xu, A. Malko, J. A. Hollingsworth, C. A. Leatherdale, H.-J. Eisler and M. G. Bawendi, *Science*, 2000, **290**, 314–317.
122. V. C. Sundar, H.-J. Eisler and M. G. Bawendi, *Adv. Mater.*, 2002, **14**, 739–743.
123. M. Bruchez Jr., M. Moronne, P. Gin, S. Weiss and A. P. Alivisatos, *Science*, 1998, **281**, 2013–2016.
124. W. C. W. Chan and S. Nie, *Science*, 1998, **281**, 2016–2018.
125. X. Michalet, F. F. Pinaud, L. A. Bentolila, J. M. Tsay, S. Doose, J. J. Li, G. Sundaresan, A. M. Wu, S. S. Gambhir and S. Weiss, *Science*, 2005, **307**, 538–544.
126. C. B. Murray, D. J. Norris and M. G. Bawendi, *J. Am. Chem. Soc.*, 1993, **115**, 8706–8715.
127. Z. A. Peng and X. Peng, *J. Am. Chem. Soc.*, 2001, **123**, 183–184.
128. L. Qu, Z. A. Peng and X. Peng, *Nano Lett.*, 2001, **1**, 333–336.
129. D. V. Talapin, A. L. Rogach, A. Kornowski, M. Haase and H. Weller, *Nano Lett.*, 2001, **1**, 207–211.
130. H. Qian, X. Qiu, L. Li and J. Ren, *J. Phys. Chem. B*, 2006, **110**, 9034–9040.
131. Y. Zhao, J.-M. Hong and J.-J. Zhu, *J. Crystal Growth*, 2004, **270**, 438–445.
132. J. Zhu, O. Palchik, S. Chen and A. Gedanken, *J. Phys. Chem. B*, 2000, **104**, 7344–7347.
133. W. Schumacher, A. Nagy, W. J. Waldman and P. K. Dutta, *J. Phys. Chem. C*, 2009, **113**, 12132–12139.
134. H. Qian, L. Li and J. Ren, *Mater. Res. Bull.*, 2005, **40**, 1726–1736.
135. (a) A. T. Bell, *Science*, 2003, **299**, 1688; (b) R. Schlgl and S. B. Abd Hamid, *Angew. Chem., Int. Ed.*, 2004, **43**, 1628; (c) D. Astruc, F. Lu and J. R. Aranzaes, *Angew. Chem., Int. Ed.*, 2005, **44**, 7852; (d) G. A. Somorjai, F. Tao and J. Y. Park, *Top. Catal.*, 2008, **47**, 1; (e) R. A. Van Santen, *Acc. Chem. Res.*, 2009, **42**, 57.
136. (a) A. Hu, G. T. Yee and W. Lin, *J. Am. Chem. Soc.*, 2005, **127**, 12486; (b) N. T. S. Phan, C. S. Gill, J. V. Nguyen, Z. J. Zhang and C. W. Jones, *Angew. Chem., Int. Ed.*, 2006, **45**, 2209; (c) R. Abu-Reziq, H. Alper, D. Wang and M. L. Post, *J. Am. Chem. Soc.*, 2006, **128**, 5279; (d) C. O. Dalaigh, S. A. Corr, Y. Gunko and S. Connon, *Angew. Chem., Int. Ed.*, 2007, **46**, 4329; (e) V. Polshettiwar and R. S. Varma, *Org. Biomol. Chem.*, 2009, **7**, 37; (f) F. Shi, M. K. Tse, S. Zhou, M.-M. Pohl, J. Radnik, S. Hubner, K. Jahnisch, A. Bruckner and M. Beller, *J. Am.*

Chem. Soc., 2009, **131**, 1775; (g) V. Polshettiwar, B. Baruwati and R. S. Varma, *Green Chem.*, 2009, **11**, 127.
137. (a) M. T. Reetz, *Size-selective synthesis of nanostructured metal and metal oxide-colloids and their use as catalysts*, in *Nanoparticles and Catalysis*, ed. D. Astruc, Wiley-VCH Verlag GmbH, Weinheim, 2008, pp. 255–279; (b) V. Polshettiwar, M. N. Nadagouda and R. S. Varma, *Chem. Commun.*, 2008, 6318.
138. S. H. Joo, J. Y. Park, C.-K. Tsung, Y. Yamada, P. Yang and G. A. Somorjai, *Nat. Mater.*, 2009, **8**, 132.
139. X. Xie, Y. Li, Z.-Q. Liu, M. Hartua and W. Shem, *Nature*, 2009, **458**, 746.
140. I. Lee, F. Delbecq, R. Morales, M. A. Albiter and F. Zaera, *Nat. Mater.*, 2009, **8**, 132.
141. B. D. Chandler and J. D. Gilbertson, *Top. Organomet. Chem.*, 2006, **20**, 97.
142. H. Lang, A. May, B. L. Iversen and B. D. Chandler, *J. Am. Chem. Soc.*, 2003, **125**, 14832.
143. C. J. Crump, J. D. Gilbertson and B. D. Chandler, *Top. Catal.*, 2008, **49**, 233.
144. C. G. Long, J. D. Gilbertson, G. Vijayaraghavan, K. J. Stevenson, C. J. Pursell and B. D. Chandler, *J. Am. Chem. Soc.*, 2008, **130**, 10103.
145. J. P. M. Niederer, A. B. J. Arnold, W. F. Hölderich, B. Spliethof, B. Tesche, M. Reetz and H. Bönnemann, *Top. Catal.*, 2002, **18**, 265.
146. L. D. Pachon, M. B. Thathagar, F. Hartl and G. Rothenberg, *Phys. Chem. Chem. Phys.*, 2006, **8**, 151.
147. P.-H. Phua, L. Lefort, J. A. F. Boogers, M. Tristany and J. G. de Vries, *Chem. Commun.*, 2009, 3747.
148. X. Zong, Y. Na, F. Wen, G. Ma, J. Yang, D. Wang, Y. Ma, M. Wang, L. Sunbc and C. Li, *Chem. Commun.*, 2009, 4536.
149. A.-H. Lu, E. L. Salabas and F. Schuth, *Angew. Chem. Ind., Ed.*, 2007, **46**, 1222.
150. C. Xu, K. Xu, H. Gu, R. Zheng, H. Liu, X. Zhang, Z. Guo and B. Xu, *J. Am. Chem. Soc.*, 2004, **126**, 9938.
151. L. X. Chen, T. Liu, M. C. Thurnauer, R. Csencsits and T. Rajh, *J. Phys. Chem. B*, 2002, **106**, 8539.
152. T. Rajh, L. X. Chen, K. Lukas, T. Liu, M. C. Thurnauer and D. M. Tiede, *J. Phys. Chem. B*, 2002, **106**, 10543.
153. (a) B. Baruwati, D. Guin and S. V. Manorama, *Org. Lett.*, 2007, **9**, 5377; (b) D. Guin, B. Baruwati and S. V. Manorama, *Org. Lett.*, 2007, **9**, 1419.
154. T. Hirakawa, S. Tanaka, N. Usuki, H. Kanzaki, M. Kishimoto and M. Kitamura, *Eur. J. Org. Chem.*, 2009, 789.
155. O. Gleeson, R. Tekoriute, Y. K. Gunko and S. J. Connon, *Chem. Eur. J.*, 2009, **15**, 5669.
156. V. Cadierno, J. Francos and J. Gimeno, *Chem. Eur. J.*, 2008, **14**, 6601.
157. V. Polshettiwar and R. S. Varma, *Chem. Eur. J.*, 2009, **15**, 1582.
158. L. D. Pachon and G. Rothenberg, *Appl. Organomet. Chem.*, 2008, **22**, 288.

159. V. Polshettiwar and R. S. Varma, *Curr. Opin. Drug Discovery Dev.*, 2007, **10**, 723.
160. (a) A. P. Brogan, T. J. Dickerson and K. D. Janda, *Angew. Chem. Int. Ed.*, 2006, **45**, 8100; (b) Y. Hayashi, S. Samanta, H. Gotoh and H. Ishikawa, *Angew. Chem. Int. Ed.*, 2008, **47**, 6634; (c) J. Huang, X. Zhang and D. W. Armstrong, *Angew. Chem. Int. Ed.*, 2007, **46**, 9073.
161. D. G. Blackmond, A. Armstrong, V. Coombe and A. Wells, *Angew. Chem. Int. Ed.*, 2007, **46**, 3798.
162. V. Polshettiwar, B. Baruwati and R. S. Varma, *Chem. Commun.*, 2009, **20**, 1837.

Subject Index

A^3 coupling reactions, 16–17
acetals, 44–45, 75, 84, 115
acetonitrile, 14, 135
　aqueous, 66, 132, 138
acetylcholinesterase, 126, 128
acridines, 59–60
acrylamide
　emulsion polymerization, 155
　graft co-polymerization, 148, 150
　N, N'-methylenebis- (MBA), 149
　N-isopropyl- (NiPAM), 152–154, 156
　nanocomposites with calcium phosphate, 155, 171
　radical homogeneous polymerization, 146–147
acrylic acid polymerization, 146–147, 149
acrylonitrile polymerization, 149–150
addition reactions
　1,3-dipolar cycloadditions, 31–36, 47, 160, 164–168, 172
　metal-catalyzed, 11–17, 45–46
(S)-adenosylhomocysteine hydrolase, 127, 129
adenylate kinase, 126–129
alcohol dehydrogenase, 126, 128
alcohols
　allyl alcohols, 75
　deuteration, 19–20
　isomerization reactions, 17–19, 46
　oxidation to ketones, 42
　propargylic alcohols, 16, 19, 46

aldehydes
　A^3 coupling reactions, 16–17
　crotonoaldehydes, 44, 48–49
　cyanation, 78–79
　heterocycles from, 92–93, 96–102, 104–112, 118–119
aldol condensations, 109–112
Aliquat (tricaprylmethylammonium chloride), 161, 172
alkenes
　deuteration of cycloalkenes, 20–21
　1,3-dipolar cycloadditions, 31
　Heck reaction with aryl halides, 71
　hydrogenation, 40, 48, 196
alkyl dihalides, 94
alkyne-enol ethers, 44
alkynes
　arylation, 75, 84
　cycloadditions, 29, 31–32, 35, 47
　hydration to carbonyl compounds, 14–17
　vinylphosphine oxides from, 15
allylation-isomerization processes, 18
amides
　aminocarbonylation, 116
　dehydration to nitriles, 14
　halide carbonylation, 80
　intramolecular amidation, 29, 47, 105
　N-arylation, 114
　nitrile hydration to, 11–14, 45–46
　synthesis of polyamides, 158–160
amination
　polysaccharides, 164–166
　retro-reductive, 40–42, 115

amines
    aminocarbonylation, 81, 85, 116
    N-arylation, 24–26, 114
    diamines, substituted, 113–114
    heterocycles from, by double N-alkylation, 94
    ketones from, 40, 115–116
    secondary, from primary, 41, 48
    triazines from, 93
amino acids
    arylation, 24–25, 118
    reducing agents, 182, 206
    spiro, 95–96
aminocarbonylation reactions, 81, 85, 116
AMPS (2-acrylamido-2-methylpropanesulfonic acid), 147
γ-amylase, 127, 129, 137
anions, in ionic liquids, 134
anisole, 4-iodo-, 85–86
anticancer drugs, 77, 105–106
aqueous media *see* water
aryl alkynes, 75, 84
aryl azides, 27–28
aryl halides
    carbonylation reactions, 80–83
    conversion into phenols and azides, 27–28, 46–47
    cyanation reactions, 79–80
    Heck reaction, 71–75
    Hiyama reaction, 78
    N-arylation of amino acids, 118
    Suzuki-Miyaura reaction, 56–71
aryl nitriles, 79–80, 85, 92
aryl silanes, deuteration, 21, 46
aryl triflates
    carbonylation reactions, 80–83
    Hiyama reaction, 78
    Suzuki-Miyaura reaction, 67–68, 164
arylation of amines and amino acids, 24–26, 114
arylboronic acids, 56–60, 63, 65–71, 83, 164
atom transfer radical polymerization (ATRP), 156

azides
    1,3-dipolar cycloadditions, 31–36, 47, 164–168, 172
    aryl, from halides, 27–28

Barbier-type propargylations, 16
Beckmann rearrangement, 135
benzamides, 80–81, 85, 116
benzimidazoles, 93, 97
benzo*[b]*pyrans, 96, 106
benzoic acids, 22, 82–84
benzo[4,5]imidazo[1,2-*a*]pyrimidines, 102
benzonitrile, 4-methoxy-, 85
benzopyrano[4,3-*c*]pyrazoles, 96
benzothiazepinones, 111
benzothiazoles, 69, 111, 119
benzoxazines, 109
2-benzylidenemalononitrile, 198
biaryls, Suzuki-Miyaura reaction, 56, 83–84
bicyclo[2.2.2]oct-7-enes, 101
Biginelli reactions, 6, 97
bio-solvents, 3
biocatalysis *see* enzymes
biphasic media, 3, 7, 10, 16
    see also solvents
Biphenomycin B, 61–62
biphenyl, 4,5-dimethoxy-2-vinyl-2'-pivaloylamino-, 83
bond migrations, 18–20
boron compounds
    arylboronic acids, 56–60, 63, 65–71, 83, 164
    Suzuki-Miyaura reaction, 66–67, 160–161
buflavine analogues, 63
butyl methacrylate, 154

cadmium selenide and telluride, 194
calcium phosphate nanocomposites, 155, 171
carbohydrates
    see also starch; sugars
    dendrimer coupling, 37, 47
    deuteration, 20
    Ferrier carbocyclization, 31
    oligonucleotide anchoring, 35

Subject Index

carbon-carbon bonds
  C=C bond migrations, 18–20
  cross-coupling polymerizations, 160–162, 172
  cross-coupling reactions, 55–56, 63, 71, 79
carbon monoxide, 80, 82–83
carbon nanotubes, polymer-grafted, 155
carbon-nitrogen bonds, 94
carbonyl compounds
  see also ketones
  β-amino-, 112–113
  hydration of alkynes to, 14–17
  oxime hydrolysis to, 12
  preparation of α,β-unsaturated, 19, 46
  propargylation, 16–17
  reduction of α,β-unsaturated, 38
carbonylation reactions, 80–83, 85–86, 116
carboxylic acids
  aryl triflates, 81–82
  decarboxylation, 112
  from nitriles, 11
carboxylmethylcellulose (CMC), 184–185
κ-carrageenan, 149, 169
catalysts
  see also metal-catalyzed reactions; phase-transfer catalysis
  co-catalysis, 26, 76, 133
  heterogeneous, 7, 55, 68
  leach-proof, 68, 198
  nano-catalysis, 4, 7, 195–204, 208
  organocatalysis, 203–204
  photocatalysis, 188
  as susceptors, 7
cationic doping of TiO_2, 188
cationic polymerization, 162, 172
cationization of polysaccharides, 163–164
cellulases, 126, 128, 130
cellullar adhesion, 66
cellulose
  carboxymethyl-, 184–185
  co-polymerization, 148
cephem analogues, 25–26
cerium compounds, 148–150

cetyl (trimethyl)ammonium bromide (CTAB), 179, 205–206
chain transfer agents, 147, 156
chitosan, 69, 149–150, 161, 164, 166, 180
chlorine dioxide, 43–44
chromanediones, 96
chromenes, 98, 106, 118
chromium-based nanoparticles, 209
α-chymotrypsin, 125, 131
cinnamic acid and esters
  decarboxylation, 112
  Heck reaction, 74–75, 84
  transfer hydrogenation, 40, 48
'click reactions,' 31–32, 34–35, 160, 164–168, 172
CMC (carboxylmethylcellulose), 184–185
CMUI see ultrasound
cobalt-based nanoparticles, 196, 208–209
colloids
  polymer, 151–152, 154
  semiconductor, 194
  TiO_2, 187
colour change, nanoparticle formation, 206–207
controlled/living radical polymerizations, 155–157
copper-based catalysts
  addition reactions, 16
  cyclization reactions, 28, 32–37, 47, 106
  iron-copper co-catalysis, 26
  palladium-copper co-catalysis, 76
  redox reactions, 43–44
  substitution reactions, 24–28, 46, 118
copper selenides, 195
core-shell co-polymers, 156
core-shell nanoparticles, 154, 195–196
coumarin, 4-hydroxy-, 98, 106–108
coupling reactions see A^3 coupling reactions; cross-coupling reactions
creatine phosphokinase, 126, 128
cross-coupling reactions
  aldol condensations, 111–112
  carbonylation reactions, 80–83, 85–86
  cyanation reactions, 79–80, 85

220 Subject Index

Heck reaction, 71–75, 84
Hiyama reaction, 78–79
introduced, 55–56
representative syntheses, 83–86
Sonogashira reaction, 75–78, 84
Stille reaction, 60, 78
Suzuki-Miyaura reaction, 56–71, 83–84, 160–162, 164
cross-metathesis, 44
crotonaldehyde, 44
crystallization, 58, 155, 179, 183–184, 187–191, 194
CTAB (cetyl (trimethyl)ammonium bromide), 179, 205–206
cyanation reactions, 79–80, 85
cyclization reactions
  see also heterocyclic compounds
  1,3-dipolar cycloadditions, 31–36, 47, 160, 164–168, 172
  amine double *N*-alkylation, 94
  metal-catalyzed, 28–38, 47–48
  peptide esters, 103–104
  polytriazoles, 160
  ureas, 30, 47, 93
cycloalkane deuteration, 20
cyclodextrins, 160
cyclohexa-2,5-dien-1-one, 4-hydroxy-, 77–78
cyclohexane-1,3-dione, 93
cyclohexane derivatives
  cyclohexanones from sugars, 31
  sulfides and sulfoxides, 112
  tricyclohexylphosphine, 59
cyclopentenones, 29–30
cytochrome *c* reductase, 126, 128

de-allylation, 200–201
dendrimers
  azide cycloadditions and, 36–38, 47, 164–168, 172
  encapsulation of nanostructures, 196
dendritic nanostructures, 193–194, 198
deprotection, 25, 44–45, 115
deuteration, 19–21, 46, 84
2,7-DHN (2,7-dihydroxynaphthalene), 179, 182, 205–206

dielectric constants, 134–136
dielectric heating, 123, 177–178
Diels-Alder-type cycloadditions, 31
diffusion limited reactions, 5
dioxanes, 109, 118
dioxoles, 203
dipole moments, 137, 152, 178
dipole orientation, 4
dispersion polymerization, 151
DMSO (dimethylsulfoxide) oxidation, 43
dodecanoic acid, 12-amino, 158–159
dopamine, 198, 202
DOTA dendrimers, 38
drug delivery, 150, 178, 191
drug discovery
  anticancer drugs, 77, 105–106
  antiviral nucleosides, 65, 114–115
  aromatic amines, 117–118
  green chemistry and, 1–2
  nitrogen-containing heterocycles, 92–93, 98–106
  oxygen- and sulfur-containing heterocycles, 106–112
  synthetic heterocycles generally, 91–92, 203

electrophoresis, 131
emulsion polymerization, 151–157, 170–171
enaminoketones, 95
enantioselectivity, 30, 132, 137, 198, 201
energy efficiency
  green chemistry and, 11
  microwave heating, 124, 177, 187
enzymes
  anhydrous and hydrophobic conditions, 124–125
  low-temperature reactions, 137
  microwave acceleration, 124–125
  microwave inactivation, 124, 126–129, 133
  microwave non-thermal effects, 124, 126–127, 136–137
  protein digestion, 130–132
  reactions in aqueous organic solvents, 132–133

Subject Index   221

reactions in ionic liquids, 133–136
reactions in water or aqueous
 buffer, 125–132
representative procedures, 137–138
simultaneous cooling, 137
esters
 arylation of amino, 25
 cinnamic, 40
 thioesterification of polymers, 167–168
 trifluoromethanesulfonic acid p-nitrophenyl ester, 85
ethanone, 1-(1H-benzimidazol-2-yl)-2-(3- or 4- hydroxyphenyl), 97
ethers
 4-iodoanisole, 85
 alkyne-enol ethers, 44, 48
 isomerization to ketones, 18
 polyether synthesis, 157–158, 171
europium, 155

Ferrier carbocyclization, 31
ferrite nanoparticles, 191–194, 198, 208–209
fluorene, 2,7-dibromo-9,9-dioctyl-, 172
fluorous chemistry, 70–71
fluorous liquids, 3
free radicals see radical polymerization
furo[2,3-b]pyrazines, 34
furo[3,4-e]pyrazolo[3,4-b]pyridines, 102
furo[3',4':5,6]pyrido[2,3-d]pyrimidines, 102

α-galactosidase, 130
β-galactosidase, 127, 129, 133, 138
glucose 6-phosphate dehydrogenase, 126, 128
β-glucosidase, 130
glutamic oxaloacetic transaminase, 126, 128
glutathione, 183, 203–204, 206–207
glycopeptidomimetics, 34
gold-based catalysts
 addition reactions, 12, 14–15, 45–46
 Au/Ag co-catalysis, 30, 47
gold nanoparticles, 178–182, 196, 205

graft co-polymerization, 147–150, 169
granular activated carbon (GAC), 43
grape pomace see wine and grape pomace
green chemistry
 energy efficiency and, 11
 nanoparticle production, 182, 184, 201–203
 principles, 1–3
 water as a green solvent, 3–4, 10, 49, 92, 99

H/D-exchanges, 19–20, 46
Hanzsch reaction, 99
heat capacity of solvents, 123
heating see thermal effects
Heck reaction, 71–75, 84, 134–135
 Sonogashira-Heck alkynylation, 76
heteroaromatic compounds
 deuteration, 22
 from intermolecular amination, 30
heterocyclic compounds
 miscellaneous, 112–118
 nanocatalytic synthesis, 203–204
 nitrogen-containing, 92–106
 oxygen-containing, 106–111, 118
 palladacycles, 74–75, 79, 81, 85
 pharmaceutical importance, 91–92
 pyrido-fused, 100
 representative procedures, 118–119
 sulfur-containing, 111–112
heterogeneous catalysts, 7, 12, 21
 nano-particles as, 195, 198, 201, 203
 supported metals as, 55, 68, 70, 76
hexamethylenetetramine (HMT), 188–189, 207
HIV-1 integrase inhibitors, 97
HIV-1 protease inhibitors, 63, 81
Hiyama reaction, 78–79, 196
homopropargylic alcohol, 16
horseradish peroxidase, 126, 128
Huisgen reaction, 32, 35, 38, 164
hydration reactions
 alkynes to carbonyl compounds, 14–17
 nitriles to amides, 11–14, 202–203

hydrazine and its derivatives
  bicyclo[2.2.2]oct-7-enes from, 101
  cyclization, 94, 96, 204
  as reducing agent, 185
hydrazones, 24, 44, 110
hydrodechlorination, 23–24
hydrogenation
  nanocatalysis, 196
  transfer hydrogenations, 38–41
hydrolases, 127, 129, 133
hydrophobic solvents, 10
  enzyme activity, 125, 135–136
hydroxycarbonylation reactions, 81–82, 85–86

imidazo[1,2-*a*]pyridines, 65
imidazoles, 44, 93, 97 112–113
imidazolium compounds, 31, 133–135
imides
  N-arylation, 114
  synthesis of polyimides, 158–160, 171–172
indeno[1,2-*b*]quinolines, 101–102
indium-based catalysts, 16, 46
integrin antagonists, 66
ionic liquids (ILs), 132–136
iridium-based catalysts, 196
iron-based catalysts
  C-N coupling, 30
  iron-copper co-catalysis, 26
iron/ferrite nanoparticles, 182, 191–194, 196, 198, 208–209
  Fe_3O_4-PS-co-acrylamide, 155
isochromenones, 28, 106
isomerization reactions, 17–19, 30, 46, 196
isotopic exchange *see* deuteration
isoxazolo[5,4-*b*]pyridines, 102

ketals, 115
ketones
  alcohol oxidation to, 42
  from amines, 40, 115–116
  β-amino-, 112–113, 119
  cyclization of 1,3-diketones, 28, 204
  dioxanes from, 109–110

enaminoketones, 95
  from hydration of alkynes, 15
  from isomerization of alcohols and ethers, 17–18, 46
ketopantolactone, 42–43
Kröhnke reaction, 112

β-lactam arylation, 25–26, 114
lactate dehydrogenase, 126, 128–129
lead selenide, 195
lipases, 125, 127, 130, 135
living radical polymerizations, 155–157
low-temperature enzymatic reactions, 137
lysozyme, 126, 128, 131–132, 138

macrocycles, 61–63
macromolecular chain grafting, 147–150
magnetic nanoparticles, 155, 191–192, 197–198, 202
magnetically separable catalysts, 12, 198, 201
malonates, 135, 185
malononitriles, 96, 99–100, 106, 118, 198
manganese-based nanoparticles, 208–209
Mannich reaction, 112–113
Markovnikov additions, 15
MBA (*N,N'*-methylenebisacrylamide), 149
MCT (microwave coagulation therapy), 124
metal carbonyls, 80, 82, 116
metal-catalyzed reactions
  see also nano-catalysis; *individual metals*
  addition reactions, 11–17, 45–46
  advantages, 10–11
  cyclization reactions, 28–38, 47–48
  isomerization reactions, 17–19, 46
  redox reactions, 38–45, 48
  substitution reactions, 19–28, 46–47
metal nanoparticles, 178–186, 196
  see also individual metals
metal oxide nanoparticles, 186–195
methacrylamide, *N*-isopropyl-, 156
5'-methylthioadenosine phosphorylase, 127, 129

Subject Index

Meyer-Schuster rearrangement, 19
micelles, 151, 191
molybdenum-based nanoparticles, 197, 209
molybdenum hexacarbonyl, 80, 82, 116
MPA (3-mercaptopropionic acid), 167–168, 195

nano-catalysis, 4, 7, 195–204, 208
nanoferrite, 12–13, 201
nanoparticles
  carbon nanotubes, 155
  crosslinked, 154
  inorganic-organic composites, 154–155, 171, 188
  magnetic, 155, 191–192, 197–198
  metal nanoparticle synthesis, 178–186, 196
  metal oxide nanoparticle synthesis, 186–195
  morphology and colour, 206
  morphology and its control, 179–180, 184, 186, 189–190
  morphology and temperature, 180, 187
  particle size, 182, 187, 191–192, 196, 198
  PMMA, 171
  polysiloxane, 162
  polystyrene, 153
  quantum dots, 194–195
  representative syntheses, 205–209
  surface-to-volume ratio, 177
nanotechnology as interdisciplinary, 176
naphthalene derivatives
  1,8-bis[3-(3,5-dimethylphenyl)- 9-acridyl]-, 59–60
  2,7-dihydroxy- (2,7-DHN), 179, 182, 205–206
  ethyl 3-(6-methoxy-2-naphthyl) propanoate, 84
nickel catalysts, 20–22, 196, 208
NiPAM (N-isopropylacrylamide), 152–154, 156
nitriles
  aryl nitriles, 79–80, 85, 92
  cyanation reactions, 79–80
  from dehydration of amides, 14
  hydration to amides, 11–14, 202
malononitriles, 96, 99–100, 106, 118, 198
nitrogen-containing cations, 133–134
nitrogen-containing heterocycles, 92–106, 118
nitroxide-mediated polymerization (NMP), 156, 171
non-thermal effects of microwaves, 5–6, 124, 128–130, 136–137
nucleoside analogues, 34, 114–115
nucleosides, acyclo-, 65–66

oligonucleotides, 166–167
organosilanes, 21, 46, 78–79
organostannanes, 78
organotrifluoroborates, 67
oriented attachment, 191
oxetanes, 157
oxidations, metal-catalyzed, 41–44, 48
oxime hydrolysis, 12
oxindoles, 29, 105
oxygen-containing heterocycles, 106–111, 118

Paal-Knorr reaction, 204
palladium-based catalysts
  carbon supported, 57, 94, 134–135
  co-catalysis, 76
  cross-coupling reactions, 55, 71–78, 134–135, 161, 164
  cyanation reactions, 79–80
  cyclization reactions, 29
  heterocyclic complexes, 68–69
  intramolecular amidation, 105
  palladacycles, 74–75, 79, 81, 85
  palladium acetate (non-Suzuki reactions), 73, 79–80, 82–83, 85
  palladium acetate (Suzuki reaction), 57–59, 61, 63, 68–71, 164
  palladium dichloride, 72–73, 76
  Pd(PPh$_3$)$_4$, 56, 60, 63–67, 72, 78
  polymer supported, 68–70, 72–73
  POPd$_2$, 57
  redox reactions, 38–40, 48, 196, 198
  retro-reductive aminations, 40–42, 115
  substitution reactions, 22–23

224

Subject Index

palladium contamination, 68, 77
palladium nanoparticles, 182, 185–186, 196, 198
pantolactone, 42–43
particle size
  see also nanoparticles
  cationic polymerization, 162
  emulsion polymerization, 153–154
  surface-to-volume ratio, 177
Pauson-Khand reaction, 29–30
PCy$_3$ (tricyclohexylphosphine), 59
PEG see polyethylene glycol
penem analogues, 25–26
peptides
  Biphenomycin B, 61–61
  cyclization, 103
  dendrimeric, 36, 164–167
  functionalization, 164, 166–167
  glutathione, 183, 203–204, 206–207
  solid-phase synthesis, 169
  spiro-amino acids, 95–96
perfluorinated organics, 70–71
pharmaceuticals, deuteration, 23
  see also drug discovery
phase-transfer catalysis (PTC), 4, 5, 16, 67
  see also tetrabutylammonium bromide
  absence of, 109
  polyethylene glycol, 70–71
phenanthroline derivatives, 98
phenethylamines, 2-substituted-, 63–64
phenols, from aryl halides, 27, 46–47
phenylalanines, 58, 66
phenylboronic acids see arylboronic acids
phosphatases, 128
phosphines
  diarylphosphinopropane sulfonates, 61
  diphosphines, 30
  Pd(PPh$_3$)$_4$, 56, 60, 63–67, 72, 78
  phosphine oxides, 15–16, 194
  tricyclohexyl-, 59
phosphorus-containing cations, 134
photocatalysis, 187–188
photonics/photoluminescence, 182, 194–195

piperazines, 2,5-diketo-, 96, 103
platinum-based catalysts
  nanoparticles, 195–196
  redox reactions, 41, 48
  substitution reactions, 21–22, 46
platinum nanoparticles, 182, 185–186
PMI (pulsed microwave irradiation), 151–152, 154, 170
PMMA see poly(methylmethacrylate)
podophyllotoxin, 4-aza-, 105
polarity of solvents, 6–7, 10
pollutants, treatment, 23–24, 43
pollution avoidance, 1, 3
polyamides, 158–160
polyanilines, 169
polyaspartic acid, 158
polychlorinated aromatics, 24
polyethers, step-growth synthesis, 157–158, 171
polyethylene glycol (PEG), 113, 167
  ATRP co-polymers with PS, 156
  cross-coupling reactions, 70–71, 74, 77, 164
polyethylene oxide co-polymers, 188
polyimides, 158–160, 171–172
polymer modifications, 162–169, 172
polymerization
  see also the corresponding monomers
  advantages of aqueous media, 145–146
  cationic polymerization, 162, 172
  experimental procedures, 170–172
  graft co-polymerization, 147–150, 170
  polymer modifications, 162–169, 172
  radical polymerization, 146–157
  step-growth polymerization, 157–162
poly(methylmethacrylate) (PMMA)
  coated microparticles and nanotubes, 154–155
  emulsion polymerization, 153–154, 170
  grafting onto polysaccharides, 147, 149–150, 170
  nanoparticles, 171
poly(p-phenylene)s, 160
polyphenols, 182, 184
poly[rosin-(2-acryloyloxy)ethyl ester] (PRAEE), 149–150

Subject Index

polysaccharides
  see also chitosan; starch
  cationization, 163–164
  macromolecular chain grafting, 147–149, 169–170
  O-methylation, 163
  polymer modifications, 163–169
polysiloxanes, 162
polystyrene (PS)
  ATRP co-polymers, 156
  emulsion polymerization, 154–155, 171
  microspheres and nanoparticles, 151, 153–155
  nanoparticles, 153
  solvent effects on particle size, 153
polystyrenesulfonic acid (PSSA), 97–98, 109–110, 113–114, 118
polysuccinimide (PSI), 158
polythiophenes, 160
polyvinylpyrrolidone (PVP), 151, 169, 180, 191
pomace see wine and grape pomace
POPd$_2$ (dihydrogen di-μ-chloro-dichlorobis(di-*tert*-butyl-phosphinito-kP)dipalladate), 57
porphyrin precursors, 97
PRAEE (poly[rosin-(2-acryloyloxy)-ethyl ester]), 149–150
precatalysts, 68–69, 74, 76, 84
pronase E, 131
propanesulfonic acid, 2-acrylamido-2-methyl- (AMPS), 147
propargylic alcohols, 16, 19, 46
propionic acid, 3-mercapto- (MPA), 167–168, 195
proteins
  enzymatic digestion, 130–133, 138
  functionalization, 63
  microwave susceptibility, 137
  separation, 198
PS/PSI see polystyrene; polysuccinimide
PTC see phase-transfer catalysis
pulsed microwave irradiation (PMI), 151–152, 154, 170
purines, 58, 114–115

PVP (polyvinylpyrrolidone), 151, 169, 180, 191
4*H*-pyrano[2,3-*c*]pyrazoles, 100
pyrazine, 2,5-diethoxy-3-isopropyl-, 96
2(1*H*)-pyrazinones, 32–34, 78
1*H*-pyrazoles, 94–95, 203
pyrazolidine, 1-phenyl, 94–95
pyrazolo[4',3':5,6]pyrido[2,3-*d*]pyrimidines, 102
pyridazinones, 66–67
pyridine
  5-chloro-2-phenylethynyl-, 84
  1,4-dihydro- and derivatives, 99, 104
  6-ferrocenyl- derivatives, 99
  4-*N,N*-dimethylamino- analogues, 201
  fused heterocycles from, 100
  terpyridines from 2-acetyl-, 112
pyridinealdoximes, 68, 73, 76
2-pyridones, 99
di(2-pyridyl)methylamine-PdCl$_2$ complexes, 72–73, 76
pyrimidinones, 97, 111–112, 118
pyrranes, tri-, tetra- and penta-, 97
pyrroles, 30, 97, 204
pyrrolidines, 76, 93, 100
  see also PVP
pyrrolidinium ions, 133–134

quantum dots, 194–195
quinazolines, 30, 106–107, 116
quinoline, 6-amino, 98

radical polymerization
  acrylamide and acrylic acid, 146–147
  controlled/living polymerizations, 155–157
  in dispersed media, 151–155
  macromolecular chain grafting, 147–150
RAFT (reversible addition-fragmentation chain transfer), 156–157
reaction rates
  water enhancement, 10
red wine see wine and grape pomace
reductions
  see also hydrogenation
  metal-catalyzed reactions, 38–41, 48
  retro-reductive amination, 40–42, 115

residence times, 7
reversible addition-fragmentation chain transfer (RAFT), 156–157
rhodanine derivatives, 111
rhodium-based catalysts
  addition reactions, 15
  cyclization reactions, 29
  redox reactions, 42, 196
  substitution reactions, 22–23
ring-closing metathesis, 6, 63
room temperature ionic liquids (RTILs), 133
ruthenium-based catalysts
  addition reactions, 11–12
  cross-metathesis, 44
  redox reactions, 43, 48, 196, 198, 200
  Ru-Fe$_3$O$_4$, 201
  substitution reactions, 20, 22
ruthenium hydroxide nanocatalyst, 202–203

salicylaldehyde N(4)-hexamethylene-iminylthiosemicarbazone, 59–60
semiconductors, 160–161, 189, 194
SERS (surface enhanced Raman spectroscopy), 178, 182
silica-coated nanoparticles, 195–198, 200
silicon compounds
  N,N-diethylamino-propylated silica, 106
  organosilanes, 21, 46, 78–79
  polysiloxanes, 162, 172
silver nanoparticles, 182–185, 206–207
soapless emulsion polymerization, 152, 154–155
sodium bicarbonate decomposition, 6
sodium hypophosphite, 24
solvent replacement, 2–3
solvents
  biphasic systems, 3, 7, 10, 16
  dielectric constants, 134–135
  dipole moments, 178
  enzyme activity in organic, 124–125
  enzyme reactions in aqueous organics, 132–133
  heat capacity, 123

  hydrophobic, 125, 135
  ionic liquids as, 133–136
  polarity, 123, 134, 153
  superheating effect, 124
Sonogashira reaction, 75–78, 84
spectrometry-spectroscopy, 130–131, 178, 182
spin coating, 189
starch
  see also polysaccharides
  amination, 164–165
  graft co-polymers, 149–150
  hydrolysis, 137
  stabilizing agent, 182, 206
  superabsorbent materials based on, 147
step-growth polymerization, 157–162
stereoselective reactions
  additions, 15
  cyclizations, 30–31
  enantioselectivity, 30, 132, 137, 198, 201
steric effects
  cross-coupling reactions, 59, 61, 64
  Mannich reactions, 113
steroids, 15
stilbenes, 112
Stille reaction, 60, 78
styrene, Heck reaction, 73
substitution reactions
  deuteration, 19–21
  metal-catalyzed, 19–28, 46–47
sugars, 31, 34, 166, 180–181
  see also nucleosides; polysaccharides
sulfides
  β-hydroxy-, 112
  diaryl-, 27
sulfonamides, 201
sulfophthalein dyes, 70
sulfoxides
  β-hydroxy-, 112
  DMSO oxidation, 43
sulfur-containing heterocycles, 111–112, 119
superabsorbent materials, 147, 149
supercritical CO_2, 3

Subject Index

superheating effect, 14–15, 124, 136
surface-to-volume ratio, 177
susceptors, 7
Suzuki-Miyaura reaction, 56–71, 83–84
  intramolecular, 61
  polymers, 160–162, 164

tandem bis-aldol reactions, 109–110
tandem bis-aza-Michael reactions, 113–114
TBAB/TBAI *see* tetrabutyl-
terazoles, 92–93
terpyridines, 112
tetrabutylammonium bromide (TBAB)
  other cross-coupling ractions, 72–77, 79–80, 84
  Suzuki reaction, 56–58, 60–70, 83
  synthesis of bioactive heterocyclics, 104–106, 111–112
  synthesis of gold nanoparticles, 180
  synthesis of polyethers, 157
tetrabutylammonium fluoride, 197
tetrabutylammonium iodide (TBAI), 57
tetraphenylborate, 68
tetronic acid, 102, 105
thermal effects of microwaves, 4–5, 137, 177–178
thiazoles, 65
thiazolidinones, 111
thiobarbituric acid, 111–112
thioesterification of polymers, 167–168
3-thiophenemalonic acid, 185
tighter ion-pair effect, 6
tin-based catalysts
  addition reactions, 16
  isomerization reactions, 18
tin compounds (organostannanes), 78
titanium dioxide (TiO_2) nanoparticles, 154, 187–189, 196
titanium-ruthenium co-catalyst, 18
toluene, 4-phenyl, 83–84
TON (turnover number), 69, 198
toxic substances
  carbon monoxide, 80
  halogenated organics, 23–24
  hydrodechlorination, 23–24
  metals, 196
  nanoparticles, 184
  solvents, 2–3, 145, 178, 202
  tin compounds, 78–79
transfer hydrogenations, 38–41
transglycosylation, 133, 138
transition metals *see* metal-catalyzed reactions; *individual metals*
triaza-benzo[b]flouren-6-one, 93–94
1,3,5-triazines, 92–93
1,2,3-triazoles, 32–33, 35–37, 47, 160
1,2,4-triazoles, 65–66
tricyclohexylphosphine (PCy_3), 59
trypsin, 126, 128, 131–132, 138
tungsten-based catalysts, 31
turnover number (TON), 69, 198
two-dimensional electrophoresis (2-DE), 131
two-phase systems *see* biphasic media

Ullmann-type reactions, 24–26
ultrasound, 60, 99–100
ureas, cyclization, 30, 47, 93

vinyl monomer grafting polymerization, 147–149
vinylation
  Heck reaction, 71–72
  Hiyama reaction, 78–79
vinylphosphine oxides, 15

wastewater treatment, 43
water
  see also solvents
  effectiveness of metal catalysts, 11
  enzyme reactions in, 124–132
  green chemistry and, 3–4, 10, 49, 92, 99
  microwave selectivity towards, 6–7, 178
  organic solvents in, 132
  polymer synthesis in, 145–146
  reaction rates and, 10
  superheated, 14–15, 136
  thermal and non-thermal effects of microwaves, 4–6

wine and grape pomace, 181–182, 184, 186, 205

xyloglucan, 150

zinc acetate, 189, 194, 208
zinc chloride, 14
zinc nitrate, 189, 207
zinc oxide (ZnO) nanoparticles, 189–191
zinc selenide and sulfide, 194